Natural Selection and Social Theory

Evolution and Cognition

General Editor: Stephen Stich, Rutgers University

Published in the Series

Simple Heuristics That Make Us Smart
Gerd Gigerenzer, Peter M. Todd, and the ABC Research Group

Adaptive Thinking
Gerd Gigerenzer

Natural Selection and Social Theory:
Selected Papers of Robert Trivers
Robert Trivers

Natural Selection and Social Theory

Selected Papers of Robert Trivers

Robert Trivers

OXFORD
UNIVERSITY PRESS
2002

OXFORD
UNIVERSITY PRESS

Oxford   New York
Auckland   Bangkok   Buenos Aires   Cape Town   Chennai
Dar es Salaam   Delhi   Hong Kong   Istanbul   Karachi   Kolkata
Kuala Lumpur   Madrid   Melbourne   Mexico City   Mumbai   Nairobi
São Paulo   Shanghai   Singapore   Taipei   Tokyo   Toronto

and an associated company in Berlin

Published by Oxford University Press, Inc.
198 Madison Avenue, New York, New York 10016

www.oup.com

Oxford is a registered trademark of Oxford University Press

Library of Congress Cataloging-in-Publication Data
Trivers, Robert.
Natural selection and social theory : selected papers of Robert Trivers / by Robert Trivers.
p.   cm.—(Evolution and cognition)
A collection of 10 papers, 5 published in scholarly journals between 1971–1976
and 5 between 1982–2000.
Reciprocal altruism—Parental investment and reproductive success—The
Trivers-Willard effect—Parent-offspring conflict—Haplodiploidy and the social insects
—Size and reproductive success in a lizard—Selecting good genes for daughters—
Self-deception in service of deceit—Genomic imprinting—Fluctuating asymmetry
and 2nd:4th digit ratio in children.
Includes bibliographical references and index.
ISBN 13 978-0-19-513062-1 (pbk.)
ISBN 0-19-513061-8; 0-19-513062-6 (pbk.)
1. Sociobiology.   2. Evolution (Biology)   3. Social evolution in animals.
I. Title.   II. Series.
GN365.9.T762002
304.5—dc21   2001036767

3 5 7 9 8 6 4 2

Printed in the United States of America
on acid free paper

For my mother,
Mildred Raynolds Trivers,
author of seven children and
seven books of poetry,
born July 4, 1912

with love

PREFACE

This book is a collection of some of my scientific papers, along with accounts of how they were written and short postscripts that attempt to bring the reader up-to-date. I have naturally concentrated on my early papers, the five between 1971 and 1976 that had such a strong influence. In addition to these I have included an empirical paper on lizard size and reproductive success. I have also included a paper (with Jon Seger) showing a bias in female choice toward the interests of daughters, a widely neglected paper that I hope this volume will help resurrect. My thinking on self-deception is scattered in various places, including chapter 16 of my book *Social Evolution*. Here I reproduce a paper with Huey Newton analyzing the contribution of self-deception to the crash of an airplane, along with a paper just published that summarizes my current understanding of the subject. Since finishing my book on *Social Evolution* in 1985, I have largely been occupied trying to understand the genetics and evolution of selfish genetic elements, a vast subject from which I reproduce here only a chapter with Austin Burt on genomic imprinting, itself one of the most interesting discoveries in mammalian genetics in the past twenty years. I close with a brief summary of the results of the only long-term study of fluctuating asymmetry (and the 2nd : 4th digit ratio) in humans.

The postscripts are meant to lead the reader to some useful work published after the paper itself. I have not kept up in depth with many of these topics, but when I know of important work I try to draw the reader's attention to it, and I also try to give a set of recent references that can quickly lead the reader to the larger subject. Having said this, I must emphasize that with modern computer-driven search engines (such as Web of Science) the interested reader has only to find the "cited reference search" and type in the reference to my paper, and the computer will spew out very recent references on the topic.

Acknowledgments

This book was completed under a John Simon Guggenheim Memorial Fellowship, as well a Smithkline-Beecham Fellowship at the School of Law at Arizona State University. I am most grateful for these fellowships. I am also grateful to the Ann and Gordon Getty Foundation and the Biosocial Research Foundation for support of my work. I thank my editor, Kirk Jensen, for his continual encouragement and help. I thank Amy Jacobson for a truly outstanding copyediting of the entire manuscript.

I thank the publishers for permission to reproduce the copyrighted material listed below:

Chapter 1: Trivers, Robert L. 1971. The evolution of reciprocal altruism. *The Quarterly Review of Biology* 46(Mar.): 35–57.

Chapter 2: Trivers, Robert L. 1972. Parental investment and sexual selection. In *Sexual Selection and the Descent of Man 1871–1971*, ed. Bernard Campbell, 136–179. Chicago: Aldine Publishing Company.

Chapter 3: Trivers, Robert L., and Dan E. Willard. 1973. Natural selection of parental ability to vary the sex ratio of offspring. *Science* 179(Jan.): 90–92.

Chapter 4: Trivers, Robert L. 1974. Parent–offspring conflict. *American Zoologist* 14: 249–264.

Chapter 5: Trivers, Robert L., and Hope Hare. 1976. Haplodiploidy and the evolution of the social insects: The unusual traits of the social insects are uniquely explained by Hamilton's kinship theory. *Science* 191(Jan.): 249–263.

Chapter 6: Trivers, Robert L. 1976. Sexual selection and resource-accruing abilities in *Anolis garmani*. *Evolution* 30(Jul.): 253–269.

Chapter 7: Seger, Jon, and Robert Trivers. 1986. Asymmetry in the evolution of female mating preferences. *Nature* 319: 771–773.

Chapter 8: Trivers, Robert, and Huey P. Newton. 1982. The crash of Flight 90: Doomed by self-deception? *Science Digest* (Nov.): 66–67, 111. Trivers, Robert. 2000. The elements of a scientific theory of self-deception. *Annals of the New York Academy of Sciences* 907(Apr.): 114–131.

Chapter 9: Burt, Austin, and Robert Trivers. 1998. Genetic conflicts in genomic imprinting. *Proceedings of the Royal Society of London (B)* 265: 2393–2397.

Chapter 10: Manning, J. T., R. L. Trivers, R. Thornhill, and D. Singh. 2000. The 2nd : 4th digit ratio and asymmetry of hand performance in Jamaican children. *Laterality* 5(2): 121–132.

CONTENTS

Natural Selection and Social Theory

1

RECIPROCAL ALTRUISM

It was the spring semester of 1969 and the population geneticist Richard Lewontin had come to Harvard to deliver a talk on a new methodology that promised to revolutionize the field of population genetics. Lewontin was then at nearly the height of his powers, and 250 people, myself included, were crammed into a lecture hall to hear his eagerly anticipated talk. It was my devout wish that he would fall flat on his face.

The reason for this is that I had just been introduced to him by E. O. Wilson, Harvard's fabulous social insect man, at the tea preceding his talk. Lewontin had at once proceeded to dump on me, a mere first-year graduate student. The previous fall I had written a very negative paper attacking the work of two theoretical ecologists, Robert MacArthur and Richard Levins, and Lewontin was a personal friend of Levins at the University of Chicago. A fellow graduate student had carried the manuscript to a conference and allowed someone to make a photocopy, and like a true pathogen it had spread rapidly around the globe. I even received a glowing letter from a well-known Australian ecologist. When I realized the manuscript had gone public I sent copies to MacArthur and Levins. MacArthur wrote a very nice letter in return; Levins did not deign to reply. Lewontin began by saying, "Oh yes, you're the fellow that wrote that wrong-headed paper on Mac-Arthur and Levins," and dismissed it in a few sentences. He pointed to some equations on the board that he had apparently shown to Ed Wilson and told me that one of my criticism was easily handled if you used these equations. As Wilson squinted myopically toward the board (and with only one good eye at that!) I thought I could see at a glance that Lewontin had merely complexified the problem—thereby hiding the error more deeply—but that the same problem remained in his formulation as in the work of Levins himself. I have learned in my life that my memory of what I regard as odious

3

behavior is always more negative than the actual behavior itself, but in any case Lewontin had a somewhat arrogant and condescending style, and his treatment of me, I later learned, was by no means out of character. I took an immediate dislike to him.

Ernst Mayr, Harvard's venerated evolutionary biologist, introduced Lewontin at the talk, and I remember Lewontin thanked "Ernst" for his introduction, which drew an almost audible gasp from the audience, who had never before heard Professor Mayr referred to in public by only his first name. Lewontin then went on to deliver a masterful talk. He began by describing a problem that he said lay at the heart of evolutionary biology, namely, the degree of genetic variation present in natural populations. There were two competing schools of thought: one, that there was a lot of variation; the other, that there was very little. Each view had important implications for other matters. He then described a hopeless methodological conundrum, one method biased to an unknown degree in one direction, and the alternative methodology biased to an unknown degree in the opposite. Enter from stage left Lewontin with his new methodology, gel electrophoresis. This method allowed a large number of loci to be sampled relatively easily for genetic variants. He described the method and then gave the first wave of results, which clearly supported one school of thought: there was a lot of genetic variation in natural populations. He then, as I remember it, presented a second wave of data that showed parallel geographical patterns of variation in closely related species, thus suggesting the ongoing action of natural selection (since random forces, acting independently, should not produce *parallel* distributions).

Some people claim that at the end of the lecture he shot his chalk twenty feet into the air and caught it in his suit pocket, but if this occurred I did not see it. The point is that he put on quite a show, both in content and in presentation. I remember halfway through the lecture feeling some intense internal pain. This organism—odious cretin though he was—was not about to fall on his face. This contradiction led me to some soul-searching about my paper on MacArthur and Levins. I realized that the paper was entirely negative; that is, I had nothing positive with which to replace their incorrect views—I had, at best, only a bag full of *their* errors. While this is not without its use in science, its long-term value is strictly limited, for the criticism will soon sink out of sight along with the views it criticizes. Did I have anything positive to say on the subject of theoretical ecology? I knew at once that I did not. To make progress in this area you needed mathematical skills and discipline that I was not about to develop, and you would also benefit greatly from twenty or thirty years of running around in the woods. That is, if you could combine intuition and real knowledge of nature with the kind of mathematics required to handle complex interactions, then perhaps you could make some headway. I decided, in the space of a few minutes, not to try to publish this attack or to do any more work along those lines.

What positive thoughts did I have? Two immediately came to mind. One was reciprocal altruism—you scratch my back and I'll scratch yours—as an evolutionary problem. And the other was what became parental investment and sexual selection. As I recount in chapter 2, I had been watching pigeons for some time and knew that male pigeons practiced the same type of sexual double standard and had the same kind of attendant psychological problems (or at least behavioral ones) as did human males. This suggested some novel thoughts about the evolution of sex differences. I decided at once to concentrate on both of these problems and to tackle reciprocal altruism first, because the solution to that problem was more straightforward, but to continue my pigeon observations and associated thoughts.

I left Lewontin's talk, as I remember it, in high, good spirits, a burden lifted from my shoulders and my eyes firmly set on the prize ahead. For one thing, it was such a relief to kiss mathematical ecology good-bye! It was a small burden then but one that could easily have grown with time. I wonder how many academics happen to get into areas they neither really enjoy nor are particularly suited for, yet stay in those areas for many years. At the same time, by turning to what I really cared about and throwing the same kind of energy and effort into it as I had thrown into attacking MacArthur and Levins, I would really have something to show for my time.

It later often seemed ironic to me that Lewontin should have helped put me so firmly on a path that he came to detest so much himself. He was to return to Harvard as a full professor when I was an assistant professor, and he was to become chairman of Organismic and Evolutionary Biology when I came up for tenure. During that bitter dispute I heard by the grapevine that he had disparaged me to one group of students as an "intellectual opportunist." Of course I was an intellectual opportunist! What else made sense in this short life? The inability of biologists to think clearly on matter of social behavior and evolution for over a hundred years had left a series of important problems untackled. And it was a wonderful opportunity, especially if you had some social and psychological insight (my rough analog to twenty years in the woods). The area was so underdeveloped you hardly needed mathematics. Logic plus fractions would get you through most situations!

I had started to focus on reciprocal altruism largely because of the wonderful work of W. D. Hamilton on kinship and social behavior. Hamilton had shown that you could define a variable—degree of relatedness (or coefficient of relatedness) to others—that would have a strong effect on selection acting on social behavior between related individuals. In particular, an individual could sacrifice personal reproductive success (or fitness, as it was usually called then) and still be favored by natural selection as long as this benefited relatives such that, when the benefit was devalued by the degree of relatedness, the resulting number was still larger than the cost suffered. I take up kinship theory in much greater depth in chapters 4 and 5.

Hamilton did something else in his famous 1964 paper that was deceptively simple—he defined the four major categories of social interaction in terms of their effects on the reproductive success of the two individuals involved. Thus, "altruistic" behavior was behavior that caused a loss to the actor and a benefit to the recipient where these were defined in terms of effects on their reproductive success. Selfish behavior was the reverse, while in cooperative behavior both parties benefited and in spiteful behavior neither party benefited: each suffered a cost. This fourfold classification of behavior, or social traits, more broadly, had the benefit of immediately stating how natural selection was acting on the interaction from the standpoint of each of the two individuals.

This was a use of language that social scientists and others would come to detest. When they were aroused in full antipathy toward "sociobiology," you often read that this was a perversion of language. True altruism had to refer to pure internal motivations or other-directed internal motivations without thought or concern for self. To an evolutionist, this seemed absurd. You begin with the *effect* of behavior on actors and recipients; you deal with the problem of internal motivation, which is a secondary problem, afterward. If you made this point to some of these naysayers, they would often argue back that, if this were the case, then Hamilton should have chosen words that had no connotations in everyday language. This was also a very short-sighted view. In the extreme case, I suppose, Hamilton could have called the behaviors x, y, w, and z, so as to avoid any but alphabetical connotations. But this, of course, meant that you would always have to be translating those symbols into some verbal system that made sense to you before you could think clearly. Incidentally, when the great sociobiology controversy did roll forth, I soon came to see that the real function of these counterarguments was to slow down your work and, if possible, stop it cold. If you start with motivation, you have given up the evolutionary analysis at the outset. If you are forced to use arbitrary symbols, progress will be slowed for no good reason. Even the invitation to argue with them seemed to me to benefit them by wasting your time!

In any case, once you have conceptualized certain behaviors or traits as "altruistic" yet can show that they exist in nature, then you immediately have a problem, and a severe one, because in classic Darwinian theory, natural selection removes genes that tend to induce altruistic acts. The phrase "a gene for altruism" was also later subject to bitter attack. How could a single gene cause altruism? And again, the primary function of this objection was to make it very difficult for people to talk and think about the analysis in question. Lewontin, incidentally, who was one of those who raised this and related objections, I always felt fought sociobiology tooth and nail, because he had committed himself already to an alternative, marxist or pseudo-marxist system of social interpretation, and he would rather keep his evolutionary theory for population genetics and the kinds of problems he tackled in his office, while retaining marxist thinking for handling social interactions. In any case, by simply calling certain behaviors and acts altruistic and defining them the way he did, Hamilton brought into clear focus the problem involved. In his own kinship theory he solved many such cases in nature while at the same time producing much deeper implications as well. Again, I will leave this subject for later chapters.

When I came into biology at age twenty-two, never having had a course in biology and knowing next to nothing bout animal behavior, my knowledge was almost entirely restricted to our own species. In adult humans it was obvious that, though kinship was a very important factor—blood being thicker than water—it could not explain all phenomena. We had strong positive feelings toward friends, and we were willing to act altruistically toward them and others. Kinship could not explain this. What could?

Well, reciprocity, in some form, could obviously do the trick—that is, you scratch my back and I'll scratch yours—but reciprocity required some thinking to get the argument right. When we are each scratching each other's backs we are simultaneously trading benefits and suffering costs. That does not create much of an evolutionary problem. But what about when we act nicely toward an individual and the return benefit, if any, must come later? This raised some interesting evolutionary problems. So, I saw that what in the human species was obviously a major area of life involving deep and complex behaviors was not explained by Hamilton's theory, and required some new explanation. Note that the use of the term "altruism" helped immediately in thinking about reciprocity or reciprocal altruism. Reciprocity, after all, can be negative—reciprocal spite—as Frans deWaal is fond of emphasizing. Anthropologists, I soon learned, were fond of talking about "reciprocity" but by using that term they usually bypassed the theoretical problem at once. They were even able to dream up formulations where parental care was an example of "reciprocity" between the generations, because the offspring would later invest parental care in its own offspring.

So one day I sat down and wrote out a short manuscript on the evolution of reciprocal altruism. I began with the evolutionary problem and tried to formulate the matter mathematically. I remember I had a locus with genes that affected altruism but I realized that I had better have a second locus in the other individual directing the return effort so that my analysis would not become confused with kinship complications at that same locus. I wrote out a short and, I might add, feeble "mathematical-genetic" section. I had two examples in mind from nature, neither of which, in fact, was a good example of reciprocal altruism, though both were good examples of "return effect altruism." That is, you act altruistic and a return benefit reaches you but not because another individual chooses to reciprocate your original altruism.

The first example was cleaning symbioses in fish. I used to read very widely in those days, and I remembered seeing an advertisement for a book entitled *The Biology of the Mouth*. I thought to myself, Who wouldn't want to understand the biology of the mouth? and I sent off for my copy. I was thoroughly disappointed in most of the articles, which gave no evolutionary understanding or overview of the mouth but dealt only with problems related to dentistry. One chapter, though, caught my eye, a chapter on cleaning symbioses in the ocean. These are symbioses in which a member of one species, the cleaner species, cleans ectoparasites from the body and sometimes the mouth of a member of the host species. Both parties benefit instantaneously. That is, with each bite, the cleaner gets a meal and the host loses an ectoparasite: evolution had clearly favored warm interactions between the two. But in the article, I read of behavior not so easily explained. In some cases, when a cleaner is cleaning the inside of a host fish's mouth, the host will spot a predator, and instead of (as I imagine I would have done) simply swallowing the cleaner and getting the hell out of there, the host closed its mouth and then opened it—as a warning to the cleaner to depart— and then took off running itself. Ahhh, here was a delay requiring explanation. Was there a return benefit? As you will see if you read the paper that follows, I did indeed gather evidence that the hosts often return to the same cleaners and that they probably benefit from doing so, and therefore, a concern for the welfare of the life of the cleaner, even at some cost to your own life, may pay its way. The individuals involved were members of different species, so kinship could be ruled out.

The second problem from nature was warning cries in birds. In many, many species of birds, individuals who spot a predator give an alarm call, which warns other individuals of a predator at some presumed cost, however small, to the individual giving the warning call. Indeed, Peter Marler (1955) had written a beautiful paper showing that the sound characteristics of warning calls were very different than those of territorial calls in a couple of species of birds: the latter had a wide frequency variation while the warning calls tended to be pure tone. Wide frequency variation permits listeners to locate, easily and quickly, the direction from which the sound is coming, while pure tones are almost *ventriloquial*. So, it was nice for the bird that its warning call did not reveal much about its location while its territorial assertion call did. But, from an evolutionary stand-

point, those pure tones had to evolve. They had to begin as calls with frequency variation. So that beautiful pure-tone picture that Marler published connoted to me so many dead birds to get to that pure pitch. That
was indirect evidence of cost.

Now, there was in principle no reason not to explain all of bird calls as
kin directed. On the other hand, nothing was known about the kin structure
of any bird species, so there was no direct evidence in favor of this interpretation. How might one explain bird calls in my system? The obvious
would be reciprocal altruism itself: I warn you when I spot the predator,
you warn me when you spot the predator, and we both show a net gain over
time. I saw a problem with this: it seemed a hopeless system to police. How
are you going to handle cheaters? To put it in human terms, you can imagine
a bird spotting a predator late, almost being eaten itself, and then going to
its neighbor Fred and saying, "Why the hell didn't you warn me?" Fred
throws up his wings and says, "I was as surprised as you, brother!" There
was no way to identify cheaters, no way to punish them, except not calling
yourself when the shoe was on the other foot. All I could see selection doing
was silencing the birds throughout their range.

But there were plenty of opportunities for return-effect altruism to explain warning calls. I knew from my teacher William Drury that predators
had been shown in some cases to form specific search images, in which
experience killing one member of a species greatly increased their chance
of spotting and killing other members of the same species. Imagine a highly
cryptic (i.e., camouflaged) moth resting against the bark of a tree. You over-
look many, but when you spot the first one as edible food and consume it,
you become aware of their existence, rapidly learn characteristics to discriminate them from bark, and start cleaning up on moths resting on bark.
In short, my neighbor not getting eaten might decrease the chance of a
predator learning useful things when it turned to attack me. But some small
familiarity with predator–prey interactions in animals immediately suggested more direct possibilities. When predators are sneaking up on prey
and are spotted, and a warning cry is subsequently issued and everyone
dashes to safety, often the predators then move to some other area where
they have not yet been spotted. So, warning your neighbor that a predator
is nearby may be the quickest way to get the predator to move on elsewhere.
I tried to think through as many ways as I could of natural selection favoring
alarm calls in birds through return effects. Since there was no direct evidence
on any of these possibilities, my thinking at least had the virtue of generating
alternative explanations to the kinship logic.

I then wrote, if I remember correctly, a very short few pages on human
reciprocal altruism with no evidence cited or research results discussed.

I Meet W.D.

A little bit after this, perhaps in the fall of 1969, William Hamilton himself came to Harvard to lecture. He was coming from a "Man and Beast" symposium at the Smithsonian in Washington where he had presented some of his latest thinking, which I believe was the same talk he gave us. There were perhaps eighty or ninety people, almost filling a lecture hall, most of us with eager anticipation. Hamilton got up and gave one of the worst lectures, as a lecture, that I had ever heard. There was an emeritus Harvard professor who occasionally used to give a lecture, widely appreciated, on how to give a poor lecture. W. D. did not need any teaching in this regard and has generated some wonderful tricks of his own. I say this as a man who loved W. D., but his early troubles in this regard were sometimes very funny. For one thing, he lectured for a full fifty-five minutes without yet getting to the point. It was abstruse and technical; he often had his back to us while he was writing things on the board; you had difficulty hearing his voice; you did not get any overview of where he was going or why he was going there. When he realized that he was five minutes over time and still had not gotten to the point, or indeed very near it, he looked down at Ed Wilson, his host, and asked him if he could have some more time, perhaps ten or fifteen minutes. Of course Professor Wilson granted him some more time, but he also made a rolling, "let's-try-to-speed-this-up" motion with his arms. Hamilton then called for slides. The room went dark, and there was a rumble and a roaring sound as about 90% of the audience took this opportunity to exit the room for some fresh air. Some students were nearly trampled, I am told.

I remember walking home from the lecture with Ernst Mayr, both of us shaking our heads. It was obvious that the man was brilliant, a deep thinker, and his every thought well worth attending to, but whoa, was he bad in public! Hamilton was not unaware of this problem and once told a class we taught together that after hearing his lectures many students would doubt that he understood even his own ideas! By the way, he improved considerably in the intervening years, but he still showed the touch of a true master. He was invited to lecture to law professors in Squaw Valley, as a guest of Margaret Gruter's Institute for the Study of Law and Behavioral Sciences, and there he introduced a new trick that I had not seen before. He showed a number of interesting but complex slides on parasite–host interactions. He had a hand-held microphone but no pointer, so he used the microphone as a pointer. Often all you would hear him say was, "Here, as you can see . . ." and then the microphone would point to various parts of the slide, while his mouth continued to move. Then you might hear again, "And then in the next slide . . ." and then once again you would not hear anything about the slide, though you could see Dr. Hamilton pointing to various places in it with his microphone!

After his Harvard talk I got the chance to meet him in person. Mary Jane West-Eberhard, the celebrated student of wasps, was I believe a postdoc with Wilson at the time, and she held a small party for Hamilton. I brought along the little paper that I had written on reciprocal altruism to give him to read, and I think he expressed interest in meeting me (he had heard I had been thinking along these lines, or something like that). He was a shy man, very soft spoken. You often had to lean into him to catch what he was saying. He had a kind of horsey-looking face, as he would describe it to others (e.g., when he was meeting someone new at the train station). I remember thinking, at some point in my relationship with Bill, that if the argument ever became physical, the contest I would least like to be engaged in against Hamilton was a shoving contest. I felt that he would dig in his heels, that you would be unable to move him, and that he would lean forward and shove you slowly and stubbornly to wherever he wanted to get you.

Some time thereafter I received a letter from W. D. and I remembered liking the gentle tone with which he dealt with my efforts. He encouraged me to continue. He could see why I had chosen separate loci for actor and recipient, but he suggested (without telling me that my mathematical attempts were littered with errors, as they were) that it would perhaps be better to delete the "maths," as he called it, from the paper. I only half followed his advice. I deleted the separate loci but tried to expand the "maths" and of course introduced new errors in the process. For example, you will see that I make assertions about the evolution of reciprocal altruism genes that in fact apply only when the gene for altruism is exactly 50% in frequency. This is not a very helpful formulation from an evolutionary standpoint. The evolutionary problem is to take a gene that begins at low frequency and get it to a high frequency—what is happening at the 50/50 mark is usually an irrelevancy: But I was very much trying to mimic, however feebly, Hamilton's kind of thinking itself: try, if at all possible, to get the formulation down to the level of genes, to be more sure that you have got the argument right, and to give a quantitative form to it if possible. Hamilton also pointed out that my two nonhuman examples were not actually examples of reciprocal altruism and suggested that I rename the paper "the evolution of return-effect altruism." This I was not about to do, though later I wished I had taken the opportunity to stress more clearly the differences between the two categories.

Human Reciprocal Altruism

The key to the paper, besides its evolutionary approach, was the section on human reciprocal altruism. Could I reorganize facts about human psychology around this new argument in a coherent and interesting way? If so, there would be immediate pay-off for the argumentation itself. This required that I become familiar with a literature from social psychology that I knew nothing about. Naturally I sought shortcuts. I saw that in the Harvard catalog a man was teaching a course on moral behavior, and from the description in the catalog, this sounded just about right. So I said to myself that I would humble myself and take the entire course (as an auditor, of course) in order to learn this material, and at the end of the semester I should have the human section in hand, or at least be ready to write it. I attended, I believe, only two lectures. The problem with the course was that the teacher seemed to think it was immoral to teach a course on morality without first canvassing all the students to find out what they wanted to know about the subject. This was not what I wanted. I wanted someone to lay out the subject matter for me so that I could reinterpret it as I wished. However, I noted that his graduate teaching assistant, Dennis Krebs (now a Professor of Psychology at Simon Fraser University), was cited on the reading list as having written a review, soon to be published, of the literature on altruistic behavior. Ahh, this was perfect! As a graduate student of the professor, you could take it for granted that his review was going to be very thorough and very likely better than the professor himself would have done. I could skip the lectures entirely and simply read Krebs's paper, assuming he was willing to provide a copy. I went to Mr. Krebs and he very kindly provided me with a copy. I never saw the class or its professor again.

On the other hand, I did not immediately digest Krebs's paper, either. It was written in a different language, and I was required to master that language before I could understand exactly what he was saying. I read around in the paper to try to familiarize myself with the kinds of evidence that were available. One thing that amused me, right off the bat, was that social psychologists called altruistic behavior "pro-social behavior." Now, that seemed immediately somewhat ill-defined. Social is easily opposed to solitary, so social interaction is an interaction involving more than one organism. I can be fighting you and it is a social interaction. Am I pro-social if I am in favor of fights? An antisocial individual might be a hermit, but antisocial was being used to refer to someone whose behavior was bad for others, as the author saw it, and ought best to be curbed. Again, defining categories of behavior in this way seemed like a hopeless way to proceed if you were interested in building up a solid scientific approach to human behavior. There were, of course, other linguistic hurdles I had to overcome. Krebs had organized his paper along conventional distinctions within that discipline. For example, there were immediate precipitating variables where you were more or less to give a dollar to a beggar if snot was coming out of his nose. There were personality variables in the potential pro-socialists. There were situational variables that might affect the tendency. And, of course, all this research was guided by the goal of increasing pro-social behavior—not really understanding it, and certainly not at any deep level, but increasing its occurrence in society. What was missing was exactly what was missing from the discipline itself: any functional understanding of the behaviors that they were discussing. Why did it make sense for the organism to do it? This was, of course, what evolutionary biology, and myself in particular, was set to provide. So, all I had to do was master the literature cited by Krebs and then reorganize it appropriately.

This sounded easier than it turned out to be, and the more I comprehended Krebs's paper, the more I dreaded having to study the works that he was citing. They would be written in a bizarre language and I was afraid to learn too much about the actual methodologies used. I read a few of the papers and then decided to take another shortcut and write the section based on Krebs's review. I piously told myself that I would, of course, read all the papers whose citations I would be lifting from Krebs's review, as by good scientific and academic procedure I should, but in fact I never did.

I was visiting my parents' home in the Midwest for two or three weeks and decided to take this opportunity to write the human section. I was depressed the first week and lay around doing little more than sleeping. Then I roused myself and said, You have two boring weeks ahead of you, however you slice it, so you might as well do this work, which has to be done anyway. I spread out the few papers I had actually copied, reread Krebs's account, and simply reorganized the information around obvious psychological categories, such as sympathy, gratitude, and moralistic aggression.

How to Write a Classic Paper

In retrospect, I think my paper on reciprocal altruism can be used to illustrate how one might go about writing a classic paper. Here is my recipe:

1. Pick an important topic.
2. Try to do a little sustained thinking on the topic, always keeping close to the task at hand.
3. Generalize outward from your chosen topic.
4. Write in the language of your discipline but, of course, try to do so simply and clearly.
5. If at all possible, reorganize existing evidence around your theory.

Pick an important topic. This is perhaps easier said than done, because you must pick an important topic on which progress can be made. But it still seems remarkable to me how often people bypass what are more important subjects to work on less important ones. Constructing a scientific understanding of human psychology and social behavior is an important task, and within that subject reciprocal altruism was, to me, an important topic when I began work on it. The very fact that so many strong emotions could be associated, in humans, with friendship, with helping others, with guarding against being cheated, said to me that it was an important topic in human psychology. There was no reason, in advance, to believe that it would not be an important topic in at least some other species, certainly closely related monkeys and apes, but very likely other species as well.

Do some sustained thinking. To me this is easier than it sounds. There are no great intellectual gymnastics in my paper. I am not proving Fermat's Last Theorem or generating Goedel's Proof. I am only trying to think simply and clearly on an interesting and important subject. I was amazed when I went into academic work—and it still baffles me today—why so many people take the first available path off their main argument into trivia land. The sustained thinking must always be directed back to the key subject itself. I suppose it is easier, at first, to write a section on the semantics of discussing altruism, or sometimes to review previous failed efforts in considerable detail if these failed efforts are in front of you and easy to interpret. But this is all a waste of energy and effort. It diverts you from your main task. Get to the point and stick to the point. When I sent the paper to M. L. Roonwal, the great Indian student of termites and locusts (whom I had met on a very memorable ten-day monkey-viewing visit with Irv DeVore), he praised the paper for its "intellectual architecture." And I thought the phrase was very apt. I had constructed the paper much as you might build a house, with roughly the same kinds of mental operations required. Nothing brilliant or flashy, just the steady construction of a series of arguments and facts regarding reciprocal altruism.

Generalize outward. In most of my papers I took the road opposite the one biologists usually take, or in any case are usually accused of, which is arguing from a knowledge of animals to suppositions about humans. I usually begin with humans and then try to generalize outward to include as many other species and phenomena as I can. In this case, I took pains to formulate the argument with as few assumptions as possible. These only appear later as limitations. For example, both Charles Darwin and George C. Williams had a few words to say on human reciprocal altruism, but each presupposed that its appearance would require the kind of intellectual talents that we know exist in our species: to recognize individuals, to remember past interactions, to alter behavior appropriately, and so on. This limits the argument to human beings in advance, and in a way that is completely unnecessary. It seemed to me obviously preferable to avoid any such limitation unless absolutely forced to accept it. The mental processes they cite could easily evolve *after* the fact. If you could just get a little bit of reciprocal altruism going, selection pressures to spot cheaters, to reward especially good fellow altruists, and so on ought to evolve easily.

Write in the language of your discipline, but simply and clearly. I was trying to make an evolutionary argument within the field of evolutionary biology. You need to satisfy the criteria of your discipline to be accepted or noticed.

Of course, as I have admitted, I dressed up a pseudo-mathematical genetics section to look as if I were using the concepts and language of my discipline more than I was, but I am convinced that had the later material, on humans for example, been written without any citations or references, it could easily have been neglected. So, even though the literature that was cited was a good bit weaker than I would have liked, I think it was important for the seriousness with which people took the paper. Alas, it had very little effect, quite opposite to my supposition, on social psychology itself. I naively imagined that social psychologists would be delighted to see that their work could be given much more interest and meaning when put in an evolutionary framework, and that they would immediately see new ways of doing their work that would prove more fruitful than the paths that they were taking. For example, let us assume you show, as they did, that you are more likely to act nicely toward a person who resembles you more. As an evolutionist you immediately wonder whether this is because of kin selection, in which case the organism is unconsciously measuring degree of relatedness, or whether it might be due to a reciprocal mechanism in which the likelihood of exchanging benefits was more likely with closer resemblance. Yet this immediately raises another question since reciprocal altruism between individuals who are complementary in characteristics may give greater value in some contexts than when they are similar. Pursuing this line of thought would suggest experiments or observations that discriminated these possibilities as well. Nothing like this has, in fact, happened. I know of no social psychologists who have altered their work because of my paper (I would be very happy to hear of any) and have been very disappointed to discover that almost the only new work done is by people who first start with an evolutionary interest and then turn to trying to do something social psychological.

The virtues of writing simply and clearly should be obvious, certainly from the standpoint of the recipient, but attempting to do so has virtues for the writer as well, since it repeatedly forces you to think through your subject clearly. I was very fortunate at Harvard to have several professors who were excellent critical readers. The most useful from my standpoint was Irven DeVore, the famous Harvard "baboon man" and anthropologist. Irv is a superb stylist and through five or six rewrites of the paper he would continually improve the presentation. Similarly, my advisor, Ernest Williams, would continually force me to tighten the argument and clarify the presentation. Incidentally, when I brought Professor Williams the sixth draft of the paper I remember saying to him, "Is it finished Dr. Williams? Is it finished?" There was a small pause and then Ernest looked at me and said, "A paper is *never* finished, Bob, it is only abandoned." He then told me that he thought that this paper was ready to be abandoned.

All those drafts that the paper was put through earned a rich reward when it was actually published. There was an immediate, large, and very welcome response. I began to receive reprint requests in the mail, mostly from the United States but in fact from all over the world. I had ordered six hundred reprints (this was in the days before wholesale photocopying of papers) and soon enough the pile was exhausted. It gave me immense pleasure, especially receiving the foreign reprint requests, to know that my thinking would be studied in countries around the world. I was especially gratified at the number of reprint requests that came from Soviet bloc or communist countries because, of course, there was otherwise so little exchange across the so-called Iron Curtain. After a while, it struck me that their socialist ideology and emphasis on the possibility of naturally cooperative behavior among humans would make this an interesting topic for them and, perhaps, even one that was socially acceptable to pursue. I also learned that in their literature, Peter Kropotkin was an early pioneer whom they would have expected me to cite.

Try to reorganize or reinterpret existing information. Many, many theoretical papers in biology and elsewhere fail to show, at the end of the paper, that there is anything out there in the real world to which the argumentation just given actually applies. It applies in principle, of course, but they are unable to reinterpret any existing information. A typical effort might discuss an important topic, provide some sustained mathematical thinking on some part of the problem, usually misrepresented as being central to the problem, and then derive results that not only cannot immediately be tested against reality, but also would require an awful lot of work to do so. The value of the work, of course, depends upon how restrictive the initial assumptions are. If they are highly restrictive, then even if your mathematics are correct, you have solved a highly specialized problem and it may be of very marginal interest to test whether your statements are true, especially as this may require very detailed and difficult measurements. Of course, if you have modeled something central to a problem and your results are not easy to test, that is another matter.

In my case, I think the last section gave life to the paper in a very important way because it showed, in principle, that the simple argument given could reinterpret information gathered in social psychology and give us a deeper understanding of similar kinds of facts known from our everyday experience.

A Sense of Justice

A very agreeable feature of my reciprocal altruism argument, which I had not anticipated in advance, was that a sense of justice or fairness seemed a natural consequence of selection for reciprocal altruism. That is, you could easily imagine that a sense of fairness would evolve as a way of regulating reciprocal tendencies, a way of judging the degree to which other people were cheating you (and you them!). There seemed no plausible way in which a sense of fairness would evolve from kin interactions, for these did not require any return benefit or equality of contribution on the part of the two organisms. An older sibling could repeatedly act altruistically toward a younger sibling with little or no expectation of returned favor yet feel good about doing so. Not so, of course, for relations between unrelated individuals. I naively thought that those with a self-professed concern for justice in society would welcome this argument. After all, it suggested that a sense of fairness was not some arbitrary cultural construct or an easily changed effect of socialization. It suggested, instead, that there were deep biological roots to our sense of fairness that to me would seem to encourage a commitment to fairness or justice. Although, as I have mentioned, scholars in so-called socialist countries became especially interested in my paper, the same was not true of their leftist counterparts in the United States and England.

Indeed, while we are on the subject, I may say that I felt then that *all* of my papers were on the side of the angels! Parental investment and sexual selection gave an objective and unbiased system for thinking about the evolution of the two sexes, one that naturally gave equal weight to the two sexes and was built on a sex-blind concept that could be applied everywhere (relative parental investment). My paper with Dan Willard elevated the lower class female to special status—an otherwise counterintuitive status that has now been abundantly confirmed in humans. My paper on parent–offspring conflict elevated the offspring to an equal status with the parent in affecting their interactions. Yet none of these papers was greeted by the political left as any kind of positive contribution. Quite the contrary. Because the papers dared to take an evolutionary approach in the first place—which required natural selection working on genetic variation—they were ruled to be regressive in their implications at the very outset.

I shared the paper on reciprocal altruism, of course, with Professor Wilson, and I was a little put off by his response. He liked it, but he urged me to push the analysis further, to lay a proper mathematical foundation for my argument just as Hamilton had done for kinship theory, with its inevitable improvements in precision, logic, and testability. I responded somewhat hotly that kinship lent itself to a simple mathematical foundation, permitting Hamilton's rules to be derived, while reciprocal altruism depended upon frequency of interaction and distributions of relevant payoffs in a complex manner that did not lend itself to easy mathematical formulation. Wilson replied, as I remember it, that this was well and good and might even be true but that he believed that this represented a "failure of imagination" on my part. Even as his comments rankled me anew, I admired the phrase and have used it to good effect myself. In any case, his words were to prove prophetic. In ten short years, Hamilton himself, along with Robert Axelrod, a political scientist at the University of Michigan, were to do precisely what Ed had urged me to accomplish. Far from complexity, they isolated a simple rule of action, tit-for-tat, which was evolutionarily stable.

The Evolution of Reciprocal Altruism

ROBERT L. TRIVERS

Abstract.   A model is presented to account for the natural selection of what is termed reciprocally altruistic behavior. The model shows how selection can operate against the cheater (non-reciprocator) in the system. Three instances of altruistic behavior are discussed, the evolution of which the model can explain: (1) behavior involved in cleaning symbioses; (2) warning cries in birds; and (3) human reciprocal altruism.

Regarding human reciprocal altruism, it is shown that the details of the psychological system that regulates this altruism can be explained by the model. Specifically, friendship, dislike, moralistic aggression, gratitude, sympathy, trust, suspicion, trustworthiness, aspects of guilt, and some forms of dishonesty and hypocrisy can be explained as important adaptations to regulate the altruistic system. Each individual human is seen as possessing altruistic and cheating tendencies, the expression of which is sensitive to developmental variables that were selected to set the tendencies at a balance appropriate to the local social and ecological environment.

Altruistic behavior can be defined as behavior that benefits another organism, not closely related, while being apparently detrimental to the organism performing the behavior, benefit and detriment being defined in terms of contribution to inclusive fitness. One human being leaping into water, at some danger to himself, to save another distantly related human from drowning may be said to display altruistic behavior. If he were to leap in to save his own child, the behavior would not necessarily be an instance of "altruism"; he may merely be contributing to the survival of his own genes invested in the child.

Models that attempt to explain altruistic behavior in terms of natural selection are models designed to take the altruism out of altruism. For example, Hamilton (1964) has demonstrated that degree of relationship is an important parameter in predicting how selection will operate, and behavior which appears altruistic may, on knowledge of the genetic relationships of the organisms involved, be explicable in terms of natural selection: those genes being selected for that contribute to their own perpetuation, regardless of which individual the genes appear in. The term "kin selection" will be used in this paper to cover instances of this type—that is, of organisms being selected to help their relatively close kin.

The model presented here is designed to show how certain classes of behavior conveniently denoted as "altruistic" (or "reciprocally altruistic") can be selected for even when the recipient is so distantly related to the organism performing the altruistic act that kin selection can be ruled out. The model will apply, for example, to altruistic behavior between members of different species. It will be argued that under certain conditions natural selection favors these altruistic behaviors because in the long run they benefit the organism performing them.

The Model

One human being saving another, who is not closely related and is about to drown, is an instance of altruism. Assume that the chance of the drowning man dying is one-half if no one leaps in to save him, but that the chance that his potential rescuer will drown if he leaps in to save him is much smaller, say, one in twenty. Assume that the drowning man always drowns when his rescuer does and that he is always saved when the rescuer survives the rescue attempt. Also assume that the energy costs involved in rescuing are trivial compared to the survival probabilities. Were this an isolated event, it is clear that the rescuer should not bother to save the drowning man. But if the drowning man reciprocates at some future time, and if the survival chances are then exactly reversed, it will have been to the benefit of each participant to have risked his life for the other. Each participant will have traded a one-half chance of dying for about a one-tenth chance. If we assume that the entire population is sooner or later exposed to the same risk of drowning, the two individuals who risk their lives to save each other will be selected over those who face drowning on their own. Note that the benefits of reciprocity depend on the unequal cost/benefit ratio of the altruistic act, that is, the benefit of the altruistic act to the recipient is greater than the cost of the act to the performer, cost and benefit being defined here as the increase or decrease in chances of the relevant alleles propagating themselves in the population. Note also that, as defined, the benefits and costs depend on the age of the altruist and recipient (see *Age-dependent changes* below). (The odds assigned above may not be unrealistic if the drowning man is drowning because of a cramp or if the rescue can be executed by extending a branch from shore.)

Why should the rescued individual bother to reciprocate? Selection would seem to favor being saved from drowning without endangering one-self by reciprocating. Why not cheat? ("Cheating" is used throughout this paper solely for convenience to denote failure to reciprocate; no conscious intent or moral connotation is implied.) Selection will discriminate against the cheater if cheating has later adverse affects on his life which outweigh the benefit of not reciprocating. This may happen if the altruist responds to the cheating by curtailing all future possible altruistic gestures to this indi-vidual. Assuming that the benefits of these lost altruistic acts outweigh the costs involved in reciprocating, the cheater will be selected against relative to individuals who, because neither cheats, exchange many altruistic acts.

This argument can be made precise. Assume there are both altruists and nonaltruists in a population of size N and that the altruists are characterized by the fact that each performs altruistic acts when the cost to the altruist is well below the benefit to the recipient, where cost is defined as the degree to which the behavior retards the reproduction of the genes of the altruist and benefit is the degree to which the behavior increases the rate of repro-duction of the genes of the recipient. Assume that the altruistic behavior of an altruist is controlled by an allele (dominant or recessive), a_2, at a given locus and that (for simplicity) there is only one alternative allele, a_1, at that locus and that it does not lead to altruistic behavior. Consider three possi-bilities: (1) the altruists dispense their altruism randomly throughout the population; (2) they dispense it nonrandomly by regarding their degree of genetic relationship with possible recipients; or (3) they dispense it nonran-domly by regarding the altruistic tendencies of possible recipients.

(1)   Random dispensation of altruism

There are three possible genotypes: a_1a_1, a_2a_1, and a_2a_2. Each allele of the heterozygote will be affected equally by whatever costs and benefits are associated with the altruism of such individuals (if a_2 is dominant) and by whatever benefits accrue to such individuals from the altruism of others, so they can be disregarded. If altruistic acts are being dispensed randomly throughout a large population, then the typical a_1a_1 individual benefits by $(1/N)\Sigma b_i$, where b_i is the benefit of the ith altruistic act performed by the altruist. The typical a_2a_2 individual has a net benefit of $(1/N)\Sigma b_i - (1/N)\Sigma c_j$, where c_j is the cost to the a_2a_2 altruist of his jth altruistic act. Since $-(1/N)\Sigma c_j$ is always less than zero, allele a_1 will everywhere replace allele a_2.

(2)   Nonrandom dispensation by reference to kin

This case has been treated in detail by Hamilton (1964), who concluded that if the tendency to dispense altruism to close kin is great enough, as a function of the disparity between the average cost and benefit of an altruistic act, then a_2 will replace a_1. Technically, all that is needed for Hamilton's form of selection to operate is that an individual with an "altruistic allele" be able to distinguish between individuals with and without this allele and discriminate accordingly. No formal analysis has been attempted of the possibilities for selection favoring individuals who increase their chances of receiving altruistic acts by appearing as if they were close kin of altruists, although selection has clearly sometimes favored such parasitism (e.g., Drury and Smith, 1968).

(3)   Nonrandom dispensation by reference to the altruistic tendencies of the recipient

What is required is that the net benefit accruing to a typical a_2a_2 altruist exceed that accruing to an a_1a_1 nonaltruist, or that

$$(1/p^2) (\Sigma b_k - \Sigma c_j) > (1/q^2)\Sigma b_m,$$

where b_k is the benefit to the a_2a_2 altruist of the kth altruistic act performed toward him, where c_j is the cost of the jth altruistic act by the a_2a_2 altruist, where b_m is the benefit of the mth altruistic act to the a_1a_1 nonaltruist, and where p is the frequency in the population of the a_2 allele and q that of the a_1 allele. This will tend to occur if Σb_m is kept small (which will simultaneously reduce Σc_j). And this in turn will tend to occur if an altruist responds to a "nonaltruistic act" (that is, a failure to act altruistically toward the altruist in a situation in which so doing would cost the actor less than it would benefit the recipient) by curtailing future altruistic acts to the nonaltruist.

Note that the above form of altruism does not depend on all altruistic acts being controlled by the same allele at the same locus. Each altruist could be motivated by a different allele at a different locus. All altruistic alleles would tend to be favored as long as, for each allele, the net average benefit to the homozygous altruist exceeded the average benefit to the homozygous nonaltruist; this would tend to be true if altruists restrict their altruism to fellow altruists, regardless of what allele motivates the other individual's altruism. The argument will therefore apply, unlike Hamilton's (1964), to altruistic acts exchanged between members of different species. It is the *exchange* that favors such altruism, not the fact that the allele in question sometimes or often directly benefits its duplicate in another organism.

If an "altruistic situation" is defined as any in which one individual can dispense a benefit to a second greater than the cost of the act to himself, then the chances of selecting for altruistic behavior, that is, of keeping $\Sigma c_j + \Sigma b_m$ small, are greatest (1) when there are many such altruistic situations in the lifetime of the altruists, (2) when a given altruist repeatedly interacts with the same small set of individuals, and (3) when pairs of altruists are exposed "symmetrically" to altruistic situations, that is, in such a way that the two are able to render roughly equivalent benefits to each other at roughly equivalent costs. These three conditions can be elaborated into a set of relevant biological parameters affecting the possibility that reciprocally altruistic behavior will be selected for.

(1) Length of lifetime.   Long lifetime of individuals of a species maximizes the chance that any two individuals will encounter many altruistic situations, and all other things being equal one should search for instances of reciprocal altruism in long-lived species.

(2) Disperal rate.   Low dispersal rate during all or a significant portion of the lifetime of individuals of a species increases the chance that an individual will interact repeatedly with the same set of neighbors, and other things being equal one should search for instances of reciprocal altruism in such species. Mayr (1963) has discussed some of the factors that may affect dispersal rates.

(3) Degree of mutual dependence.   Interdependence of members of a species (to avoid predators, for example) will tend to keep individuals near each other and thus increase the chance they will encounter altruistic situations together. If the benefit of the mutual dependence is greatest when only a small number of individuals are together, this will greatly increase the chance that an individual will repeatedly interact with the same small set of individuals. Individuals in primate troops, for example, are mutually dependent for protection from predation, yet the optimal troop size for foraging is often small (Crook, 1969). Because they also meet the other conditions outlined here, primates are almost ideal species in which to search for reciprocal altruism. Cleaning symbioses provide an instance of mutual dependence between members of different species, and this mutual dependence appears to have set the stage for the evolution of several altruistic behaviors discussed below.

(4) Parental care.   A special instance of mutual dependence is that found between parents and offspring in species that show parental care. The relationship is usually so asymmetrical that few or no situations arise in which an offspring is capable of performing an altruistic act for the parents or even for another offspring, but this is not entirely true for some species (such as primates) in which the period of parental care is unusually long. Parental care, of course, is to be explained by Hamilton's (1964) model, but there is no reason why selection for reciprocal altruism cannot operate between close kin, and evidence is presented below that such selection has operated in humans.

(5) Dominance hierarchy.   Linear dominance hierarchies consist by definition of asymmetrical relationships; a given individual is dominant over another but not vice versa. Strong dominance hierarchies reduce the extent to which altruistic situations occur in which the less dominant individual is capable of performing a benefit for the more dominant which the

more dominant individual could not simply take at will. Baboons (*Papio cynocephalus*) provide an illustration of this. Hall and DeVore (1965) have described the tendency for meat caught by an individual in the troop to end up by preemption in the hands of the most dominant males. This ability to preempt removes any selective advantage that food-sharing might otherwise have as a reciprocal gesture for the most dominant males, and there is no evidence in this species of any food-sharing tendencies. By contrast, Van Lawick-Goodall (1968) has shown that in the less dominance-oriented chimpanzees more dominant individuals often do not preempt food caught by the less dominant. Instead, they besiege the less dominant individual with "begging gestures," which result in the handing over of small portions of the catch. No strong evidence is available that this is part of a reciprocally altruistic system, but the absence of a strong linear dominance hierarchy has clearly facilitated such a possibility. It is very likely that early hominid groups had a dominance system more similar to that of the modern chimpanzee than to that of the modern baboon (see, for example, Reynolds, 1966).

(6) Aid in combat.   No matter how dominance-oriented a species is, a dominant individual can usually be aided in aggressive encounters with other individuals by help from a less dominant individual. Hall and DeVore (1965) have described the tendency for baboon alliances to form which fight as a unit in aggressive encounters (and in encounters with predators). Similarly, vervet monkeys in aggressive encounters solicit the aid of other, often less dominant, individuals (Struhsaker, 1967). Aid in combat is then a special case in which relatively symmetrical relations are possible between individuals who differ in dominance.

The above discussion is meant only to suggest the broad conditions that favor the evolution of reciprocal altruism. The most important parameters to specify for individuals of a species are how many altruistic situations occur and how symmetrical they are, and these are the most difficult to specify in advance. Of the three instances of reciprocal altruism discussed in this paper only one, human altruism, would have been predicted from the above broad conditions.

The relationship between two individuals repeatedly exposed to symmetrical reciprocal situations is exactly analogous to what game theorists call the Prisoner's Dilemma (Luce and Raiffa, 1957; Rapoport and Chammah, 1965), a game that can be characterized by the payoff matrix

	A_2	C_2
A_1	R, R	S, T
C_1	T, S	P, P

where $S < P < R < T$ and where A_1 and A_2 represent the altruistic choices possible for the two individuals, and C_1 and C_2, the cheating choices (the first letter in each box gives the payoff for the first individual, the second letter the payoff for the second individual). The other symbols can be given the following meanings: R stands for the reward each individual gets from an altruistic exchange if neither cheats; T stands for the temptation to cheat; S stands for the sucker's payoff that an altruist gets when cheated; and P is the punishment that both individuals get when neither is altruistic (adapted from Rapoport and Chammah, 1965). Iterated games played between the same two individuals mimic real life in that they permit each player to respond to the behavior of the other. Rapoport and Chammah (1965) and others have conducted such experiments using human players, and some of their results are reviewed below in the discussion of human altruism.

W. D. Hamilton (pers. commun.) has shown that the above treatment of reciprocal altruism can be reformulated concisely in terms of game theory as follows. Assuming two altruists are symmetrically exposed to a series of reciprocal situations with identical costs and identical benefits, then after $2n$ reciprocal situations, each has been "paid" nR. Were one of the two a nonaltruist and the second changed to a nonaltruistic policy after first being cheated, then the initial altruist would be paid $S + (n - 1)P$ (assuming he had the first opportunity to be altruistic) and the nonaltruist would receive $T + (n - 1)P$. The important point here is that unless $T >> R$, then even with small n, nR should exceed $T + (n - 1)P$. If this holds, the nonaltruistic type, when rare, cannot start to spread. But there is also a barrier to the spread of altruism when altruists are rare, for $P > S$ implies $nP > S + (n - 1)P$. As n increases, these two total payoffs tend to equality, so the barrier to the spread of altruism is weak if n is large. The barrier will be overcome if the advantages gained by exchanges between altruists outweigh the initial losses to nonaltruistic types.

Reciprocal altruism can also be viewed as a symbiosis, each partner helping the other while he helps himself. The symbiosis has a time lag, however; one partner helps the other and must then wait a period of time before he is helped in turn. The return benefit may come directly, as in human food-sharing, the partner directly returning the benefit after a time lag. Or the return may come indirectly, as in warning calls in birds (discussed below), where the initial help to other birds (the warning call) sets up a casual chain through the ecological system (the predator fails to learn useful information) which redounds after a time lag to the benefit of the caller. The time lag is the crucial factor, for it means that only under highly specialized circumstances can the altruist be reasonably guaranteed that the casual chain he initiates with his altruistic act will eventually return to him and confer, directly or indirectly, its benefit. Only under these conditions will the cheater be selected against and this type of altruistic behavior evolve.

Although the preconditions for the evolution of reciprocal altruism are specialized, many species probably meet them and display this type of altruism. This paper will limit itself, however, to three instances. The first, behavior involved in cleaning symbioses, is chosen because it permits a clear discrimination between this model and that based on kin selection (Hamilton, 1964). The second, warning calls in birds, has already been elaborately analyzed in terms of kin selection; it is discussed here to show how the model presented above leads to a very different interpretation of these familiar behaviors. Finally, human reciprocal altruism is discussed in detail because it represents the best documented case of reciprocal altruism known, because there has apparently been strong selection for a very complex system regulating altruistic behavior, and because the above model permits the functional interpretation of details of the system that otherwise remain obscure.

Altruistic Behavior in Cleaning Symbioses

The preconditions for the evolution of reciprocal altruism are similar to those for the operation of kin selection: long lifetime, low dispersal rate, and mutual dependence, for example, tend to increase the chance that one is interacting with one's close kin. This makes it difficult to discriminate the two alternative hypotheses. The case of cleaning symbiosis is important to analyze in detail because altruistic behavior is displayed that cannot be explained by kin selection, since it is performed by members of one species for the benefit of members of another. It will be shown instead that the behavior can be explained by the model presented above. No elaborate explanation is needed to understand the evolution of the mutually advantageous cleaning symbiosis itself; it is several additional behaviors displayed by the host fish to its cleaner that require a special explanation because they meet the criteria for altruistic behavior outlined above—that is, they benefit the cleaner while apparently being detrimental to the host.

Feder (1966) and Maynard (1968) have recently reviewed the literature on cleaning symbiosis in the ocean. Briefly, one organism (e.g., the wrasse, *Labroides dimidiatus*) cleans another organism (e.g., the grouper, *Epinephelus striatus*) of ectoparasites (e.g., caligoid cope-pods), sometimes entering into the gill chambers and mouth of the "host" in order to do so. Over forty-five species of fish are known to be cleaners, as well as six species of shrimp. Innumerable species of fish serve as hosts. Stomach analyses of cleaner fish demonstrate that they vary greatly in the extent to which they depend on their cleaning habits for food, some apparently subsisting nearly entirely on a diet of ectoparasites. Likewise, stomach analyses of host fish reveal that cleaners differ in the rate at which they end up in the stomachs of their hosts, some being apparently almost entirely immune to such a fate. It is a striking fact that there seems to be a strong correlation between degree of dependence on the cleaning way of life and immunity to predation by hosts.

Cleaning habits have apparently evolved independently many times (at least three times in shrimps alone), yet some remarkable convergence has taken place. Cleaners, whether shrimp or fish, are distinctively colored and behave in distinctive ways (for example, the wrasse, *L. dimidiatus*, swims up to its host with a curious dipping and rising motion that reminds one of the way a finch flies). These distinctive features seem to serve the function of attracting fish to be cleaned and of inhibiting any tendency in them to feed on their cleaners. There has apparently been strong selection to avoid eating one's cleaner. This can be illustrated by several observations. Hediger (1968) raised a grouper (*Epinephelus*) from infancy alone in a small tank for six years, by which time the fish was almost four feet in length and accustomed to snapping up anything dropped into its tank. Hediger then dropped a small live cleaner (*L. dimidiatus*) into the grouper's tank. The grouper not only failed to snap up the cleaner but opened its mouth and permitted the cleaner free entry and exit.

> Soon we watched our second surprise: the grouper made a movement which in the preceding six years we had never seen him make: he spread the right gill-covering so wide that the individual gill-plates were separated from each other at great distances, wide enough to let the cleaner through. (translated from Hediger, 1968, p. 93)

When Hediger added two additional *L. dimidiatus* to the tank, all three cleaned the grouper with the result that within several days the grouper appeared restless and nervous, searched out places in the tank he had formerly avoided, and shook himself often (as a signal that he did not wish to be cleaned any longer). Apparently three cleaners working over him constantly was too much for him, yet he still failed to eat any of them. When Hediger removed two of the cleaners, the grouper returned to normal. There is no indication the grouper ever possessed any edible ectoparasites, and almost two years later (in December, 1968) the same cleaner continued to "clean" the grouper (pers. observ.) although the cleaner was, in fact, fed separately by its zoo-keepers.

Eibl-Eibesfeldt (1959) has described the morphology and behavior of two species (e.g., *Aspidontus taeniatus*) that mimic cleaners (e.g., *L. dimidiatus*) and that rely on the passive behavior of fish which suppose they are about to be cleaned to dart in and bite off a chunk of their fins. I cite the evolution of these mimics, which resemble their models in appearance and initial swimming behavior, as evidence of strong selection for hosts with no intention of harming their cleaners.

Of special interest is evidence that there has been strong selection not to eat one's cleaner even after the cleaning is over. Eibl-Eibesfeldt (1955) has made some striking observations on the goby, *Elacitinus oceanops*:

> I never saw a grouper snap up a fish after it had cleaned it. On the contrary, it announced its impending departure by two definite signal movements. First it closed its mouth vigorously, although not completely, and immediately opened it wide again. Upon this intention movement, all the gobies left the mouth cavity. Then the grouper shook its body laterally a few times, and all the cleaners returned to their coral. If one frightened a grouper it never neglected these forewarning movements. (translated from Eibl-Eibesfeldt, 1955, p. 208)

Randall has made similar observations on a moray eel (*Gymnothorax japonicus*) that signalled with a "sharp lateral jerk of the eel's head," after which "the wrasse fairly flew out of the mouth, and the awesome jaws snapped shut" (Randall, 1958, 1962). Likewise, Hediger's Kasper Hauser grouper shook its body when it had enough of being cleaned.

Why does a large fish not signal the end to a cleaning episode by swallowing the cleaner? Natural selection would seem to favor the double benefit of a good cleaning followed by a meal of the cleaner. Selection also operates, of course, on the cleaner and presumably favors mechanisms to avoid being eaten. The distinctive behavior and appearance of cleaners have been cited as evidence of such selection. One can also cite the distinctive behavior of the fish being cleaned. Feder (1966) has pointed out that hosts approaching a cleaner react by "stopping or slowing down, allowing themselves to assume awkward positions, seemingly in a hypnotic state." Fishes sometimes alter their color dramatically before and while being cleaned, and Feder (1966) has summarized instances of this. These forms of behavior suggest that natural selection has operated on cleaners to avoid attempting to clean fish without these behaviors, presumably to avoid wasting energy and to minimize the dangers of being eaten. (Alternatively, the behaviors, including color change, may aid the cleaners in finding ectoparasites. This is certainly possible but not, I believe, adequate to explain the phenomenon completely. See, for example, Randall, 1962.)

Once the fish to be cleaned takes the proper stance, however, the cleaner goes to work with no apparent concern for its safety: it makes no effort to avoid the dangerous mouth and may even swim inside, which as we have seen, seems particularly foolhardy, since fish being cleaned may suddenly need to depart. The apparent unconcern of the cleaner suggests that natural selection acting on the fish being cleaned does not, in fact, favor eating one's cleaner. No speculation has been advanced as to why this may be so, although some speculation has appeared about the mechanisms involved. Feder advances two possibilities, that of Eibl-Eibesfeldt (1955) that fish come to be cleaned only after their appetite has been satisfied, and one of his own, that the irritation of ectoparasites may be sufficient to inhibit hunger. Both possibilities are contradicted by Hediger's observation, cited above, and seem unlikely on functional grounds as well.

A fish to be cleaned seems to perform several "altruistic" acts. It desists from eating the cleaner even when it easily could do so and when it must go to special pains (sometimes at danger to itself) to avoid doing so. Furthermore, it may perform two additional behaviors which seem of no direct benefit to itself (and which consume energy and take time); namely, it signals its cleaner that it is about to depart even when the fish is not in its mouth, and it may chase off possible dangers to the cleaner:

> While diving with me in the Virgin Islands, Robert Schroeder watched a Spanish hogfish grooming a bar jack in its bronze color state. When a second jack arrived in the pale color phase, the first jack immediately drove it away. But later when another jack intruded on the scene and changed its pale color to dark bronze it was not chased. The bronze color would seem to mean "no harm intended; I need service." (Randall, 1962, p. 44)

The behavior of the host fish is interpreted here to have resulted from natural selection and to be, in fact, beneficial to the host because the cleaner is worth more to it alive than dead. This is because the fish that is cleaned "plans" to return at later dates for more cleanings, and it will be benefited by being able to deal with the same individual. If it eats the cleaner, it may have difficulty finding a second when it needs to be cleaned again. It may lose valuable energy and be exposed to unnecessary predation in the search for a new cleaner. And it may in the end be "turned down" by a new cleaner or serviced very poorly. In short, the host is abundantly repaid for the cost of its altruism.

To support the hypothesis that the host is repaid its initial altruism, several pieces of evidence must be presented: that hosts suffer from ectoparasites; that finding a new cleaner may be difficult or dangerous; that if one does not eat one's cleaner, the same cleaner can be found and used a second time (e.g., that cleaners are site-specific); that cleaners live long enough to be used repeatedly by the same host; and if possible, that individual hosts do, in fact, reuse the same cleaner.

(1) The cost of ectoparasites.   It seems almost axiomatic that the evolution of cleaners entirely dependent on ectoparasites for food implies the selective disadvantage for the cleaned of being ectoparasite-ridden. What is perhaps surprising is the effect that removing all cleaners from a coral reef has on the local "hosts" (Limbaugh, 1961). As Feder (1966) said in his review:

> Within a few days the number of fishes was drastically reduced. Within two weeks almost all except territorial fishes had disappeared, and many of these had developed white fuzzy blotches, swellings, ulcerated sores, and frayed fins. (p. 366)

Clearly, once a fish's primary way of dealing with ectoparasites is by being cleaned, it is quickly vulnerable to the absence of cleaners.

(2) The difficulty and danger of finding a cleaner.   There are naturally very few data on the difficulty or danger of finding a new cleaner. This is partially because, as shown below, fish tend repeatedly to return to familiar cleaners. The only observation of fish being disappointed in their search for cleaners comes from Eibl-Eibesfeldt (1955): "If the cleaners fail to appear over one coral in about half a minute, the large fishes swim to another coral and wait there a while" (translated from p. 210). It may be that fish have several alternative cleaning stations to go to, since any particular cleaning station may be occupied or unattended at a given moment. So many fish tend to be cleaned at coral reefs (Limbaugh, 1961, observed a cleaner service 300 fish in a 6-hour period), that predators probably frequent coral reefs in search of fish being cleaned. Limbaugh (1961) suggested that good human fishing sites are found near cleaning stations. One final reason why coming to be cleaned may be dangerous is that some fish must leave their element to do so (Randall, 1962):

> Most impressive were the visits of moray eels, which do not ordinarily leave their holes in the reef during daylight hours, and of the big jacks which swam up from deeper water to the reef's edge to be "serviced" before going on their way. (p. 43)

(3) Site specificity of cleaners.   Feder (1966) has reviewed the striking evidence for the site specificity of cleaners and concludes:

> Cleaning fishes and cleaning shrimps have regular stations to which fishes wanting to be cleaned can come. (p. 367)

Limbaugh, Pederson, and Chase (1961) have reviewed available data on the six species of cleaner shrimps, and say:

> The known cleaner shrimps may conveniently be divided into two groups on the basis of behavior, habitat and color. The five species comprising one group are usually solitary or paired. . . . All five species are territorial and remain for weeks and, in some cases, months or possibly years within a meter or less of the same spot. They are omnivorous to a slight extent but seem to be highly dependent upon their hosts for food. This group is tropical, and the individuals are brightly marked. They display themselves to their hosts in a conspicuous manner. They probably rarely serve as prey for fishes. A single species, *Hippolysmata californica*, comprises the second group. . . . This species is a gregarious, wandering, omnivorous animal . . . and is not highly dependent upon its host for survival. So far as is known, it does not display itself to attract fishes. (p. 238)

It is *H. californica* that is occasionally found in the stomachs of at least one of its hosts. The striking correlation of territoriality and solitariness with cleaning habits is what theory would predict. The same correlation can be found in cleaner fish. *Labroides*, with four species, is the genus most completely dependent on cleaning habits. No *Labroides* has ever been found in the stomach of a host fish. All species are highly site-specific and tend to be solitary. Randall (1958) reports that an individual *L. dimidiatus* may sometimes swim as much as 60 feet from its cleaning station, servicing fish on the way. But he notes,

> This was especially true in an area where the highly territorial damsel fish *Pomacentris nigricans* (Lepede) was common. As one damsel fish was being tended, another nearby would assume a stationary pose with fins erect and the *Labroides* would move on to the latter with little hesitation. (p. 333)

Clearly, what matters for the evolution of reciprocal altruism is that the same two individuals interact repeatedly. This will be facilitated by the site specificity of either individual. Of temperate water cleaners, the species most specialized to cleaning is also apparently the most solitary (Hobson, 1969).

(4) Lifespan of cleaners.   No good data exist on how long cleaners live, but several observations on both fish and shrimp suggest that they easily live long enough for effective selection against cheaters. Randall (1958) repeatedly checked several ledges and found that different feeding stations were occupied for "long periods of time," apparently by the same individuals. One such feeding station supported two individuals for over three years. Of one species of cleaner shrimp, *Stenopus hispidus*, Limbaugh, Pederson, and Chase (1961) said that pairs of individuals probably remain months, possibly years, within an area of a square meter.

(5) Hosts using the same cleaner repeatedly.   There is surprisingly good evidence that hosts reuse the same cleaner repeatedly. Feder (1966) summarizes the evidence:

> Many fishes spend as much time getting cleaned as they do foraging for food. Some fishes return again and again to the same station, and show a definite time pattern in their daily arrival. Others pass from station to station and return many times during the day; this is particularly true of an injured or infected fish. (p. 368)

Limbaugh, Pederson, and Chase (1961) have presented evidence that in at least one species of cleaner shrimp (*Stenopus scutellus*), the shrimp may reservice the same individuals:

> One pair was observed in the same football-sized coral boulder from May through August 1956. During that period, we changed the position and orientation of the boulder several times within a radius of approximately seven meters without disturbing the shrimp. Visiting fishes were momentarily disturbed by the changes, but they soon relocated the shrimps. (p. 254)

Randall (1958) has repeatedly observed fish swimming from out of sight directly to cleaning stations, behavior suggesting to him that they had prior acquaintance with the stations. During two months of observations at several feeding stations, Eibl-Eibesfeldt (1955) became personally familiar with several individual groupers (*Epinephelus striatus*) and repeatedly observed them seeking out and being cleaned at the same feeding stations, presumably by the same cleaners.

In summary, it seems fair to say that the hosts of cleaning organisms perform several kinds of altruistic behavior, including not eating their cleaner after a cleaning, which can be explained on the basis of the above model. A review of the relevant evidence suggests that the cleaner organisms and their hosts meet the preconditions for the evolution of reciprocally altruistic behavior. The host's altruism is to be explained as benefiting him because of the advantage of being able quickly and repeatedly to return to the same cleaner.

Warning Calls in Birds

Marler (1955, 1957) has presented evidence that warning calls in birds tend to have characteristics that limit the information a predator gets from the call. In particular, the call characteristics do not allow the predator easily to determine the location of the call-giver. Thus, it seems that giving a warning call must result, at least occasionally, in the otherwise unnecessary death of the call-giver, either at the hands of the predator that inspired the call or at the hands of a second predator formerly unaware of the caller's presence or exact location.

Given the presumed selection against call-giving, Williams (1966) has reviewed various models to explain selection for warning cries:

(1) Warning calls are functional during the breeding season in birds in that they protect one's mate and offspring. They have no function outside the breeding season, but they are not deleted then because "in practice it is not worth burdening the germ plasm with the information necessary to realize such an adjustment" (Williams, 1966, p. 206).

(2) Warning calls are selected for by the mechanism of group selection (Wynne-Edwards, 1962).

(3) Warning calls are functional outside the breeding season because there is usually a good chance that a reasonably close kin is near enough to be helped sufficiently (Hamilton, 1964; Maynard Smith, 1964). Maynard Smith (1965) has analyzed in great detail how closely related the benefited kin must be, at what benefit to him the call must be, and at what cost to the caller, in order for selection to favor call-giving.

The first is an explanation of last resort. While it must sometimes apply in evolutionary arguments, it should probably only be invoked when no other explanation seems plausible. The second is not consistent with the known workings of natural selection. The third is feasible and may explain the warning calls in some species and perhaps even in many. But it does depend on the somewhat regular nearby presence of closely related organisms, a matter that may often be the case but that has been demonstrated only as a possibility in a few species and that seems very unlikely in some. A fourth explanation is suggested by the above model:

(4) Warning calls are selected for because they aid the bird giving the call. It is disadvantageous for a bird to have a predator eat a nearby conspecific because the predator may then be more likely to eat him. This may happen because the predator will

   (i) be sustained by the meal,
  (ii) be more likely to form a specific search image of the prey species,
 (iii) be more likely to learn the habits of the prey species and perfect his predatory techniques on it,
 (iv) be more likely to frequent the area in which the birds live, or
  (v) be more likely to learn useful information about the area in which the birds live.

In short, in one way or another, giving a warning call tends to prevent predators from specializing on the caller's species and locality.

There is abundant evidence for the importance of learning in the lives of predatory vertebrates (see, for example, Tinbergen, 1960; Leyhausen, 1965; Brower and Brower, 1965). Rudebeck (1950, 1951) has presented important observations on the tendency of avian predators to specialize individually on prey types and hunting techniques. Owen (1963) and others have presented evidence that species of snails and insects may evolve polymorphisms as a protection against the tendency of their avian predators to learn their appearance. Similarly, Kuyton (1962; cited in Wickler, 1968) has described the adaptation of a moth that minimizes the chance of its predators forming a specific search image. Southern (1954), Murie (1944), and numerous others have documented the tendency of predators to specialize on certain localities within their range. Finally, Blest (1963) has presented evidence that kin selection in some cryptic saturnid moths has favored rapid, post-reproductive death to minimize predation on the young. Blest's evidence thus provides an instance of a predator gaining useful information through the act of predation.

It does not matter that in giving a warning call the caller is helping its non-calling neighbors more than it is helping itself. What counts is that it outcompetes conspecifics from areas in which no one is giving warning calls. The non-calling neighbors of the caller (or their offspring) will soon find themselves in an area without any caller and will be selected against relative to birds in an area with callers. The caller, by definition, is always in an area with at least one caller. If we assume that two callers are preferable to one, and so on, then selection will favor the spread of the warning-call genes. Note that this model depends on the concept of *open* groups, whereas "group selection" (Wynne-Edwards, 1962) depends partly on the concept of closed groups.

It might be supposed that one could explain bird calls more directly as altruistic behavior that will be repaid when the other birds reciprocate, but there are numerous objections to this. It is difficult to visualize how one would discover and discriminate against the cheater, and there is certainly no evidence that birds refrain from giving calls because neighbors are not reciprocating. Furthermore, if the relevant bird groupings are very fluid, with much emigration and immigration, as they often are, then cheating would seem to be favored and no selection against it possible. Instead, according to the model above, it is the mere fact that the neighbor survives that repays the call-giver his altruism.

It is almost impossible to gather the sort of evidence that would discriminate between this explanation and that of Hamilton (1964). It is difficult to imagine how one would estimate the immediate cost of giving a warning call or its benefit to those within earshot, and precise data on the genetic relationships of bird groupings throughout the year are not only lacking but would be most difficult to gather. Several lines of evidence suggest, however, that Hamilton's (1964) explanation should be assumed with caution:

(1) There exist no data showing a decrease in warning tendencies with decrease in the genetic relationship of those within earshot. Indeed, a striking feature of warning calls is that they are given in and out of the breeding season, both before and after migration or dispersal.

(2) There do exist data suggesting that close kin in a number of species migrate or disperse great distances from each other (Ashmole, 1962; Perdeck, 1958; Berndt and Sternberg, 1968; Dhont and Hublé, 1968).

(3) One can advance the theoretical argument that kin selection under some circumstances should favor kin dispersal in order to avoid competition (Hamilton, 1964, 1969). This would lead one to expect fewer closely related kin near any given bird, outside the breeding season.

The arguments advanced in this section may also apply, of course, to species other than birds.

Human Reciprocal Altruism

Reciprocal altruism in the human species takes place in a number of contexts and in all known cultures (see, for example, Gouldner, 1960). Any complete list of human altruism would contain the following types of altruistic behavior:

(1) helping in times of danger (e.g., accidents, predation, intraspecific ag-
      gression);
(2) sharing food;
(3) helping the sick, the wounded, or the very young and old;
(4) sharing implements; and
(5) sharing knowledge.

All these forms of behavior often meet the criterion of small cost to the giver and great benefit to the taker.

During the Pleistocene, and probably before, a hominid species would have met the preconditions for the evolution of reciprocal altruism: long lifespan; low dispersal rate; life in small, mutually dependent, stable, social groups (Lee and DeVore, 1968; Campbell, 1966); and a long period of parental care. It is very likely that dominance relations were of the relaxed, less linear form characteristic of the living chimpanzee (Van Lawick-Goodall, 1968) and not of the more rigidly linear form characteristic of the baboon (Hall and DeVore, 1965). Aid in intraspecific combat, particularly by kin, almost certainly reduced the stability and linearity of the dominance order in early humans. Lee (1969) has shown that in almost all Bushman fights which are initially between two individuals, others have joined in. Mortality, for example, often strikes the secondaries rather than the principals. Tool use has also probably had an equalizing effect on human dominance relations, and the Bushmen have a saying that illustrates this nicely. As a dispute reaches the stage where deadly weapons may be employed, an individual will often declare: "We are none of us big, and others small; we are all men and we can fight; I'm going to get my arrows," (Lee, 1969). It is interesting that Van Lawick-Goodall (1968) has recorded an instance of strong dominance reversal in chimpanzees as a function of tool use. An individual moved from low in dominance to the top of the dominance hierarchy when he discovered the intimidating effects of throwing a metal tin around. It is likely that a diversity of talents is usually present in a band of hunter-gatherers such that the best maker of a certain type of tool is not often the best maker of a different sort or the best user of the tool. This contributes to the symmetary of relationships, since altruistic acts can be traded with reference to the special talents of the individuals involved.

To analyze the details of the human reciprocal-altruistic system, several distinctions are important and are discussed here.

(1) Kin selection. The human species also met the preconditions for the operation of kin selection. Early hominid hunter-gatherer bands almost certainly (like today's hunter-gatherers) consisted of many close kin, and kin selection must often have operated to favor the evolution of some types of altruistic behavior (Haldane, 1955; Hamilton, 1964, 1969). In general, in attempting to discriminate between the effects of kin selection and what

might be called reciprocal-altruistic selection, one can analyze the form of the altruistic behaviors themselves. For example, the existence of discrimination against non-reciprocal individuals cannot be explained on the basis of kin selection, in which the advantage accruing to close kin is what makes the altruistic behavior selectively advantageous, not its chance of being reciprocated. The strongest argument for the operation of reciprocal-altruistic selection in man is the psychological system controlling some forms of human altruism. Details of this system are reviewed below.

(2) Reciprocal altruism among close kin. If both forms of selection have operated, one would expect some interesting interactions. One might expect, for example, a lowered demand for reciprocity from kin than from nonkin, and there is evidence to support this (e.g., Marshall, 1961; Balikci, 1964). The demand that kin show some reciprocity (e.g., Marshall, 1961; Balikci, 1964) suggests, however, that reciprocal-altruistic selection has acted even on relations between close kin. Although interactions between the two forms of selection have probably been important in human evolution, this paper will limit itself to a preliminary description of the human reciprocally altruistic system, a system whose attributes are seen to result only from reciprocal-altruistic selection.

(3) Age-dependent changes. Cost and benefit were defined above without reference to the ages, and hence reproductive values (Fisher, 1958), of the individuals involved in an altruistic exchange. Since the reproductive value of a sexually mature organism declines with age, the benefit to him of a typical altruistic act also decreases, as does the cost to him of a typical act he performs. If the interval separating the two acts in an altruistic exchange is short relative to the lifespans of the individuals, then the error is slight. For longer intervals, in order to be repaid precisely, the initial altruist must receive more in return than he himself gave. It would be interesting to see whether humans in fact routinely expect "interest" to be added to a long overdue altruistic debt, interest commensurate with the intervening decline in reproductive value. In humans reproductive value declines most steeply shortly after sexual maturity is reached (Hamilton, 1966), and one would predict the interest rate on altruistic debts to be highest then. Selection might also favor keeping the interval between act and reciprocation short, but this should also be favored to protect against complete non-reciprocation. W. D. Hamilton (pers. commun.) has suggested that a detailed analysis of age-dependent changes in kin altruism and reciprocal altruism should show interesting differences, but the analysis is complicated by the possibility of reciprocity to the kin of a deceased altruist (see *Multiparty interactions* below).

(4) Gross and subtle cheating. Two forms of cheating can be distinguished, here denoted as gross and subtle. In *gross cheating* the cheater fails to reciprocate at all, and the altruist suffers the costs of whatever altruism he has

dispensed without any compensating benefits. More broadly, gross cheating may be defined as reciprocating so little, if at all, that the altruist receives less benefit from the gross cheater than the cost of the altruist's acts of altruism to the cheater. That is, $\Sigma_i C_{ai} > \Sigma_j b_{aj,}$ where c_{ai} is the cost of the ith altruistic act performed by the altruist and where b_{aj} is the benefit to the altruist of the jth altruistic act performed by the gross cheater; altruistic situations are assumed to have occurred symmetrically. Clearly, selection will strongly favor prompt discrimination against the gross cheater. *Subtle cheating*, by contrast, involves reciprocating, but always attempting to give less than one was given, or more precisely, to give less than the partner would give if the situation were reversed. In this situation, the altruist still benefits from the relationship but not as much as he would if the relationship were completely equitable. The subtle cheater benefits more than he would if the relationship were equitable. In other words,

$$\sum_{i,j}(b_{ai} - c_{qj}) > \sum_{i}(b_{qi} - c_{ai}) > \sum_{i,j}(b_{aj} - c_{ai})$$

where the ith altruistic act performed by the altruist has a cost to him of c_{ai} and a benefit to the subtle cheater of b_{qi} and where the jth altruistic act performed by the subtle cheater has a cost to him of c_{qi} and a benefit to the altruist of b_{aj}. Because human altruism may span huge periods of time, a lifetime even, and because thousands of exchanges may take place, involving many different "goods" and with many different cost/benefit ratios, the problem of computing the relevant totals, detecting imbalances, and deciding whether they are due to chance or to small-scale cheating is an extremely difficult one. Even then, the altruist is in an awkward position, symbolized by the folk saying, "half a loaf is better than none," for if attempts to make the relationship equitable lead to the rupture of the relationship, the altruist, assuming other things to be equal, will suffer the loss of the substandard altruism of the subtle cheater. It is the subtlety of the discrimination necessary to detect this form of cheating and the awkward situation that ensues that permit some subtle cheating to be adaptive. This sets up a dynamic tension in the system that has important repercussions, as discussed below. *(5) Number of reciprocal relationships.*   It has so far been assumed that it is to the advantage of each individual to form the maximum number of reciprocal relationships and that the individual suffers a decrease in fitness upon the rupture of any relationship in which the cost to him of acts dispensed to the partner is less than the benefit of acts dispensed toward him by the partner. But it is possible that relationships are partly exclusive, in the sense that expanding the number of reciprocal exchanges with one of the partners may necessarily decrease the number of exchanges with another. For example, if a group of organisms were to split into subgroups for much of the day (such as breaking up into hunting pairs), then altruistic exchanges will be more likely between members of each subgroup than between members of different subgroups. In that sense, relationships may be partly exclusive, membership in a given subgroup necessarily decreasing exchanges with others in the group. The importance of this factor is that it adds further complexity to the problem of dealing with the cheater and it increases competition within a group to be members of a favorable subgroup. An individual in a subgroup who feels that another member is subtly cheating on their relationship has the option of attempting to restore the relationship to a completely reciprocal one or of attempting to join another subgroup, thereby decreasing to a minimum the possible exchanges between himself and the subtle cheater and replacing these with exchanges between a new partner or partners. In short, he can switch friends. There is evidence in hunter-gatherers that much movement of individuals from one band to another occurs in response to such social factors as have just been outlined (Lee and DeVore, 1968).

(6) Indirect benefits or reciprocal altruism? Given mutual dependence in a group it is possible to argue that the benefits (nonaltruistic) of this mutual dependence are a positive function of group size and that altruistic behaviors may be selected for because they permit additional individuals to survive and thereby confer additional indirect (nonaltruistic) benefits. Such an argument can only be advanced seriously for slowly reproducing species with little dispersal. Saving an individual's life in a hunter-gatherer group, for example, may permit nonaltruistic actions such as cooperative hunting to continue with more individuals. But if there is an optimum group size, one would expect adaptations to stay near that size, with individuals joining groups when the groups are below this size, and groups splitting up when they are above this size. One would only be selected to keep an individual alive when the group is below optimum and not when the group is above optimum. Although an abundant literature on hunter-gatherers (and also nonhuman primates) suggests that adaptations exist to regulate group size near an optimum, there is no evidence that altruistic gestures are curtailed when groups are above the optimum in size. Instead, the benefits of human altruism are to be seen as coming directly from reciprocity—not indirectly through nonaltruistic group benefits. This distinction is important because social scientists and philosophers have tended to deal with human altruism in terms of the benefits of living in a group, without differentiating between nonaltruistic benefits and reciprocal benefits (e.g., Rousseau, 1954; Baier, 1958).

The Psychological System Underlying Human Reciprocal Altruism

Anthropologists have recognized the importance of reciprocity in human behavior, but when they have ascribed functions to such behavior they have done so in terms of group benefits, reciprocity cementing group relations and encouraging group survival. The individual sacrifices so that the group may benefit. Recently psychologists have studied altruistic behavior in order to show what factors induce or inhibit such behavior. No attempt has been made to show what function such behavior may serve, nor to describe and interrelate the components of the psychological system affecting altruistic behavior. The purpose of this section is to show that the above model for the natural selection of reciprocally altruistic behavior can readily explain the function of human altruistic behavior and the details of the psychological system underlying such behavior. The psychological data can be organized into functional categories, and it can be shown that the components of the system complement each other in regulating the expression of altruistic and cheating impulses to the selective advantage of individuals. No concept of group advantage is necessary to explain the function of human altruistic behavior.

There is no direct evidence regarding the degree of reciprocal altruism practiced during human evolution nor its genetic basis today, but given the universal and nearly daily practice of reciprocal altruism among humans today, it is reasonable to assume that it has been an important factor in recent human evolution and that the underlying emotional dispositions affecting altruistic behavior have important genetic components. To assume as much allows a number of predictions.

(1) A complex, regulating system. The human altruistic system is a sensitive, unstable one. Often it will pay to cheat: namely, when the partner will not find out, when he will not discontinue his altruism even if he does find out, or when he is unlikely to survive long enough to reciprocate adequately. And the perception of subtle cheating may be very difficult. Given this unstable character of the system, where a degree of cheating is adaptive, natural selection will rapidly favor a complex psychological system in each individual regulating both his own altruistic and cheating tendencies and his responses to these tendencies in others. As selection favors subtler forms of cheating, it will favor more acute abilities to detect cheating. The system that results should simultaneously allow the individual to reap the benefits of altruistic exchanges, to protect himself from gross and subtle forms of cheating, and to practice those forms of cheating that local conditions make adaptive. Individuals will differ not in being altruists or cheaters but in the degree of altruism they show and in the conditions under which they will cheat.

The best evidence supporting these assertions can be found in Krebs' (1970) review of the relevant psychological literature. Although he organizes it differently, much of the material supporting the assertions below is taken from his paper. All references to Krebs below are to this review. Also, Hartshorne and May (1928–1930) have shown that children in experimental situations do not divide bimodally into altruists and "cheaters" but are distributed normally; almost all the children cheated, but they differed in how much and under what circumstances. ("Cheating" was defined in their work in a slightly different but analogous way.)

(2) Friendship and the emotions of liking and disliking.   The tendency to like others, not necessarily closely related, to form friendships and to act altruistically toward friends and toward those one likes will be selected for as the immediate emotional rewards motivating altruistic behavior and the formation of altruistic partnerships. (Selection may also favor helping strangers or disliked individuals when they are in particularly dire circumstances.) Selection will favor a system whereby these tendencies are sensitive to such parameters as the altruistic tendencies of the liked individual. In other words, selection will favor liking those who are themselves altruistic.

Sawyer (1966) has shown that all groups in all experimental situations tested showed more altruistic behavior toward friends than toward neutral individuals. Likewise, Friedrichs (1960) has shown that attractiveness as a friend was most highly correlated among undergraduates with altruistic behavior. Krebs has reviewed other studies that suggest that the relationship between altruism and liking is a two-way street: one is more altruistic toward those one likes and one tends to like those who are most altruistic (e.g., Berkowitz and Friedman, 1967; Lerner and Lichtman, 1968).

Others (Darwin, 1871); Williams, 1966; and Hamilton, 1969) have recognized the role friendship might play in engendering altruistic behavior, but all have viewed friendship (and intelligence) as prerequisites for the appearance of such altruism. Williams (1966), who cites Darwin (1871) on the matter, speaks of this behavior as evolving.

> in animals that live in stable social groups and have the intelligence and other mental qualities necessary to form a system of personal friendships and animosities that transcend the limits of family relationships. (p. 93)

This emphasis on friendship and intelligence as prerequisites leads Williams to limit his search for altruism to the Mammalia and to a "minority of this group." But according to the model presented above, emotions of friendship (and hatred) are not prerequisites for reciprocal altruism but may evolve after a system of mutual altruism has appeared, as important ways of regulating the system.

(3) Moralistic aggression.   Once strong positive emotions have evolved to motivate altruistic behavior, the altruist is in a vulnerable position because cheaters will be selected to take advantage of the altruist's positive emotions. This in turn sets up a selection pressure for a protective mechanism. Moralistic aggression and indignation in humans was selected for in order

(a)   to counteract the tendency of the altruist, in the absence of any reciprocity, to continue to perform altruistic acts for his own emotional rewards;

(b) to educate the unreciprocating individual by frightening him with immediate harm or with the future harm of no more aid; and

(c) in extreme cases, perhaps, to select directly against the unreciprocating individual by injuring, killing, or exiling him.

Much of human aggression has moral overtones. Injustice, unfairness, and lack of reciprocity often motivate human aggression and indignation. Lee (1969) has shown that verbal disputes in Bushmen usually revolve around problems of gift-giving, stinginess, and laziness. DeVore (pers. commun.) reports that a great deal of aggression in hunter-gatherers revolves around real or imagined injustices—inequities, for example, in food-sharing (see, for example, Thomas, 1958; Balikci, 1964; Marshall, 1961). A common feature of this aggression is that it often seems out of all proportion to the offenses committed. Friends are even killed over apparently trivial disputes. But since small inequities repeated many times over a lifetime may exact a heavy toll in relative fitness, selection may favor a strong show of aggression when the cheating tendency is discovered. Recent discussions of human and animal aggression have failed to distinguish between moralistic and other forms of aggression (e.g., Scott, 1958; Lorenz, 1966; Montague, 1968; Tinbergen, 1968; Gilula and Daniels, 1969). The grounds for expecting, on functional grounds, a highly plastic developmental system affecting moralistic aggression are discussed below.

(4) Gratitude, sympathy, and the cost/benefit ratio of an altruistic act.   If the cost/benefit ratio is an important parameter in determining the adaptiveness of reciprocal altruism, then humans should be selected to be sensitive to the cost and benefit of an altruistic act, both in deciding whether to perform one and in deciding whether, or how much, to reciprocate. I suggest that the emotion of gratitude has been selected to regulate human response to altruistic acts and that the emotion is sensitive to the cost/benefit ratio of such acts. I suggest further that the emotion of sympathy has been selected to motivate altruistic behavior as a function of the plight of the recipient of such behavior; crudely put, the greater the potential benefit to the recipient, the greater the sympathy and the more likely the altruistic gesture, even to strange or disliked individuals. If the recipient's gratitude is indeed a function of the cost/benefit ratio, then a sympathetic response to the plight of a disliked individual may result in considerable reciprocity.

There is good evidence supporting the psychological importance of the cost/benefit ratio of altruistic acts. Gouldner (1960) has reviewed the sociological literature suggesting that the greater the need state of the recipient of an altruistic act, the greater his tendency to reciprocate; and the scarcer the resources of the donor of the act, the greater the tendency of the recipient to reciprocate. Heider (1958) has analyzed lay attitudes on altruism and finds that gratitude is greatest when the altruistic act does good. Tesser, Gatewood, and Driver (1968) have shown that American undergraduates thought they would feel more gratitude when the altruistic act was valuable and cost the benefactor a great deal. Pruitt (1968) has provided evidence that humans reciprocate more when the original act was expensive for the benefactor. He shows that under experimental conditions more altruism is induced by a gift of 80 per cent of $1.00 than 20 per cent of $4.00. Aronfreed (1968) has reviewed the considerable evidence that sympathy motivates altruistic behavior as a function of the plight of the individual arousing the sympathy.

(5) Guilt and reparative altruism. If an organism has cheated on a reciprocal relationship and this fact has been found out, or has a good chance of being found out, by the partner and if the partner responds by cutting off all future acts of aid, then the cheater will have paid dearly for his misdeed. It will be to the cheater's advantage to avoid this, and, providing that the cheater makes up for his misdeed and does not cheat in the future, it will be to his partner's benefit to avoid this, since in cutting off future acts of aid he sacrifices the benefits of future reciprocal help. The cheater should be selected to make up for his misdeed and to show convincing evidence that he does not plan to continue his cheating sometime in the future. In short, he should be selected to make a reparative gesture. It seems plausible, furthermore, that the emotion of guilt has been selected for in humans partly in order to motivate the cheater to compensate his misdeed and to behave reciprocally in the future, and thus to prevent the rupture of reciprocal relationships.

Krebs has reviewed the evidence that harming another individual publicly leads to altruistic behavior and concludes:

> Many studies have supported the notion that public transgression whether intentional or unintentional, whether immoral or only situationally unfortunate, leads to reparative altruism. (p. 267)

Wallace and Sadalla (1966), for example, showed experimentally that individuals who broke an expensive machine were more likely to volunteer for a painful experiment than those who did not, but only if their transgression had been discovered. Investigators disagree on the extent to which guilt feelings are the motivation behind reparative altruism. Epstein and Hornstein (1969) supply some evidence that guilt is involved, but on the assumption that one feels guilt even when one behaves badly in private, Wallace and Sadalla's (1966) result contradicts the view that guilt is the only motivating factor. That private transgressions are not as likely as public ones to lead to reparative altruism is precisely what the model would predict, and it is possible that the common psychological assumption that one feels guilt even when one behaves badly in private is based on the fact that many transgressions performed in private are *likely* to become public knowledge. It should often be advantageous to confess sins that are likely to be discovered before they actually are, as evidence of sincerity (see below on detection of mimics).

(6) Subtle cheating: the evolution of mimics.   Once friendship, moralistic aggression, guilt, sympathy, and gratitude have evolved to regulate the altruistic system, selection will favor mimicking these traits in order to influence the behavior of others to one's own advantage. Apparent acts of generosity and friendship may induce genuine friendship and altruism in return. Sham moralistic aggression when no real cheating has occurred may nevertheless induce reparative altruism. Sham guilt may convince a wronged friend that one has reformed one's ways even when the cheating is about to be resumed. Likewise, selection will favor the hypocrisy of pretending one is in dire circumstances in order to induce sympathy-motivated altruistic behavior. Finally, mimicking sympathy may give the appearance of helping in order to induce reciprocity, and mimicking gratitude may mislead an individual into expecting he will be reciprocated. It is worth emphasizing that a mimic need not necessarily be conscious of the deception; selection may favor feeling genuine moralistic aggression even when one has not been wronged if so doing leads another to reparative altruism.

Instances of the above forms of subtle cheating are not difficult to find. For typical instances from the literature on hunter-gatherers see Rasmussen (1931), Balikci (1964), and Lee and DeVore (1968). The importance of these forms of cheating can partly be inferred from the adaptations to detect such cheating discussed below and from the importance and prevalence of moralistic aggression once such cheating is detected.

(7) Detection of the subtle cheater: Trust-worthiness, trust, and suspicion.   Selection should favor the ability to detect and discriminate against subtle cheaters. Selection will clearly favor detecting and countering sham moralistic aggression. The argument for the others is more complex. Selection may favor distrusting those who perform altruistic acts without the emotional basis of generosity or guilt because the altruistic tendencies of such individuals may be less reliable in the future. One can imagine, for example, compensating for a misdeed without any emotional basis but with a calculating, self-serving motive. Such an individual should be distrusted because the calculating spirit that leads this subtle cheater now to compensate may in the future lead him to cheat when circumstances seem more advantageous (because of unlikelihood of detection, for example, or because the cheated individual is unlikely to survive). Guilty motivation, insofar as it evidences a more enduring commitment to altruism, either because guilt teaches or because the cheater is unlikely not to feel the same guilt in the future, seems more reliable. A similar argument can be made about the trustworthiness of individuals who initiate altruistic acts out of a calculating rather than a generous-hearted disposition or who show either false sympathy or false gratitude. Detection on the basis of the underlying psycho-

logical dynamics is only one form of detection. In many cases, unreliability may more easily be detected through experiencing the cheater's inconsistent behavior. And in some cases, third party interactions (as discussed below) may make an individual's behavior predictable despite underlying cheating motivations.

The anthropological literature also abounds with instances of the detection of subtle cheaters (see above references for hunter-gatherers). Although I know of no psychological studies on the detection of sham moralistic aggression and sham guilt, there is ample evidence to support the notion that humans respond to altruistic acts according to their perception of the motives of the altruist. They tend to respond more altruistically when they perceive the other as acting "genuinely" altruistic, that is, voluntarily dispatching an altruistic act as an end in itself, without being directed toward gain (Leeds, 1963; Heider, 1958). Krebs (1970) has reviewed the literature on this point and notes that help is more likely to be reciprocated when it is perceived as voluntary and intentional (e.g., Goranson and Berkowitz, 1966; Lerner and Lichtman, 1968) and when the help is appropriate, that is, when the intentions of the altruist are not in doubt (e.g., Brehm and Cole, 1966; Schopler and Thompson, 1968). Krebs concludes that, "When the legitimacy of apparent altruism is questioned, reciprocity is less likely to prevail." Lerner and Lichtman (1968) have shown experimentally that those who act altruistically for ulterior benefit are rated as unattractive and are treated selfishly, whereas those who apparently are genuinely altruistic are rated as attractive and are treated altruistically. Berscheid and Walster (1967) have shown that church women tend to make reparations for harm they have committed by choosing the reparation that approximates the harm (that is, is neither too slight nor too great), presumably to avoid the appearance of inappropriateness.

Rapoport and Dale (1967) have shown that when two strangers play iterated games of Prisoner's Dilemma in which the matrix determines profits from the games played there is a significant tendency for the level of cooperation to drop at the end of the series, reflecting the fact that the partner will not be able to punish for "cheating" responses when the series is over. If a long series is broken up into subseries with a pause between subseries for totaling up gains and losses, then the tendency to cheat on each other increases at the end of each subseries. These results, as well as some others reported by Rapoport and Chammah (1965), are suggestive of the instability that exists when two strangers are consciously trying to maximize gain by trading altruistic gestures, an instability that is presumably less marked when the underlying motivation involves the emotions of friendship, of liking others, and of feeling guilt over harming a friend. Deutsch (1958), for example, has shown that two individuals playing iterated games of Prisoner's Dilemma will be more cooperative if a third individual, disliked by both, is present. The perceived mutual dislike is presumed to create a bond between the two players.

It is worth mentioning that a classic problem is social science and philosophy has been whether to define altruism in terms of motives (e.g., real vs. "calculated" altruism) or in terms of behavior, regardless of motive (Krebs, 1970). This problem reflects the fact that, wherever studied, humans seem to make distinctions about altruism partly on the basis of motive, and this tendency is consistent with the hypothesis that such discrimination is relevant to protecting oneself from cheaters.

(8) Setting up altruistic partnerships.   Selection will favor a mechanism for establishing reciprocal relationships. Since humans respond to acts of altruism with feelings of friendship that lead to reciprocity, one such mechanism might be the performing of altruistic acts toward strangers, or even enemies, in order to induce friendship. In short, do unto others as you would have them do unto you.

The mechanism hypothesized above leads to results inconsistent with the assumption that humans always act more altruistically toward friends than toward others. Particularly toward strangers, humans may initially act more altruistically than toward friends. Wright (1942) has shown, for example, that third grade children are more likely to give a more valuable toy to a stranger than to a friend. Later, some of these children verbally acknowledged that they were trying to make friends. Floyd (1964) has shown that, after receiving many trinkets from a friend, humans tend to *decrease* their gifts in return, but after receiving many trinkets from a neutral or disliked individual, they tend to *increase* their gifts in return. Likewise, after receiving few trinkets from a friend, humans tend to increase their gifts in return, whereas receiving few trinkets from a neutral or disliked individual results in a decrease in giving. This was interpreted to mean that generous friends are taken for granted (as are stingy non-friends). Generosity from a non-friend is taken to be an overture to friendship, and stinginess from a friend as evidence of a deteriorating relationship in need of repair. (Epstein and Hornstein, 1969, provide new data supporting this interpretation of Floyd, 1964.)

(9) Multiparty interactions.   In the close-knit social groups that humans usually live in, selection should favor more complex interactions than the two-party interactions so far discussed. Specifically, selection may favor learning from the altruistic and cheating experiences of others, helping others coerce cheaters, forming multiparty exchange systems, and formulating rules for regulated exchanges in such multiparty systems.

(i) Learning from others.   Selection should favor learning about the altruistic and cheating tendencies of others indirectly, both through observing interactions of others and, once linguistic abilities have evolved, by hearing about such interactions or hearing characterizations of individuals (e.g., "dirty, hy-

pocritical, dishonest, untrustworthy, cheating louse"). One important result of this learning is that an individual may be as concerned about the attitude of onlookers in an altruistic situation as about the attitude of the individual being dealt with.

(ii) Help in dealing with cheaters.   In dealing with cheaters selection may favor individuals helping others, kin or non-kin, by direct coercion against the cheater or by everyone refusing him reciprocal altruism. One effect of this is that an individual, through his close kin, may be compensated for an altruistic act even after his death. An individual who dies saving a friend, for example, may have altruistic acts performed by the friend to the benefit of his offspring. Selection will discriminate against the cheater in this situation, if kin of the martyr, or others, are willing to punish lack of reciprocity.

(iii) Generalized altruism.   Given learning from others and multiparty action against cheaters, selection may favor a multiparty altruistic system in which altruistic acts are dispensed freely among more than two individuals, an individual being perceived to cheat if in an altruistic situation he dispenses less benefit for the same cost than would the others, punishment coming not only from the other individual in that particular exchange but from the others in the system.

(iv) Rules of exchange.   Multiparty altruistic systems increase by several-fold the cognitive difficulties in detecting imbalances and deciding whether they are due to cheating or to random factors. One simplifying possibility that language facilitates is the formulation of rules of conduct, cheating being detected as infraction of such a rule. In short, selection may favor the elaboration of norms of reciprocal conduct.

There is abundant evidence for all of the above multiparty interactions (see the above references on hunter-gatherers). Thomas (1958), for example, has shown that debts of reciprocity do not disappear with the death of the "creditor" but are extended to his kin. Krebs has reviewed the psychological literature on generalized altruism. Several studies (e.g., Darlington and Macker, 1966) have shown that humans may direct their altruism to individuals other than those who were hurt and may respond to an altruistic act that benefits themselves by acting altruistically toward a third individual uninvolved in the initial interaction. Berkowitz and Daniels (1964) have shown experimentally, for example, that help from a confederate leads the subject to direct more help to a third individual, a highly dependent supervisor. Freedman, Wallington, and Bless (1967) have demonstrated the surprising result that, in two different experimental situations, humans engaged in reparative altruism only if it could be directed to someone other than the individual harmed, or to the original individual only if they did not expect to meet again. In a system of strong multiparty interactions it is possible that in some situations individuals are selected to demonstrate generalized altruistic tendencies and that their main concern when they have harmed another is to show that they are genuinely altruistic, which they best do by acting altruistic without any apparent ulterior motive, e.g., in the experiments by acting altruistic toward an uninvolved third party. Alternatively, A. Rapoport (pers. commun.) has suggested that the reluctance to direct reparative altruism toward the harmed individual may be due to unwillingness to show thereby a recognition of the harm done him. The re-direction serves to allay guilt feelings without triggering the greater reparation that recognition of the harm might lead to.

(10) Developmental plasticity.   The conditions under which detection of cheating is possible, the range of available altruistic trades, the cost/benefit ratios of these trades, the relative stability of social groupings, and other relevant parameters should differ from one ecological and social situation to another and should differ through time in the same small human population. Under these conditions one would expect selection to favor developmental plasticity of those traits regulating altruistic and cheating tendencies and responses to these tendencies in others. For example, developmental plasticity may allow the growing organism's sense of guilt to be educated, perhaps partly by kin, so as to permit those forms of cheating that local conditions make adaptive and to discourage those with more dangerous consequences. One would not expect any simple system regulating the development of altruistic behavior. To be adaptive, altruistic behavior must be dispensed with regard to many characteristics of the recipient (including his degree of relationship, emotional makeup, past behavior, friendships, and kin relations), of other members of the group, of the situation in which the altruistic behavior takes place, and of many other parameters, and no simple developmental system is likely to meet these requirements.

Kohlberg (1963), Bandura and Walters (1963), and Krebs have reviewed the developmental literature on human altruism. All of them conclude that none of the proposed developmental theories (all of which rely on simple mechanisms) can account for the known diverse developmental data. Whiting and Whiting (in prep.) have studied altruistic behavior directed towards kin by children in six different cultures and find consistent differences among the cultures that correlate with differences in child-rearing and other facets of the cultures. They argue that the differences adapt the children to different adult roles available in the cultures. Although the behavior analyzed takes place between kin and hence Hamilton's model (1964) may apply rather than this model, the Whitings' data provide an instance of the adaptive value of developmental plasticity in altruistic behavior. No careful work has been done analyzing the influence of environmental factors on the development of altruistic behavior, but some data exist. Krebs has reviewed the evidence that altruistic tendencies can be increased by the effects of warm, nurturant models, but little is known on how long such effects endure. Rosenhan (1967) and Rettig (1956) have shown a correlation between altruism in parents and altruism in their college-age children, but these studies do not separate genetic and environmental influences. Class differences in altruistic behavior (e.g., Berkowitz, 1968; Ugurel-Semin, 1952; Almond and Verba, 1963) may primarily reflect environmental influences. Finally, Lutzker (1960) and Deutsch (1958) have shown that one can predict the degree of altruistic behavior displayed in iterated games of Prisoner's Dilemma from personality typing based on a questionnaire. Such personality differences are probably partly environmental in origin.

It is worth emphasizing that some of the psychological traits analyzed above have applications outside the particular reciprocal altruistic system being discussed. One may be suspicious, for example, not only of individuals likely to cheat on the altruistic system, but of any individual likely to harm oneself; one may be suspicious of the known tendencies toward adultery of another male or even of these tendencies in one's own mate. Likewise, a guilt-motivated show of reparation may avert the revenge of someone one has harmed, whether that individual was harmed by cheating on the altruistic system or in some other way. And the system of reciprocal altruism may be employed to avert possible revenge. The Bushmen of the Kalahari, for example, have a saying (Marshall, 1959) to the effect that, if you wish to sleep with someone else's wife, you get him to sleep with yours, then neither of you goes after the other with poisoned arrows. Likewise, there is a large literature on the use of reciprocity to cement friendships between neighboring groups, now engaged in a common enterprise (e.g., Lee and DeVore, 1968).

The above review of the evidence has only begun to outline the complexities of the human altruistic system. The inherent instability of the Prisoner's Dilemma, combined with its importance in human evolution, has led to the evolution of a very complex system. For example, once moralistic aggression has been selected for to protect against cheating, selection favors sham moralistic aggression as a new form of cheating. This should lead to selection for the ability to discriminate the two and to guard against the latter. The guarding can, in turn, be used to counter real moralistic aggression: one can, in effect, *impute* cheating motives to another person in order to protect one's own cheating. And so on. Given the psychological and cognitive complexity the system rapidly acquires, one may wonder to what extent the importance of altruism in human evolution set up a selection pressure for psychological and cognitive powers which partly contributed to the large increase in hominid brain size during the Pleistocene.

ACKNOWLEDGMENTS

I thank W. H. Drury, E. Mayr, I. Nisbet, E. E. Williams and E. O. Wilson for useful comments on earlier drafts of this paper, and I thank especially I. DeVore and W. D.

Hamilton for detailed comment and discussion. I thank A. Rapoport and D. Krebs for access to unpublished material. This work was completed under a National Science Foundation pre-doctoral fellowship and partially supported by grant number NIMH 13156 to I. DeVore.

LIST OF LITERATURE

Almond, G. A., and S. Verba. 1963. *The Civic Culture*. Princeton University Press, Princeton, N.J.

Aronfreed, J. 1968. *Conduct and Conscience*. Academic Press, N.Y.

Ashmole, M. 1962. Migration of European thrushes: a comparative study based on ringing recoveries. *Ibis*, 104: 522–559.

Baier, K. 1958. *The Moral Point of View*. Cornell University Press, Ithaca, N.Y.

Balikci, A. 1964. Development of basic socio-economic units in two Eskimo communities. National Museum of Canada Bulletin No. 202, Ottawa.

Bandura, A., and R. H. Walters. 1963. *Social Learning and Personality Development*. Holt, Rhinehart and Winston, N.Y.

Berkowitz, L. 1968. Responsibility, reciprocity and social distance in help-giving: an experimental investigation of English social class differences. *J. Exp. Soc. Psychol.*, 4: 664–669.

Berkowitz, L., and L. Daniels. 1964. Affecting the salience of the social responsibility norm: effects of past help on the response to dependency relationships. *J. Abnorm. Soc. Psychol.*, 68: 275–281.

Berkowitz, L., and P. Friedman. 1967. Some social class differences in helping behavior. *J. Personal. Soc. Psychol.*, 5: 217–225.

Berndt, R., and H. Sternberg. 1968. Terms, studies and experiments on the problems of bird dispersion. *Ibis*, 110: 256–269.

Berscheid, E. and E. Walster. 1967. When does a harm-doer compensate a victim? *J. Personal. Soc. Psychol.*, 6: 435–441.

Blest, A. D. 1963. Longevity, palatability and natural selection in five species of New World Saturuiid moth. *Nature*, 197: 1183–1186.

Brehm, J. W., and A. H. Cole, 1966. Effect of a favor which reduces freedom. *J. Personal. Soc. Psychol.*, 3: 420–426.

Brower, J.V.Z., and L. P. Brower. 1965. Experimental studies of mimicry. 8. *Am. Natur.*, 49: 173–188.

Campbell, B. 1966. *Human Evolution*. Aldine, Chicago.

Crook, J. H. 1969. The socio-ecology of primates. In J. H. Crooke (ed.), *Social Behavior in Birds and Mammals*, p. 103–166. Academic Press, London.

Darlington, R. B., and C. E. Macker. 1966. Displacement of guilt-produced altruistic behavior. *J. Personal. Soc. Psychol.*, 4: 442–443.

Darwin, C. 1871. *The Descent of Man and Selection in Relation to Sex*. Random House, N.Y.

Deutsch, M. 1958. Trust and suspicion. *J. Conflict Resolution*, 2: 267–279.

Dhont, A. A., and J. Hublé. 1968. Fledging-date and sex in relation to dispersal in young Great Tits. *Bird Study*, 15: 127–134.

Eibl-Eibesfeldt, I. 1955. Über Symbiosen, Parasitismus and andere besondere zwischenartliche Bezichungen tropischer Meeresfische. *Z. f. Tierpsychol.*, 12: 203–219.

———. 1959. Der Fisch *Aspidontus taeniatus* als Nachahmer des Putzers *Labroides dimidiatus*. *Z. f. Tierpsychol.*, 16: 19–25.

Epstein, Y. M., and H. A. Horstein. 1969. Penalty and interpersonal attraction as factors influencing the decision to help another person. *J. Exp. Soc. Psychol.*, 5: 272–282.

Feder, H. M. 1966. Cleaning symbioscs in the marine environment. In S. M. Henry (ed.), *Symbiosis*, Vol. 1, p. 327–380. Academic Press, N.Y.

Fisher, R. A. 1958. *The Genetical Theory of Natural Selection.* Dover, N.Y.

Floyd, J. 1964. Effects of amount of award and friendship status of the other on the frequency of sharing in children. Unpublished doctoral dissertation. University of Minnesota. (Reviewed in Krebs, 1970).

Freedman, J. L., S. A. Wallington, and E. Bless. 1967. Compliance without pressure: the effect of guilt. *J. Personal. Soc. Psychol.*, 7: 117–124.

Friedrichs, R. W. 1960. Alter versus ego: an exploratory assessment of altruism. *Am. Sociol. Rev.*, 25: 496–508.

Gilula, M. F., and D. N. Daniels. 1969. Violence and man's struggle to adapt. *Science*, 164: 395–405.

Goranson, R., and Berkowitz. 1966. Reciprocity and responsibility reactions to prior help. *J. Personal. Soc., Psychol.* 3: 227–232.

Gouldner, A. 1960. The norm of reciprocity: a preliminary statement. *Am. Sociol. Rev.*, 47: 73–80.

Haldane, J.B.S. 1955. Population genetics. *New Biology*, 18: 34–51.

Hall, K.R.L., and I. DeVore. 1965. Baboon social behavior. In I. DeVore (ed.), *Primate Behavior: Field Studies of Monkeys and Apes*, p. 53–110. Holt, Rhinehart and Winston, N.Y.

Hamilton, W. D. 1964. The genetical evolution of social behavior. *J. Theoret. Biol.*, 7: 1–52.

———. 1966. The moulding of senescence by natural selection. *J. Theoret. Biol.*, 12: 12–45.

———. 1969. Selection of selfish and altruistic behavior in some extreme models. Paper presented at "Man and Beast Symposium" (in press, Smithsonian Institution).

Hartshorne, H., and M. A. May. 1928–1930. *Studies in the Nature of Character. Vol. 1, Studies in Deceit; Vol. 2, Studies in Self-Control; Vol. 3, Studies in the Organization of Character.* Macmillan, N.Y.

Hediger, H. 1968. Putzer-fische im aquarium. *Natur und Museum*, 98: 89–96.

Heider, F. 1958. *The Psychology of Interpersonal Relations.* Wiley, N.Y.

Hobson, E. S. 1969. Comments on certain recent generalizations regarding cleaning symbioses in fishes. *Pacific Science*, 23: 35–39.

Kohlberg, L. 1963. Moral development and identification. In H. W. Stevenson (ed.), *Yearbook of the National Society for the Study of Education. Part 1. Child Psychology*, p. 277–332. University of Chicago Press, Chicago.

Krebs, D. 1970. Altruism—an examination of the concept and a review of the literature. *Psychol. Bull.*, 73: 258–302.

Kuyton, P. 1962. Verhalten-beobachtungen an der Raupe des Kaiseratlas. *Z. d. Entomol.*, 72: 203–207.

Lee, R. 1969. !Kung Bushman violence. Paper presented at meeting of American Anthropological Association, Nov. 1969.

Lee, R., and I. DeVore, 1968. *Man the Hunter.* Aldine, Chicago.

Leeds, R. 1963. Altruism and the norm of giving. *Merrill-Palmer Quart.*, 9: 229–240.

Lerner, M. J., and R. R. Lichtman. 1968. Effects of perceived norms on attitudes and altruistic behavior toward a dependent other. *J. Personal. Soc. Psychol.*, 9: 226–232.

Leyhausen, P. 1965. Über die Funktion der relativen Stimmungshierarchie (dasgestellt am Beispiel der phylogenetischen und ontogenetischen Entwicklung des Beutefangs von Raubtieren. *Z. f. Tierpsychol.*, 22: 412–494.

Limbaugh, C. 1961. Cleaning symbioses. *Scient. Am.*, 205: 42–49.

Limbaugh, C., H. Pederson, and F. Chase, 1961. Shrimps that clean fishes. *Bull. Mar. Sci. Gulf Caribb.*, 11: 237–257.

Lorenz, K. 1966. *On Aggression.* Harcourt, Brace and World, N.Y.

Luce, R. D., and H. Raiffa. 1957. *Games and Decisions.* Wiley, N.Y.

Lutzker, D. 1960. Internationalism as a predictor of cooperative game behavior. *J. Conflict Resolution*, 4: 426–435.

Marler, P. 1955. The characteristics of certain animal calls. *Nature*, 176: 6–7.

———. 1957. Specific distinctiveness in the communication signals of birds. *Behavior*, 11: 13–39.

Marshall, L. K. 1959. Marriage among !Kung Bushmen. *Africa*, 29: 335–365.

———. 1961. Sharing, talking and giving: relief of social tension among !Kung Bushmen. *Africa*, 31: 231–249.

Maynard, E.C.L. 1968. Cleaning symbiosis and oral grooming on the coral reef. In P. Person (ed.), *Biology of the Month*, p. 79–88. Philip Person, American Association for the Advancement of Science, Wash., D.C.

Maynard Smith, J. 1964. Kin selection and group selection. *Nature*, 201: 1145–1147.

———. 1965. The evolution of alarm calls. *Amer. Natur.*, 99: 59–63.

Mayr, E. 1963. *Animal Species and Evolution.* Belknap Press, Cambridge.

Montagu, F.M.A. 1968. *Man and Aggression.* Oxford University Press, N.Y.

Murie, A. 1944. *The Wolves of Mount McKinley.* Fauna of National Parks, Fauna Series #5; Wash., D.C.

Owen, D. F. 1963. Similar polymorphisms in an insect and a land snail. *Nature*, 198: 201–203.

Perdeck, A. 1958. Two types of orientation in migrating starlings, *Starnus vulgaris* L., and chaffishes, *Fringilla coeleus* L., as revealed by displacement experiments. *Ardea*, 46: 1–35.

Pruitt, D. G. 1968. Reciprocity and credit building in a laboratory dyad. *J. Personal. Soc. Psychol.*, 8:143–147.

Randall, J. E. 1958. A review of the Labrid fish genus *Labroides* with descriptions of two new species and notes on ecology. *Pacific Science*, 12: 327–347.

———. 1962. Fish service stations. *Sea Frontiers*, 8: 40–47.

Rapoport, A., and A. Chammah. 1965. *Prisoner's Dilemma.* University of Michigan Press, Ann Arbor.

Rapoport, A., and P. Dale. 1967. The "end" and "start" effects in iterated Prisoner's Dilemma. *J. Conflict Resolution*, 10: 363–366.

Rasmussen, K. 1931. The Netsilik Eskimos: social life and spiritual culture. Report of the Fifth Thule Expedition 1921–1924, Vol. 8(1,2). Gyldendalske Boghandel, Copenhagen.

Rettic, S. 1956. An exploratory study of altruism. *Dissert. Abstr.*, 16: 2220–2230.

Reynolds, L. 1966. Open groups in hominid evolution. *Man*, 1: 441–452.

Rosenhan, D. 1967. *The Origins of Altruistic Social Autonomy.* Educational Testing Service, Princeton, N.J.

Rousseau. J. J. 1954. *The Social Contract.* Henry Regnery Co., Chicago.

Rudebeck, G. 1950. The choice of prey and modes of hunting of predatory birds with special reference to their selective effort. *Oikos*, 2: 65–88.

———. 1951. The choice of prey and modes of hunting of predatory birds, with special reference to their selective effort *(cont.) Oikos*, 3: 200–231.

Sawyer, J. 1966. The altruism scale: a measure of cooperative, individualistic, and competitive interpersonal orientation. *Am. J. Social.* 71: 407–416.

Schopler, J., and V. T. Thompson. 1968. The role of attribution process in mediating amount of reciprocity for a favor. *J. Personal. Soc. Psychol.*, 10: 243–250.

Scott, J. P. 1958. *Aggression.* University of Chicago Press, Chicago.

Southern, H. N. 1954. Tawny owls and their prey. *Ibis*, 96: 384–410.

Struhsaker, T. 1967. Social structure among vervet monkeys *(Cercopithecus aethiops). Behavior*, 29: 83–121.

Tesser, A., R. Gatewood, and M. Driver. 1968. Some determinants of gratitude. *J. Personal. Soc. Psychol.*, 9: 232–236.

Thomas, E. M. 1958. *The Harmless People.* Random House, N.Y.

Tinbergen, L. 1960. The dynamics of insect and bird populations in pine woods. 1. Factors influencing the intensity of predation by song birds, *Arch. Neerl. de Zool.*, 13: 265–343.

Tinbergen, N. 1968. On war and peace in animals and man. *Science*, 160: 1411–1418.

Ugurel-Semin, R. 1952. Moral behavior and moral judgment of children. *J. Abnorm. Soc. Psychol.*, 47: 463–474.

Van Lawick-Goodall, J. 1968. A preliminary report on expressive movements and communication in the Gombe Stream chimpanzees. In P. Jay (ed.), *Primates*, p. 313–374 Holt, Rinehart and Winston, N.Y.

Wallace, J., and E. Sadalla. 1966. Behavioral consequences of transgression: the effects of social recognition. *J. Exp. Res. Personal.*, 1: 187–194.

Wickler, W. 1968. *Mimicry in Plants and Animals.* McGraw-Hill, N.Y.

Williams, G. C. 1966. *Adaptation and Natural Selection.* Princeton University Press, Princeton, N.J.

Wright, B. 1942. Altruism in children and the perceived conduct of others. *J. Abnorm. Soc. Psychol.*, 37:218–233.

Wynne-Edwards, V. C. 1962. *Animal Dispersion in Relation to Social Behavior.* Hafner, N.Y.

Postscript

About the time that my reciprocal altruism paper was published I ran into a friend at the Museum of Comparative Biology at Harvard, Allen Greer, an Australian herpetologist who was a couple of years ahead of me as a graduate student. He asked me what I was up to now that my reciprocal altruism paper was done. I told him parental investment and sexual selection. He said, "No, no—that's not the way science is done. You should now perform an experiment or two on reciprocal altruism and then later write a review paper, maybe hold a conference. You should make the area your own and stay on top of it, so to speak." My reply was that I was not really interested in reciprocal altruism per se but in social theory based on natural selection, of which reciprocal altruism was one part. In any case, I am certainly glad that I never took the road he suggested, because I believe that I have not had a fresh thought on the subject since! On the other hand, I have enjoyed a response to my paper that has greatly enriched the subject both empirically and theoretically.

One of the most agreeable features of writing a paper that is widely read by others is that, soon enough, you get an empirical and/or theoretical response, which enlarges your own understanding of the subject. The empirical response was almost immediate in coming. I had wanted, in my original paper, to include the possibility of baboon reciprocal altruism because I had learned from my mentor, Professor DeVore, about his famous "central hierarchy." In baboons it was known that adult males in a group arrived separately from elsewhere and thus are unlikely to be closely related, yet three or four adult males often seem to act as a unit against other adult males in certain kinds of interactions. Support is especially dramatic when one of the central males is in consort with a single female and this relationship is challenged by a male who may be individually more dominant but who is outside the so-called central hierarchy. In this case another member of the central group may rush in and help the male in consort retain the consortship. But there was, in fact, no evidence of reciprocity, only the supposition that such reciprocity was likely in nature, so the situation seemed too weak to merit even comment in my paper. But in 1977 Craig Packer published evidence that in baboon troops one male's frequency of soliciting another male's help in such situations was positively correlated with the other male's tendency to solicit help from him. This did not quite demonstrate reciprocal altruism, but rather reciprocal solicitation toward altruism, but that was a very welcome start. (For the most recent work on reciprocal altruism in primates, see de Waal 1997a, b, and de Waal and Berger 2000; for revenge in primates, see Aureli et al 1992; for reconciliation, see Aureli and van Schaik 1991.)

Similar advances came soon enough as well. Leigh (1984) showed that hermaphroditic sea bass were reciprocal in their couplings. When they mate one acts as a female, releasing eggs, and the other acts as a male, releasing sperm. Since the latter is much less expensive than the former, selection would favor hermaphrodites that spend most of their time being males, if they can get away with it. Instead, reciprocal egg trading has evolved. Fish that have as many as two hundred eggs to spawn in an afternoon release them not all at once but instead in small numbers, say, three to five at a time, and wait for the reciprocal act before continuing the relationship. A nice feature of this work was that two species were compared that differed in their degree of reciprocity. One had a high tendency to reverse roles, and the other had a significant tendency to reverse roles but not as often as the first. Reciprocal egg trading permits the hermaphrodite over evolutionary time to save energy on investment in testes and, sure enough, both species have smaller testis/ovary size ratios than do mass spawners, and the more reciprocal species has the smallest testis/ovary size ratio.

Axelrod and Hamilton (1981) Arrives by Mail

In 1981 I was a professor at the University of California and somewhat out of it in more ways than one. I was not keeping up with the new journals, and so it came as a welcome surprise to receive one day in the mail a reprint of Axelrod and Hamilton (1981) from Hamilton. The paper was inscribed, "To Bob, Tit-for-Tat or Hamilton's Revenge, Bill." The paper described how a strategy for playing iterated games of Prisoner's Dilemma could win out against competing strategies under very robust conditions. This strategy was called tit-for-tat. In a computer tournament that Axelrod had run, with strategies submitted by invitation, the simplest strategy had turned out to be the winner. Cooperate on the first move and do whatever your partner did on the previous move. So, if your partner also cooperated you would continue to cooperate, but if your partner failed to cooperate you would then cease cooperating yourself, tit-for-tat. My first reaction to the inscription was that this paper was the tit for the tat of Trivers and Hare (1976) (see chap. 5). That is, I had stolen a portion of Hamilton's thunder on the haplodiploid Hymenoptera and he was now returning the favor on reciprocal altruism. But Bill has a very subtle mind and my first wife, Lorna, was the first to draw my attention to the frequency with which Bill's statements had double and even triple meanings. In this regard, his mental operations reminded me of the dreaded knight on the chess board, which moves in a forking motion, simultaneously attacking to the left and the right. Earlier I had on the one hand prevented Bill from publishing his first formulation of game theory applied to reciprocal altruism (when he asked permission to include it in his "Men and Beast" paper), and then had turned around and stolen it back (to be sure while citing Hamilton), and this was my tat for that tit, as well. In any case, I sat down at around eight o'clock at night, turned on some classical music, read the paper, and, as I later wrote Bill, "my heart soared." For one wild moment, I kidded him, I actually believed that there was progress in science! What Axelrod and Hamilton had done was to *prove* that in these repeated games of the Prisoner's Dilemma, there were only two evolutionary stable strategies, tit-for-tat and perpetual defection, or noncooperation. To me the paper had almost biblical proportions. That is, you could see how a kind of social heaven and social hell could evolve right here on earth. The social hell was perpetual isolation, perpetual inability to link up with others in a positive way, never being cheated by others to be sure, but at the cost of eternal loneliness. The social heaven was not heavenly in some naïve way, dancing around the mulberry bush together without regard for selfish possibilities. Instead, cooperation required perpetual vigilance to enjoy its fruits, but tit-for-tat, a very simple mechanism that could apply even to bacteria, could bring about this cooperative world. Even the first sentence of the paper set the subsequent tone: "The benefits of life are disproportionately available to cooperating creatures."

It was the simplicity of the rule that was so beguiling. When Axelrod had first held his computer tournament, he solicited an entry from me. That is, I was offered the opportunity to send in my own strategy, which would then compete against strategies submitted by others, the computer doing the dirty work. I am embarrassed to say that my first reaction was that the matter would be complicated and that some kind of complex formula for responding to the play of your partner would succeed. And, since at the time I also lacked any ability to work on a computer, the invitation to submit my entry in computer form also stood as a barrier. So I sent in nothing.

Axelrod and Hamilton, of course, had the same theoretical bias that I had, which was to try to state the argument in as general a form as possible and to apply it as broadly in nature as one could. The simplicity of the tit-for-tat strategy bypassed, in one step, the cognitive complexity that was often assumed to be required to get reciprocity going in our own species. A single-celled organism, as they pointed out, that could respond to chemicals produced by neighbors was, in principle, in a position to perform a tit-for-tat strategy, producing a cooperative chemical on the first move and an uncooperative one in response to an uncooperative one from the neighbor. There was, however, one moment in the paper where they pushed the argument one step beyond what I could then envision. Since this involved genetics, it almost certainly came from Hamilton (so I will now write as if it did). He argued that the paired chromosomes in germ cells such as oocytes might be able to respond to a selfish maneuver by returning a selfish maneuver in kind. In particular, he was thinking about meiotic drive in which one chromosome gains an advantage in reproduction at the expense of its paired chromosome. In this case, one chromosome may cause itself, for example, to land in the egg cell instead of the polar body at the relevant division, improving over the 50/50 chance that the fair rules of meiosis were supposed to produce. If I understood them right, they were arguing that if this defection occurred in one oocyte, a paired chromosome in a neighboring oocyte might be "aware" of it and do the same itself. This could easily result in both chromosomes ending up in the egg cell with the unfortunate result that the offspring would be trisomic, that is, have three copies of the chromosome (such as is found in Down syndrome, trisomy 21). I could not imagine that the neighboring oocyte could get the information that this argument required.

I remember the sensation vividly as if it were yesterday. It was as if Hamilton and I were exploring unknown territory together. I was, once again, following Hamilton through some tangled undergrowth, and he was confidently hacking his way through with a machete and assuring me that he knew just what lay ahead. But the terrain was becoming steeper and more dangerous as we descended, Bill still confidently telling me that the water hole or the river, or whatever it was that we were looking for, was straight ahead. But finally I was too frightened to continue and called out, "This far, Bill, and not *one* step farther." Many, many people have had that sensation far earlier in their explorations with Bill and have turned back, to their own disadvantage. But that is the only time that I can remember, in reading any of Bill's works, where I drew back and said, in effect, "I am not going there with you, Bill, I am turning back." (For a recent treatment of the possibility of meiotic drive, see Day and Taylor 1998.) I remember once being startled in 1986 when I heard a group of graduate students at the University of Michigan discussing Bill. He was on the faculty then but out of town. One fellow said, "The thing with Hamilton is that 20% of what he says is brilliant and 80%, off the wall, but you don't know which is which so you have to pay attention to all of it." I thought, "More like 99% and 1% and be real careful about the 1%!" (Bill, on the other hand, agreed with the student and pointed out that the remark was not based on his published papers but on his everyday comments.)

Axelrod and Hamilton had made a key advance. In fact, they had done exactly what Ed Wilson proposed that I should do at the beginning, namely, put the subject on a firm mathematical foundation. But this paper, to me, was also just a start. That is, the Prisoner's Dilemma is a very simplified version of natural interactions, and one still needs to model the kind of complexity that naturally evolves in such systems in species like ourselves and our close relatives. It has been gratifying to see, especially recently, that a series of papers have come out relaxing the rules of the Prisoner's Dilemma, allowing punishment, for example (Frank 1995; see also Clutton-Brock and Parker 1995), and forgiveness (Godfray 1992), permitting a gradation of responses so as to model the phenomenon of subtle cheaters (Roberts and Sherratt 1999), and modeling three-party interactions involving observer effects (Nowak and Sigmund 1998a,b). Additional subtleties can be found in Frean (1994), Roberts (1998), McNamara, Gasson, and Houston (1999), Bendor and Swistak (1995, 1997), de Vos and Zeggelink (1997), and Binmore (1998). Linking reciprocal altruism to game theory has the additional virtue of linking the problem to a whole series of more complicated games. For possible reciprocal altruism in whales, see Connor, Heithaus, and Barre (1999). For application to predator inspection in guppies, see Godin and Davis (1995); to evolution of an RNA virus, Turner and Chao (1999).

2

PARENTAL INVESTMENT AND REPRODUCTIVE SUCCESS

I had a wonderful teacher, a man named William Drury. I met him when he was hired to oversee some booklets I was to write on animals and animal behavior for a fifth grade course of study. In effect, for two years I had a private tutor in biology, before I ever took a formal course, paid by the hour (and at a good rate) by my employer. No need to feel any guilt as you consumed yet another hour of your teacher's time! I had graduated from college and was working for a curriculum company, putting together the new social sciences for fifth graders, just as the new mathematics and new physics had swept their curricula. We were generating a brand new kind of curriculum, including the latest footage on baboon behavior from East Africa, and the behavior of modern hunter-gatherers, such as the Netsilic Eskimos, and the !Kung of the Kalahari Desert in southern Africa. When, after six weeks working for the company, they discovered that I knew nothing about either animals or humans (in the sense of anthropology or sociology), they assigned me to write about animals, since they cared less about that material. Dr. Drury was hired as a professional biologist to help me with references and to sign off on the quality of my booklets. He was at that time the research director at the Massachusetts Audubon Society in Lincoln, Massachusetts.

Bill Drury soon taught me that natural selection referred to *individual* reproductive success, and that thinking along the lines of species advantage and group selection had little going for it. He also introduced me to animal behavior and taught me many facts about the social and psychological lives of other creatures. He taught me his biases, and they were, by and large, biases I was only too happy to be taught. I was once watching a herring gull through binoculars side by side with him. In those days, a herring gull could not scratch itself without one of us asking why natural selection favored *that*

behavior. In any case, I offered as an explanation for the ongoing gull behavior something that was nonfunctional and had the hidden supposition
that the animal was not capable of acting in its own self-interest. Bill said
quietly, "Never assume the animal you are studying is as stupid as the one
studying it." That had a very bracing effect on a young mind. On another
occasion we were discussing racial prejudice and possible biological components thereof, and he said to me, "Bob, once you've learned to think of a
herring gull as an equal, the rest is easy." What a welcome approach to the
problem, especially from within biology!

I well remember the thrill when I first learned the whole system of evolutionary logic as applied to biology from Dr. Drury. It was similar to the
feeling I had when I first fell in love with astronomy as a twelve-year-old.
Astronomy gave you inorganic creation and evolution over a 15-billion-year
period of time. Evolutionary logic gave you the comparable story over 4
billion years and evolutionary logic applied to *life*. In both cases, I felt a sense
of religious awe. Astronomy spoke of the vastness of time and space while
evolutionary biology did the same thing for living creatures. The living world
was not created 6 thousand years ago, in one blinding flash of creation, or
in seven days, perhaps. Living creatures have been forming over a 4-billion-
year period of time, with natural selection knitting together adaptive traits
over time. Living creatures are expected to be organized functionally in
exquisite and even counterintuitive forms. In no way did this perspective
diminish my sense of awe, nor did it argue, one way or another, for the
existence of an omnipotent force to which personal attention was suggested.

I do not think I fully appreciated the immense value Bill Drury had been
to me until he was dead. Indeed, in a maudlin act of self-pity I cried out
one night, "Where are you now, Bill, when *I* need you most?" and wept
bitter tears—tears for a great and wonderful teacher, lovingly remembered,
and tears for myself, forced now, at last, to navigate biology's waters alone,
or at least without a trusted guide who had a deep knowledge of animal
behavior, ecology, botany, and geology, a man who lacked the linguistic
talents I had and probably some of the logical abilities, as well. You merely
had to mine him for all he knew and then reorganize whenever possible
along appropriate lines of logic. The logic might be pretty obvious to Drury,
yet he certainly had a more limited capacity to organize multiple lines of
logic toward a single end. He indeed may have been dyslexic in the formal
sense. I know I once traveled to Lincoln on the train while reading one of
his ecological chapters, trying to reorganize the fragmented material into
coherent wholes, and sailed right on through to Concord before I realized
I had missed my stop. He was a superb artist, naturalist, and bird mimic.
(And, he had a cephalic index that must have approached exactly 1, that
is, his head was as wide as it was long and, indeed, seemed perfectly round.)
Drury was also a wonderful teacher, one on one and in small groups, but

he had great difficulty lecturing to larger groups. I describe him in this chapter because I think his effect on my thinking was probably greatest where sex differences, sexual selection, and courtship were concerned. He had a large knowledge of bird behavior, both personal and acquired through a study of the works of others, a rich trove of information against which to test my developing ideas (for his thinking in ecology, see Drury 1998.)

Go Thou to the Pigeon

After I had spent a year or so watching herring gulls and other sea birds with Bill Drury, I wanted to start a project on a species of my own, a species that I could study on land. I believe I suggested the lesser marshwren, so to speak, that is, a species whose social behavior and ecology had not yet been studied, though something would have been known about a closely related species. Drury immediately batted down that idea. He said it would take me eighteen months to find the speices on a regular basis and another eighteen months to acclimate individuals sufficiently to me to permit detailed behavioral observations. That it had not yet been studied, he said, might better be taken as a warning than as an invitation. He suggested I go in the other direction—study the pigeon, he said. They were everywhere in Cambridge and too ugly (due in part to earlier domestication) and common to attract any ornithologist since Whitman's 1919 monograph. The variability in feather patterns that helped make them ugly also made individuals easy to identify, so behavioral observations of known individuals could begin right away without the need to capture and handle the birds. As it turned out, at the North Cambridge third-floor apartment where I was living, there were in fact pigeons that roosted on the roof of the house next door that could provide a steady stream of behavioral observations right through the night!

What soon became clear in this monogamous species was that males were sexually much more insecure than were females, and males acted to deprive their mates of what they would be perfectly happy to indulge in themselves, that is, an extra-pair copulation. For example, the group outside my window began with four pigeons—two mated couples. They slept next to each other in the gutter of the roof of the house next door. They often settled on the roof any time after four o'clock in the afternoon. When spending the night together as a foursome, the two males, although they were the more aggressive sex, always sat next to each other with each one's mate on the outside. By sitting next to each other, the males could ensure that each one was sitting between his mate and the other male.

Then, for a period of several days, a new male arrived and was regularly attacked by each of the two resident males and driven off. Finally, after four or five days of persistence, such a male might still be sleeping twenty yards down the gutter from the other four pigeons, and subject to attack without notice. The very day he arrived with his own mate, however, the distance to the other birds was cut in half, suggesting that male concern about male visitors might be associated with some sexual threat or increased chance that his mate would indulge in an extra-pair copulation. More striking still, when the third couple managed to join the other two, it was no longer possible for each male to sit between his own mate and all other males. What happened then was that the outermost males kept their mates on the outside, thus sitting between their mate and the other two males, but the innermost male forced his female onto the sloping roof in front of them, rather than allow her to sit between him and his neighbor to the right! The female was not happy with this situation and would return to the more comfortable (and warmer) gutter, only to be forced back onto the sloping roof. Sometimes she would wait for him to fall asleep and would slip down beside him unnoticed, but I would soon hear roo-koo-kooing out my bathroom window and would rush to see her pushed back onto the roof. This, for me, was a surprising observation because it put the lie to the notion, so common in ornithology and evolutionary thinking at the time, that the monogamous relationship was one without internal conflict. Here was a male willing to force his own mate, mother of his offspring-to-be, up onto the sloping roof all night long because of his sexual insecurities. This suggested relatively strong selection pressures.

Whitman (1919) reported a sex difference in behavior upon viewing the partner in adultery that I thought was instructive along these same lines. Whitman said that when a male pigeon saw his female about to begin copulating with another male, he flew straight at the second male, attempting to knock him off her, that is, he interrupted the copulation as soon as possible. By contrast, a female seeing her own mate involved in the same behavior would not attempt to stop the copulation but would intervene immediately afterward, separate the couple, and act to keep the other female away from her mate. What was going on here? The obvious answer was suggested by the relative investment of the two sexes in the offspring, certainly at the time of copulation. The male's investment at copulation is trivial, or relatively minor, but the female's investment may be associated with a year's worth of reproductive effort. Thus, males chosen as extra-pair partners by females enjoy the possibility of a large immediate benefit (paternity of offspring who will be reared by the female with the help of another male) and similarly inflict a large cost on the "cuckolded" or genetically displaced male. These large potential selective effects would explain both a male's eagerness to indulge in such extra-pair copulations and his anxiety that his own mate might act similarly!

Ernst Mayr Gives Me the Key Reference

The key reference in my 1972 paper on parental investment and sexual selection was a paper by A. J. Bateman published in 1948 and, until my paper, completely overlooked in the literature on sexual selection. Even the very best, G. C. Williams, for example, was unaware of Bateman's paper. I became aware of it in the following way. I was taking a reading course in genetics from Professor Mayr. When you are admitted as a graduate student at Harvard, you take a series of tests to find out what subdisciplines you do not know, and I was found deficient in, among other things, genetics. Professor Mayr, however, detested the Harvard course on genetics at the time as being purely molecular, teaching no population genetics, for example, and thus being of limited utility to evolutionists. As chairman of my prescription committee, he therefore got everyone to agree to only "prescribe" me a "knowledge of genetics" instead of insisting that I take Harvard's genetics course.

Thus, when I wondered who might teach me a course in genetics, Professor Mayr himself loomed as an attractive candidate, very busy, to be sure, but someone who might have a difficult time ducking my request given his prior actions. In any case, Dr. Mayr agreed to teach me a reading course in genetics, assigned a book by Whitehouse called *An Introduction to Genetics*, and told me to come see him whenever I had questions or wished to discuss what I was reading. I remember that the first time I showed up I actually talked genetics with him. But on the next visit, having done no genetics reading but being consumed by my pigeon observations, I came to his office and decided to tell him pigeon stories instead.

The first session on pigeons went well. I kept Dr. Mayr's interest with my pigeon anecdotes and with my then, I think, still primitive efforts at developing a coherent theory based on relative parental investment. In any case, we had chatted for twenty or thirty minutes when Professor Mayr said, "Have you read Bateman '48 in *Heredity*?" I did not know what he was referring to. It was typical of Mayr's memory, by the way, that he had the date, the year, and the journal for you. With that big a lead, anybody can find the paper! He said that Bateman was working on concepts similar to those I was talking about and that I should read his paper. Well, I went off and continued my pigeon observations and parental investment thinking, such as it was, and I dared to show up a second time to Dr. Mayr's office with no genetics to report, only elaborations on my pigeon stories. He suffered this for only a few minutes when he leaned forward and cut me off, saying, "Have you *yet* read Bateman '48 in *Heredity*?" Well, of course, I had not read it, nor—until that moment—remembered his instruction to do so! When I confessed, he did something I always loved him later for and said,

"Well, I will not continue this conversation until you have," and sent me on my way. I left Professor Mayr's office with one goal in life, a burning desire to read Bateman (1948) in *Heredity*!

Later that night when I read the paper itself—mercifully short and to the point—the proverbial scales fell from my eyes. They almost seemed to do so literally, for Bateman also had relative parental investment, but he had something no one else had—variance in reproductive success analyzed by sex. It was his demonstration of higher variance in male than in female reproductive success and his argument linking this to low male parental investment that were key to understanding sexual selection. Thus, relative parental investment was expected to control relative variation in reproductive success by sex. Monogamous species (i.e., those with high male parental investment) would be expected to show similar variance in reproductive success in the two sexes, while species with greater male parental investment ought to show greater variance in *female* reproductive success.

The importance of sex role–reversed species was first suggested by George Williams (1966), and I certainly learned it there, but I forgot where I had learned it and imagined that I had made this extension myself, until having finished nearly the final draft of the paper. I was rereading Williams's book, in preparation for lecturing in the fall of 1971, and I was amazed to see not only that Williams had clearly outlined the application to sex role–reversed species and given evidence—from sea horses and pipefish—but also that those and other sections related to sexual selection were heavily underlined by me, and annotated in the margin. The style of applying functional reasoning down to every detail of the relationship between a male and a female was something I very much liked about Williams's work. Incidentally, I immediately sat down to write Professor Williams a letter, enclosing my paper, and saying (which did not turn out to be true) that I would revise extensively the introduction to my paper to draw greater attention to his own contribution. George wrote back a very nice letter, saying that my paper had rendered obsolete a chapter in his own new book (*Sex and Evolution*, published in 1974). In this chapter he suggested that differential male mortality in many species might be explained by sexual selection, much along the lines I did in my 1972 paper, though without the graphs (whose value he was later to praise). In any case, I always regarded the differential mortality by sex sections as the most original of my paper, and perhaps appropriately enough, they have been the most neglected. Whereas I might easily have imagined a comparative literature growing up on sex differential mortality linked to other features of sexual selection, this has not happened yet (but see Clutton Brock, Albon and Guinness 1985).

Mayr Arranges an Invited Chapter for My Work

My article on parental investment and sexual selection appeared as a chapter in a book commemorating the hundred-year anniversary of Darwin's *Sexual Selection and the Descent of Man*. Mayr himself was invited to contribute a chapter, which he did, but he also asked the editor, Bernard Campbell, Why not get a young turk in the book and not just elder statesmen such as himself? I believe Professor DeVore also spoke up on my behalf. In any case, I was invited to contribute a chapter on sexual selection in amphibians and reptiles. I immediately set out to copy the literature and spent many evenings in the Museum of Comparative Biology (when you could copy for free) bathed in the odious green glow of the copying machines (for posterity's sake, my balls pressed tightly against the side of the machine, should the light be mutagenic) copying a large number of papers on population studies of salamanders, lizards, snakes, and so on. I always knew that I wanted to make a general argument, but I made a valiant effort to collect the relevant herpetological literature, for example, on adult sex rations in nature, which in turn could give you information on differential mortality by sex, a topic I was interested in developing. But I soon discovered that very little of the literature had any value in this regard. To know whether your adult sex ratio is a valid one, you must know to what degree your capture techniques are biased in favor of one sex or another. If males wander farther, as they do in many species, they may more often become exposed to your traps, and you may be measuring not the true sex ratio but a sex difference in the tendency to wander into your traps. Most studies had no information with which to make the relevant corrections. As I continued to develop my paper, the relative proportion of material on amphibians and reptiles continued to shrink. One day I went to Professor DeVore and explained my problem. He suggested that I write Dr. Campbell and say that my paper had "drifted" somewhat from its original focus and that I was in the process of making more general arguments, as well as trying, of course, to cover amphibians and reptiles as best I could. Campbell wrote back to say that he was delighted to hear of this development but hoped, nonetheless, that I would retain as much on amphibians and reptiles as I could. I was very pleased to receive his letter and promptly struck almost all remaining references to amphibians and reptiles, with the exception of my own lizard work!

Incidentally, I might never have been given the opportunity to contribute this chapter if Professor Mayr had not suffered a rare theft from his office. Like Darwin, he liked to start folders on important subjects, collecting articles, information, and thoughts in the folder until such time as he was ready to write on the subject. I believe it was his usual procedure to permit student access to these folders but not to allow them to leave his office, or perhaps this rule only developed after the incident I am describing. In any case, a student showed up and asked to see his material on sexual selection and, as I remember it, the material was so voluminous that it was held in two or three folders and could not easily be studied in one place. So Dr. Mayr allowed the student to borrow the folders, and for once his famous memory failed him. When the folders did not return, he could not remember who had borrowed them, and he never got them back. As a consequence, although he had the key reference (Bateman 1948) and a few others, he did not have the large mass of material that the topic deserved. Otherwise, he might have attempted a much more ambitious treatment himself and left me to fend for myself. Of course, it was a great advantage to me to have my chapter invited. Absent a poor performance, this guaranteed publication of my efforts, set a deadline for its production, and then presented the paper in a context that was bound to draw attention to it: a major commemorative volume on the subject.

I remember writing the paper in about five or six weeks, that is, breaking the backbone of the first version of the paper. As was my style in those days, I would read over a period of months, up until the time that I would write the paper, and then I would sit down, reread everything, and with the various articles spread out around me, in my bachelor quarters, write the paper. This was, of course, in the old days before computers, when you put it through typewritten drafts, initially your own and then finally that of a professional typist. The creation of a clear and simple system for thinking about the evolution of sex differences, which went beyond "male" and "female" to an underlying variable, relative parental investment in the offspring, was very satisfying to me as both a public and private scientist. It provided a logic that has withstood the test of time and easily permitted additions and changes. A dream I had, while I was finishing the paper, is perhaps emblematic of this. I dreamt I was looking down a long corridor and seeing a whole bunch of male and female animals mixed together. These were, I think in retrospect, mostly mammals—ungulates, elephants, primates, and so on—two of each. And then, as if in response to some unseen signal, the males all moved to the left, and the females all moved to the right, and all entered rooms or cubicles off of this main hall, leaving it empty. When I woke up the next morning I thought that the dream symbolized what this parental investment perspective gives you, which is a clear logic by which to segregate and understand the two sexes.

Words and Acronyms: RS and PI

My paper helped to popularize two terms that were then not widely in use, "reproductive success" as a synonym for what was then called "fitness" and "parental investment" for what was sometimes called "parental care" or, in

Fisher's sex ratio argument, "parental expenditure." Let me take each of these terms in turn. Fitness (W) was measured by number of surviving offspring, yet the word "fitness" itself suggested physical fitness or moral fitness or, in the way the term actually originated, how "fitted" the organism was to its environment. None of these I thought was very precise, and the term easily invited tautological phrases such as "survival of the fittest" without the tautology being obvious—if you said "survival of those who survive," the tautology would be more apparent. Fitness was measured by number of surviving offspring, so why not call it what it is, namely, reproductive success. But what to use as a symbol for reproductive success? r would be the obvious symbol but it was already in use as the rate of population growth and, separately, r was used for degree of relatedness. So I settled on RS. But here was an immediate disadvantage. We had gone from a single symbol to an acronym of two symbols. I suppose I might have been better advised to keep W but make it mean "reproductive success."

A second problem soon emerged because Hamilton had used the term "inclusive fitness" to refer to reproductive success plus effects on the reproductive success of relatives where these effects are devalued by the relevant degrees of relatedness, and the parallel term "inclusive reproductive success" did not seem quite to capture Hamilton's formulation. Sometimes I use the term "genetic success" as a synonym for inclusive fitness, but again, the language is not quite parallel: reproductive success and genetic success.

I do not remember what thinking, if any, went into the choice of the term "parental investment." But I know that when I "rediscovered" Fisher's term "parental expenditure" I thought I saw a reason to prefer investment. "Expenditure" referred to a cost, the downside, so to speak, while "investment" invoked the future with a suggestion of compound interest when high reproductive success is repeated. "Parental care" suffered, of course, in being too limited, excluding investment in the egg or offspring itself and tending to restrict attention to behaviorally mediated benefits. It certainly was useful to isolate a term that clearly separated out the *work* involved in reproduction from those other sex differences that sexual disparities in work (relative parental investment) would be expected to generate. All too often, there was a vague sense that the sexes split up the labor involved in reproduction in some kind of complementary way: the male took care of sexual arousal and the sexual act itself, the female took care of the infant, the male then took care of both, and so on. The male was easily imagined to invest when he did not (e.g., imagined use of a male's antlers to repel attacks on the family from wolves). Again, there was the disadvantage that I made an acronym out of the term (PI) instead of a single letter. Some acronyms I readily retain and use, such as ESS for evolutionarily stable strategy and EPC for extra-pair copulation, but others seem silly, such as POCT for parent–offspring conflict theory.

───────────── ❦ ◯ ❦ ─────────────

Parental Investment and Sexual Selection

ROBERT L. TRIVERS

Charles Darwin's (1871) treatment of the topic of sexual selection was sometimes confused because he lacked a general framework within which to relate the variables he perceived to be important: sex-linked inheritance, sex ratio at conception, differential mortality, parental care, and the form of the breeding system (monogamy, polygyny, polyandry, or promiscuity). This confusion permitted others to attempt to show that Darwin's terminology was imprecise, that he misinterpreted the function of some structures, and that the influence of sexual selection was greatly overrated. Huxley (1938), for example, dismisses the importance of female choice without evidence or theoretical argument, and he doubts the prevalence of adaptations in males that decrease their chances of surviving but are selected because they lead to high reproductive success. Some important advances, however, have been achieved since Darwin's work. The genetics of sex has now been clarified, and Fisher (1958) has produced a model to explain sex ratios at conception, a model recently extended to include special mechanisms that operate under inbreeding (Hamilton 1967). Data from the laboratory and the field have confirmed that females are capable of very subtle choices (for example, Petit & Ehrman 1969), and Bateman (1948) has suggested a general basis for female choice and male-male competition, and he has produced precise data on one species to support his argument.

This paper presents a general framework within which to consider sexual selection. In it I attempt to define and interrelate the key variables. No attempt is made to review the large, scattered literature relevant to sexual selection. Instead, arguments are presented on how one might *expect* natural selection to act on the sexes, and some data are presented to support these arguments.

Variance in Reproductive Success

Darwin defined sexual selection as (1) competition within one sex for members of the opposite sex and (2) differential choice by members of one sex for members of the opposite sex, and he pointed out that this usually meant males competing with each other for females and females choosing some males rather than others. To study these phenomena one needs accurate data on differential reproductive success analyzed by sex. Accurate data on female reproductive success are available for many species, but similar data on males are very difficult to gather, even in those species that tend toward monogamy. The human species illustrates this point. In any society it is relatively easy to assign accurately the children to their biological mothers, but an element of uncertainty attaches to the assignment of children to their biological fathers. For example, Henry Harpending (personal communication) has gathered biochemical data on the Kalahari Bushmen showing that about two per cent of the children in that society do not belong to the father to whom they are commonly attributed. Data on the human species are, of course, much more detailed than similar data on other species.

To gather precise data on both sexes Bateman (1948) studied a single species, *Drosophila melanogaster*, under laboratory conditions. By using a chromosomally marked individual in competition with individuals bearing different markers, and by searching for the markers in the offspring, he was able to measure the reproductive success of each individual, whether female or male. His method consisted of introducing five adult males to five adult female virgins, so that each female had a choice of five males and each male competed with four other males.

Data from numerous competition experiments with *Drosophila* revealed three important sexual differences: (1) Male reproductive success varied much more widely than female reproductive success. Only four per cent of the females failed to produce any surviving offspring, while 21 per cent of the males so failed. Some males, on the other hand, were phenomenally successful, producing nearly three times as many offspring as the most successful female. (2) Female reproductive success did not appear to be limited by ability to attract males. The four per cent who failed to copulate were apparently courted as vigorously as those who did copulate. On the other hand, male reproductive success was severely limited by ability to attract or arouse females. The 21 per cent who failed to reproduce showed no disinterest in trying to copulate, only an inability to be accepted. (3) A female's reproductive success did not increase much, if any, after the first copulation and not at all after the second; most females were uninterested in copulating more than once or twice. As shown by genetic markers in the offspring, males showed an almost linear increase in reproductive success with increased copulations. (A corollary of this finding is that males tended not to mate with the same female twice.) Although these results were obtained in the laboratory, they may apply with even greater force to the wild, where males are not limited to five females and where females have a wider range of males from which to choose.

Bateman argued that his results could be explained by reference to the energy investment of each sex in their sex cells. Since male *Drosophila* invest very little metabolic energy in the production of a given sex cell, whereas females invest considerable energy, a male's reproductive success is not limited by his ability to produce sex cells but by his ability to fertilize eggs with these cells. A female's reproductive success is not limited by her ability to have her eggs fertilized but by her ability to produce eggs. Since in almost all animal and plant species the male produces sex cells that are tiny by comparison to the female's sex cells, Bateman (1948) argued that his results should apply very widely, that is, to "all but a few very primitive organisms, and those in which monogamy combined with a sex ratio of unity eliminated all intra-sexual selection."

Good field data on reproductive success are difficult to find, but what data exist, in conjunction with the assumption that male reproductive success varies as a function of the number of copulations,[1] support the contention that in all species, except those mentioned below in which male parental care may be a limiting resource for females, male reproductive success varies more than female reproductive success. This is supported, for example, by data from dragonflies (Jacobs 1955), baboons (DeVore 1965) common frogs (Savage 1961), prairie chickens (Robel 1966), sage grouse (Scott 1942), black grouse (Koivisto 1965), elephant seals (LeBoeuf and Peterson, 1969), dung flies (Parker 1970a) and some anoline lizards (Rand 1967 and Trivers, in preparation, discussed below). Circumstantial evidence exists for other lizards (for example, Blair 1960, Harris 1964) and for many mammals (see Eisenberg 1965). In monogamous species, male reproductive success would be expected to vary as female reproductive success, but there is always the possibility of adultery and differential female mortality (discussed below) and these factors should increase the variance of male reproductive success without significantly altering that of the female.

Relative Parental Investment

Bateman's argument can be stated in a more precise and general form such that the breeding system (for example, monogamy) as well as the adult sex ratio become functions of a single variable controlling sexual selection. I first define parental investment as *any investment by the parent in an individual offspring that increases the offspring's chance of surviving (and hence reproductive success) at the cost of the parent's ability to invest in other offspring.* So defined, parental investment includes the metabolic investment in the primary sex cells but refers to any investment (such as feeding or guarding the young) that benefits the young. It does not include effort expended in finding a member of the opposite sex or in subduing members of one's own sex in order to mate with a member of the opposite sex, since such effort (except in special cases) does not affect the survival chances of the resulting offspring and is therefore not *parental* investment.

Each offspring can be viewed as an investment independent of other offspring, increasing investment in one offspring tending to decrease investment in others. I measure the size of a parental investment by reference to its negative effect on the parent's ability to invest in other offspring: a large parental investment is one that strongly decreases the parent's ability to produce other offspring. There is no necessary correlation between the size of parental investment in an offspring and its benefit for the young. Indeed, one can show that during a breeding season the benefit from a given parental investment must decrease at some point or else species would not tend to produce any fixed number of offspring per season. Decrease in reproductive success resulting from the negative effect of parental investment on *nonparental* forms of reproductive effort (such as sexual competition for mates) is excluded from the measurement of parental investment. In effect, then, I am here considering reproductive success as if the only relevant variable were parental investment.

For a given reproductive season one can define the total parental investment of an individual as the sum of its investments in each of its offspring produced during that season, and one assumed that natural selection has favored the total parental investment that leads to maximum net reproductive success. Dividing the total parental investment by the number of individuals produced by the parent gives the typical parental investment by an individual per offspring. Bateman's argument can now be reformulated as follows. Since the total number of offspring produced by one sex of a sexually reproducing species must equal the total number produced by the other (and assuming the sexes differ in no other way than in their typical parental investment per offspring)[2] then the sex whose typical parental investment is greater than that of the opposite sex will become a limiting resource for that sex. Individuals of the sex investing less will compete among themselves to breed with members of the sex investing more, since an individual of the former can increase its reproductive success by investing successively in the offspring of several members of the limiting sex. By assuming a simple relationship between degree of parental investment and number of offspring produced, the argument can be presented graphically (Figure 1). The potential for sexual competition in the sex investing less can be measures by calculating the ratio of the number of offspring that sex optimally produces (as a function of parental investment alone, assuming the opposite sex's investment fixed at its optimal value) to the number of offspring the limiting sex optimally produces (L/M in figure 1).

What governs the operation of sexual selection is the relative parental investment of the sexes in their offspring. Competition for mates usually characterizes males because males usually invest almost nothing in their offspring.

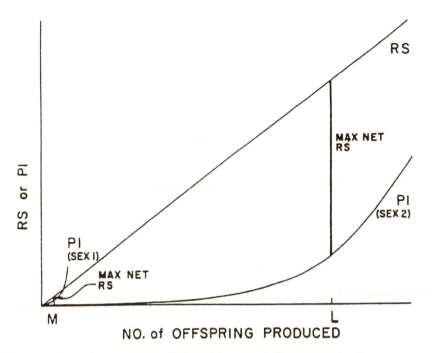

Figure 1. Reproductive success *(RS)* and decrease in future reproductive success resulting from parental investment *(PI)* are graphed as functions of the number of offspring produced by individuals of the two sexes. At *M* and *L* the net reproductive success reaches a maximum for sex 1 and sex 2 respectively. Sex 2 is limited by sex 1 (see text). The shape of the *PI* curves need not be specified exactly.

Where male parental investment per offspring is comparable to female investment one would expect male and female reproductive success to vary in similar ways and for female choice to be no more discriminating than male choice (expect as noted below). Where male parental investment strongly exceeds that of the female (regardless of which sex invests more in the sex cells) one would expect females to compete among themselves for males and for males to be selective about whom they accept as a mate.

Note that it may not be possible for an individual of one sex to invest in only part of the offspring of an individual of the opposite sex. When a male invests less per typical offspring than does a female but more than one-half what she invests (or vice-versa) then selection may not favor male competition to pair with more than one female, if the offspring of the second female cannot be parcelled out to more than one male. If the net reproductive success for a male investing in the offspring of one female is larger than that gained from investing in the offspring of two females, then the male will be selected to invest in the offspring of only one female. This argument

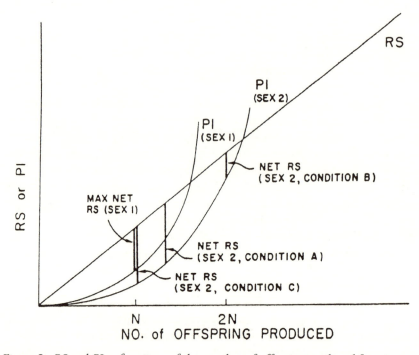

Figure 2. *RS* and *PI* as functions of the number of offspring produced for two sexes. Sex 2 invests per typical offspring more than half of what sex 1 invests. Condition A: maximum net *RS* for a member of sex 2 assuming he can invest in any number of offspring between N and 2N. Condition B: net *RS* assuming member of sex 2 invests in 2N offspring. Condition C: net RS assuming member of sex 2 invests in N offspring. If member of sex 2 must invest in an integral multiple of N offspring, natural selection favors condition C.

is graphed in Figure 2 and may be important to understanding differential mortality in monogamous birds, as discussed below.

Fisher's (1958) sex ratio model compares the parental expenditure (undefined) in male offspring with that in female offspring and suggests energy and time as measures of expenditure. Restatements of Fisher's model (for example, Kolman 1960, Willson & Pianka 1963, J. Emlen 1968, Verner 1965, Leigh 1970) employ either the undefined term, parental expenditure, or the term energy investment. In either case the key concept is imprecise and the relevant one is parental investment, as defined above. Energy investment may often be a good approximation of parental investment, but it is clearly sometimes a poor one. An individual defending its brood from a predator may expend very little energy in the process but suffer a high chance of mortality; such behavior should be measured as a large investment, not a small one as suggested by the energy involved.

Parental Investment Patterns

Species can be classified according to the relative parental investment of the sexes in their young. In the vast majority of species, the male's only contribution to the survival of his offspring is his sex cells. In these species, female contribution clearly exceeds male and by a large ratio.

A male may invest in his offspring in several ways. He may provide his mate with food as in baloon flies (Kessel 1955) and some other insects (Engelmann 1970), some spiders, and some birds (for example, Calder 1967, Royama 1966, Strokes & Williams 1971). He may find and defend a good place for the female to feed, lay eggs or raise young, as in many birds. He may build a nest to receive the eggs, as in some fish (for example, Morris 1952). He may help the female lay the eggs, as in some parasitic birds (Lack 1968). The male may also defend the female. He may brood the eggs, as in some birds, fish, frogs, and salamanders. He may help feed the young, protect them, provide opportunities for learning, and so on, as in wolves and many monogamous birds. Finally, he may provide an indirect group benefit to the young (such as protection), as in many primates. All of these forms of male parental investment tend to decrease the disparity in investment between male and female resulting from the initial disparity in size of sex cells.

To test the importance of relative parental investment in controlling sexual selection one should search for species showing greater male than female parental investment (see Williams 1966, pp. 185–186). The best candidates include the Phalaropidae and the polyandrous bird species reviewed by Lack (1968). In these species, a female's parental investment ends when she lays her eggs; the male alone broods the eggs and cares for the young after hatching. No one has attempted to assess relative parental investment in these species, but they are striking in showing very high male parental investment correlating with strong sex role reversal: females tend to be more brightly colored, more aggressive and larger than the males, and tend to court them and fight over them. In the phalaropes there is no evidence that the females lay multiple broods (Höhn 1967, Johns 1969), but in some polyandrous species females apparently go from male to male laying successive broods (for example, Beebe 1925; see also Orians 1969). In these species the female may be limited by her ability to induce males to care for her broods, and female reproductive success may vary more than male. Likewise, high male parental investment in pipefish and seahorses (Syngnathidae) correlates with female courtship and bright coloration (Fiedler 1954), and female reproductive success may be limited by male parental investment. Field data for other groups are so scanty that it is not possible to say whether there are any instances of sex role reversal among them, but available data for some dendrobatid frogs suggest at least the possibility. In these species, the male carries one or more young on his back for an unknown length of time (for example, Eaton 1941). Females tend to be more brightly colored than males (rare in frogs) and in at least one species, *Dendrobates aurata*, several females have been seen pursuing, and possibly courting, single males (Dunn 1941). In this species the male carries only one young on his back, until the tadpole is quite large, but females have been found with as many as six large eggs inside, and it is possible that females compete with each other for the backs of males. There are other frog families that show male parental care, but even less is known of their social behavior.

In most monogamous birds male and female parental investment is probably comparable. For some species there is evidence that the male invests somewhat less than the female. Kluijver (1933, cited in Coulson 1960) has shown that the male starling (*Sturnus vulgaris*) incubates the eggs less and feeds the young less often than the female, and similar data are available for other passerines (Verner & Willson, 1969). The fact that in many species males are facultative polygynists (von Haartman 1969) suggests that even when monogamous the males invest less in the young than their females. Because sex role reversal, correlating with evidence of greater male than female parental investment, is so rare in birds and because of certain theoretical considerations discussed below, I tentatively classify most monogamous bird species as showing somewhat greater female than male investment in the young.

A more precise classification of animals, and particularly of similar species, would be useful for the formulation and testing of more subtle hypotheses. Groups of birds would be ideal to classify in this way, because slight differences in relative parental investment may produce large differences in social behavior, sexual dimorphism and morality rates by sex. It would be interesting to compare human societies that differ in relative parental investment and in the details of the form of the parental investment, but the specification of parental investment is complicated by the fact that humans often invest in kin other than their children. A wealthy man supporting brothers and sisters (and their children) can be viewed functionally as a polygynist if the contributions to his fitness made by kin are devalued appropriately by their degree of relationship to him (see Hamilton 1964). There is good evidence that premarital sexual permissiveness affecting females in human societies relates to the form of parental investment in a way that would, under normal conditions, tend to maximize female reproductive success (Goethals 1971).

The Evolution of Investment Patterns

The parental investment pattern that today governs the operation of sexual selection apparently resulted from an evolutionarily very early differentiation into relatively immobile sex cells (eggs) fertilized by mobile ones (spermatozoa). An undifferentiated system of sex cells seems highly unstable: competition to fertilize other sex cells should rapidly favor mobility in some sex cells, which in turn sets up selection pressures for immobility in the others. In any case, once the differentiation took place sexual selection act ing on spermatozoa favored mobility at the expense of investment (in the form of cytoplasm). This meant that as long as the spermatozoa of different males competed directly to fertilize eggs (as in oysters) natural selection favoring increased parental investment could act only on the female. Once females were able to control which male fertilized their eggs, female choice or mortality selection on the young could act to favor some new form of male investment in addition to spermatozoa. But there exist strong selection pressures against this. Since the female already invests more than the male, breeding failure for lack of an additional investment selects more strongly against her than against the male. In this sense, her initial very great investment commits her to additional investment more than the male's initial slight investment commits him. Furthermore, male-male competition will tend to operate against male parented investment, in that any male investment in one female's young should decrease the male's chances of inseminating other females. Sexual selection, then, is both controlled by the parental investment pattern and a force that tends to mold that pattern.

The conditions under which selection favors male parental investment have not been specified for any group of animals. Except for the case of polygyny in birds, the role of female choice has not been explored; instead, it is commonly assumed that, whenever two individuals can raise more individuals together than one alone could, natural selection will favor male parental investment (Lack 1968, p. 149), an assumption that overlooks the effects of both male-male competition and female choice.

Initial Parental Investment

An important consequence of the early evolutionary differentation of the sex cells and subsequent sperm competition is that male sex cells remain tiny compared to female sex cells, even when selection has favored a total male parental investment that equals or exceeds the female investment. The male's initial parental investment, that is, his investment at the moment of fertilization, is much smaller than the female's, even if later, through parental care, he invests as much or more. Parental investment in the young can be viewed as a sequence of discrete investments by each sex. The relative investment may change as a function of time and each sex may be more or less free to terminate its investment at any time. In the human species, for example, a copulation costing the male virtually nothing may trigger a nine-month investment by the female that is not trivial, followed, if she wishes, by a fifteen-year investment in the offspring that is considerable. Although the male may often contribute parental care during this period, he need not necessarily do so. After a nine-month pregnancy, a female is more or less free to terminate her investment at any moment but doing so wastes her investment up until then. Given the initial imbalance in investment the male may maximize his chances of leaving surviving offspring by copulating and abandoning many females, some of whom, alone or with the aid of others, will raise his offspring. In species where there has been strong selection for male parental care, it is more likely that a mixed strategy will be the optimal male course—to help a single female raise young, while not passing up opportunities to mate with other females whom he will not aid.

In many birds, males defend a territory which the female also uses for feeding prior to egg laying, but the cost of this investment by the male is difficult to evaluate. In some species, as outlined above, the male may provision the female before she has produced the young, but this provisioning is usually small compared to the cost of the eggs. In any case, the cost of the copulation itself is always trivial to the male, and in theory the male need not invest anything else in order to copulate. If there is any chance the female can raise the young, either alone or with the help of others, it would be to the male's advantage to copulate with her. By this reasoning one would expect males of monogamous species to retain some psychological traits consistent with promiscuous habits. A male would be selected to differentiate between a female he will only impregnate and a female with whom he will also raise young. Toward the former he should be more eager for sex and less discriminating in choice of sex partner than the female toward him, but toward the latter he should be about as discriminating as she toward him.

If males within a relatively monogamous species are, in fact, adapted to pursue a mixed strategy, the optimal is likely to differ for different males. I know of no attempt to document this possibility in humans, but psychology might well benefit from attempting to view human sexual plasticity as an adaptation to permit the individual to choose the mixed strategy best suited to local conditions and his own attributes. Elder (1969) shows that steady dating and sexual activity (coitus and petting) in adolescent human females correlate inversely with a tendency to marry up the socioeconomic scale as adults. Since females physically attractive as adolescents tend to marry up, it is possible that females adjust their reproductive strategies in adolescence to their own assets.

Desertion and Cuckoldry

There are a number of interesting consequences of the fact that the male and female of a monogamous couple invest parental care in their offspring at different rates. These can be studied by graphing and comparing the cumulative investment of each parent in their offspring, and this is done for two individuals of a hypothetical bird species in Figure 3. I have graphed no parental investment by the female in her young before copulation, even though she may be producing the eggs before then, because it is not until the act of copulation that she commits the eggs to a given male's genes. In effect, then, I have graphed the parental investment of each individual in the other individual's offspring. After copulation, this is the same as graphing investment in their own offspring, assuming, as I do here, that the male and female copulate with each other and each other only.

To discuss the problems that confront paired individuals ostensibly cooperating in a joint parental effort, I choose the language of strategy and decision, as if each individual contemplated in strategic terms the decisions it ought to make at each instant in order to maximize its reproductive success. This language is chosen purely for convenience to explore the adaptations one might expect natural selection to favor.

At any point in time the individual whose cumulative investment is exceeded by his partner's is theoretically tempted to desert, especially if the disparity is large. This temptation occurs because the deserter loses less than his partner if no offspring are raised and the partner would therefore be more strongly selected to stay with the young. Any success of the partner will, of course, benefit the deserter. In Figure 3, for example, desertion by the male right after copulation will cost him very little, if no offspring are raised, while the chances of the female raising some young alone may be great enough to make the desertion worthwhile. Other factors are important in determining the adaptiveness of abandonment, factors such as the opportunities outside the pair for breeding and the expected shape of the deserter's investment curve if he does not desert. If the male's investment curve does not rise much after copulation, then the female's chances of raising the young alone will be greater and the time wasted by the male investing moderately in his offspring may be better spent starting a new brood.

What are the possible responses of the deserted individual? If the male is deserted before copulation, he has no choice but to attempt to start the process over again with a new female; whatever he has invested in that female is lost. If either partner is deserted after copulation, it has three choices. (1) It can desert the eggs (or eat them) and attempt to breed again with another mate, losing thereby all (or part of) the initial investment.

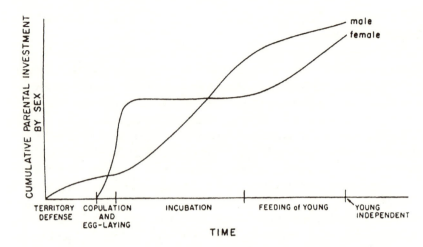

Figure 3. Hypothetical cumulative parental investment of a male and a female bird in their offspring as a function of time. *Territory defense:* Male defends area for feeding and nest building. *Copulation and egg-laying:* Female commits her eggs to male who commits his defended nest to the female. *Incubation:* Male incubates eggs while female does nothing relevant to offspring. *Feeding of young:* Each parent feeds young but female does so at a more rapid rate.

(2) It can attempt to raise the young on its own, at the risk of over-exertion and failure. Or, (3) it can attempt to induce another partner to help it raise the young. The third alternative, if successful, is the most adaptive for it, but this requires deceiving another organism into doing something contrary to its own interests, and adaptations should evolve to guard individuals from such tasks. It is difficult to see how a male could be successful in deceiving a new female, but if a female acts quickly, she might fool a male. As time goes on (for example, once the eggs are laid), it is unlikely that a male could easily be fooled. The female could thus be programmed to try the third strategy first, and if it failed, to revert to the first or second. The male deserter gains most if the female succeeds in the third strategy, nothing if she chooses the first strategy, and possibly an intermediate value if she chooses the second strategy.

If neither partner deserts at the beginning, then as time goes on, each invests more and more in the young. This trend has several consequences. On the one hand, the partner of a deserter is more capable of finishing the task alone and natural selection should favor its being more predisposed to try, because it has more to lose. On the other hand, the deserter has more to lose if the partner fails and less to gain if the partner succeeds. The balance between these opposing factors should depend on the exact form of the cumulative investment curves as well as the opportunities for further breeding outside the pair.

There is another effect with time of the increasing investment by both parents in the offspring. As the investments increase, natural selection may favor *either* partner deserting even if one has invested more in the young than the other. This is because the desertion may put the deserted partner in a cruel bind: he has invested so much that he loses considerably if he also deserts the young, even though, which should make no difference to him, the partner would lose even more. The possibility of such binds can be illustrated by an analogous situation described by Rowley (1965). Two neighboring pairs of wrens happened to fledge their young simultaneously and could not tell their young apart, so both pairs fed all six young indiscriminately, until one pair "deserted" to raise another brood, leaving their neighbors to feed all six young, which they did, even though this meant they were, in effect, being taken advantage of.

Birds should show adaptations to avoid being deserted. Females, in particular, should be able to guard against males who will only copulate and not invest subsequent parental effort. An instance of such an adaptation may be found in the red-necked phalarope, *Phalaropus lobatus*. In phalaropes the male incubates the eggs alone and alone cares for the young after hatching (Höhn 1967, Johns 1969), so that a graph of cumulative parental investment would show an initial large female investment which then remains the same through time, whereas the initial male investment is nil and increases steadily, probably to surpass the female investment. Only the female is vulnerable to being deserted and this right after copulation, since any later desertion by the male costs him his investment in incubation, the young being almost certain to perish. Tinbergen (1935) observed a female vigorously courting a male and then flying away as soon as he responded to the courtship by attempting to copulate. This coy performance was repeated numerous times for several days. Tinbergen attributed it to the "waxing and waning of an instinct," but the behavior may have been a test of the male's willingness to brood the female's eggs. The male under observation was, in fact, already brooding eggs and was courted when he left the eggs to feed on a nearby pond. In order to view a complete egg-laying sequence, Tinbergen destroyed the clutch the male was brooding. Within a half day the female permitted the male sexual access, and he subsequently brooded her eggs. The important point is that the female could apparently tell the difference between a free and an encumbered male, and she withheld sex from the latter. Courtship alternating with flight may be the test that reveals the male's true attachments: the test can show, for example, whether he is free to follow the female.

It is likely that many adaptations exist in monogamous species to guard against desertion, but despite evidence that desertion can be common (Rowley 1965) no one has attempted to analyze courtship with this danger in mind. Von Haartman (1969) has reviewed some evidence for adaptations of females to avoid being mated to a polygynous male, and being so mated is sometimes exactly equivalent to being deserted by the male (von Haartman 1951).

External fertilization requires a synchrony of behavior such that the male can usually be certain he is not attempting to fertilize previously fertilized eggs. With the evolution of internal fertilization the male cannot be so certain. For many species (for example, most mammals), the distinction is not important because the male loses so little by attempting to fertilize previously fertilized eggs. Where male parental care is involved, however, the male runs the risk of being cuckolded, of raising another male's offspring. For figure 7.1 it was assumed that the pair copulated with each other and each other only, but the male can usually not be sure that such is the case and what is graphed in such a situation is the male's investment in the *female's* offspring. Adaptations should evolve to help guarantee that the female's offspring are also his own, but these can partly be countered by the evolution of more sophisticated cuckolds.

One way a male can protect himself is to ensure that other males keep their distance. That some territorial aggression of monogamous male birds is devoted to protecting the sanctity of the pair bond seems certain, and human male aggression toward real or suspected adulterers is often extreme. Lee (1969), for example, has shown that, when the cause is known, the major cause of fatal Bushman fights is adultery or suspected adultery. In fact, limited data on other hunter-gathering groups (including Eskimos and Australian aborigines) indicate that, while fighting is relatively rare (in that organized intergroup aggression is infrequent), the "murder rate" may be relatively high. On examination, the murderer and his victim are usually a husband and his wife's real or suspected lover. In pigeons (*Columba livia*) a new male arriving alone at a nocturnal roosting place in the fall is attacked day after day by one or more resident males. As soon as the same male appears with a mate, the two are treated much more casually (Trivers, unpublished data), suggesting that an unpaired male is more threatening than a paired one.

I have argued above that a female deserted immediately after copulation may be adapted to try to induce another male to help raise her young. This factor implies adaptations on the part of the male to avoid such a fate. A simple method is to avoid mating with a female on first encounter, sequester her instead and mate with her only after a passage of time that reasonably excludes her prior impregnation by another male. Certainly males guard their females from other males, and there is a striking difference between the lack of preliminaries in promiscuous birds (Scott 1942, Kruijt & Hogan 1967) and the sometimes long lag between pair bonding and copulation in monogamous birds (Nevo 1956), a lag which usually seems to serve other functions as well.

Biologists have interpreted courtship in a limited way. Courtship is seen as allowing the individual to choose the correct species and sex, to overcome antagonistic urges and to arouse one's partner (Bastock 1967). The above analysis suggests that courtship should also be interpreted in terms of the need to guard oneself from the several possibilities of maltreatment at the hands of one's mate.

Differential Mortality and the Sex Ratio

Of special interest in understanding the effects of sexual selection are accurate data on differential morality of the sexes, especially of immature individuals. Such data are, however, among the most difficult to gather, and the published data, although important, are scanty (for example, Emlen 1940, Hays 1947, Chapman, Casida, & Cote 1938, Robinette et al. 1957, Coulson 1960, Potts 1969, Darley 1971, Myers & Krebs 1971). As a substitute one can make use of data on sex ratios within given age classes or for all age classes taken together. By assuming that the sex ratio at conception (or, less precisely, at birth) is almost exactly 50/50, significant deviations from this ratio for any age class or for all taken together should imply differential morality. Where data exist for the sex ratio at birth and where the sex ratio for the entire local population is unbalanced, the sex ratio at birth is usually about 50/50 (see above references, Selander 1965, Lack 1954). Furthermore, Fisher (1958) has shown, and others refined (Leigh 1970), that parents should invest roughly equal energy in each sex. Since parents usually invest roughly equal energy in each individual of each sex, natural selection, in the absence of unusual circumstances (see Hamilton 1967), should favor approximately a 50/50 sex ratio at conception.

It is difficult to determine accurately the sex ratio for any species. The most serious source of bias is that males and females often make themselves differentially available to the observer. For example, in small mammals sexual selection seems to have favored male attributes, such as high mobility, that tend to result in their differential capture (Beer, Frenzel, & MacLeod 1958, Myers & Krebs, 1971). If one views one's capture techniques as randomly sampling the existing population, one will conclude that males are more numerous. If one views one's capture techniques as randomly sampling the effects of mortality on the population, then one will conclude that males are more prone to mortality (they are captured more often) and therefore are less numerous. Neither assumption is likely to be true, but authors routinely choose the former. Furthermore, it is often not appreciated what a large sample is required in order to show significant deviations from a 50/50 ratio. A sample of 400 animals showing a 44/56 sex ratio, for example, does not deviate significantly from a 50/50 ratio. (Nor, although this is almost never pointed out, does it differ significantly from a 38/62 ratio.)

Mayr (1939) has pointed out that there are numerous deviations from a 50/50 sex ratio in birds and I believe it is likely that, if data were sufficiently precise, most species of vertebrates would show a significant deviation from a 50/50 sex ratio. Males and females differ in numerous characteristics relevant to their different reproductive strategies and these characters are unlikely to have equivalent effects on survival. Since it is not advantageous for the adults of each sex to have available the same number of adults of the opposite sex, there will be no automatic selective agent for keeping deviations from a 50/50 ratio small.

A review of the useful literature on sex ratios suggests that (except for birds) when the sex ratio is unbalanced it is usually unbalanced by there being more females than males. Put another way, males apparently have a tendency to suffer higher mortality rates than females. This is true for those dragonflies for which there are data (Corbet, Longfield, & Moore 1960), for the house fly (Rockstein 1959), for most fish (Beverton & Holt 1959), for several lizards (Tinkle 1967, Harris 1964, Hirth 1963, Blair 1960, Trivers, discussed below) and for many mammals (Bouliere & Verschuren 1960, Cowan 1950, Eisenberg 1965, Robinette et al. 1957, Beer, Frenzel, & MacLeod 1958, Stephens 1952, Tyndale-Biscoe & Smith 1969, Myers & Krebs 1971, Wood 1970). Hamilton (1948) and Lack (1954) have reviewed studies on other animals suggesting a similar trend. Mayr (1939) points out that where the sex ratio can be shown to be unbalanced in monogamous birds there are usually fewer females, but in polygynous or promiscuous birds there are fewer males. Data since his paper confirm this finding. This result is particularly interesting since in all other groups in which males tend to be less numerous monogamy is rare or nonexistent.

The Chromosomal Hypothesis

There is a tendency among biologists studying social behavior to regard the adult sex ratio as an independent variable to which the species reacts with appropriate adaptations. Lack (1968) often interprets social behavior as an adaptation in part to an unbalanced (or balanced) sex ratio, and Verner (1964) has summarized other instances of this tendency. The only mechanism that will generate differential mortality independent of sexual differences clearly related to parental investment and sexual selection is the chromosomal mechanism, applied especially to humans and other mammals: the unguarded X chromosome of the male is presumed to predispose him to higher mortality. This mechanism is inadequate as an explanation of differential mortality for three reasons.

1. The distribution of differential mortality by sex is not predicted by a knowledge of the distribution of sex determining mechanisms. Both sexes of fish are usually homogametic, yet males suffer higher mortality. Female birds are heterogametic but suffer higher mortality only in monogamous species. Homogametic male meal moths are outsurvived by their heterogametic female counterparts under laboratory conditions (Hamilton & Johansson 1965).

2. Theoretical predictions of the degree of differential mortality expected by males due to their unguarded X chromosome are far lower than those observed in such mammals as dogs, cattle and humans (Ludwig & Boost 1951). It is possible to imagine natural selection favoring the heterogametic sex determining mechanism if the associated differential mortality is slight and balanced by some advantage in differentiation or in the homogametic sex, but a large mortality associated with heterogamy should be counteracted by a tendency toward both sexes becoming homogametic.

3. Careful data for humans demonstrate that castrate males (who remain of course heterogametic) strongly outsurvive a control group of males similar in all other respects and the earlier in life the castration, the greater the increase in survival (Hamilton & Mestler 1969). The same is true of domestic cats (Hamilton, Hamilton & Mestler 1969), but not of a species (meal moths) for which there is no evidence that the gonads are implicated in sexual differentiation (Hamilton & Johansson 1965).

An Adaptive Model of Differential Mortality

To interpret the meaning of balanced or unbalanced sex ratios one needs a comprehensive framework within which to view life historical phenomena. Gadgil and Bossert (1970) have presented a model for the adaptive interpretation of differences between species' life histories, for example, in the age of first breeding and in the growth and survival curves. Although they did not apply this model to sexual differences in these parameters, their model is precisely suited for such differences. One can, in effect, treat the sexes as if they were different species, the opposite sex being a resource relevant to producing maximum surviving offspring. Put this way, female "species" usually differ from male species in that females compete among themselves for such resources as food but not for members of the opposite sex, whereas males ultimately compete only for members of the opposite sex, all other forms of competition being important only insofar as they affect this ultimate competition.

To analyze differential mortality by sex one needs to correlate different reproductive strategies with mortality; that is, one must show how a given reproductive strategy entails a given risk of mortality. One can do this by graphing reproductive success (RS) for the first breeding season as a function of reproductive effort expended during that season, and by graphing the diminution in future reproductive success (D) in units of first breeding season reproductive success. (Gadgil and Bossert show that the reproductive value of a given effort declines with age, hence the need to convert future reproductive success to comparable units.) For simplicity I assume that the diminution, D, results entirely from mortality between the first and second breeding seasons. The diminution could result from mortality in a later year (induced by reproductive effort in the first breeding season) which would not change the form of the analysis, or it could result from decreased ability to breed in the second (or still later) breeding season, which sometimes occurs but which is probably minor compared to the diminution due to mortality, and which does not change the analysis as long as one assumes that males and females do not differ appreciably in the extent to which they suffer this form of diminution.

Natural selection favors an individual expending in the first breeding season the reproductive effort (RE) that results in a maximum net reproductive success (RS − D). The value of D at this RE gives the degree of expected mortality between the first and second breeding seasons (see figures 7.4 and 7.5). Differences between the sexes in D will give the expected differential mortality. The same analysis can be applied to the nth breeding season to predict mortality between it and the nth + 1 breeding season. Likewise, by a trivial modification, the analysis can be used to generate differences in juvenile mortality: let D represent the diminution in chances of surviving to the first breeding season as a function of RE at first breeding. Seen this way, one is measuring the cost in survival of developing during the juvenile period attributes relevant to adult reproductive success.

Species with Little or No Male Parental Investment

In figure 4, I have graphed RS and D as functions of reproductive effort in the first breeding season for females of a hypothetical species in which males invest very little parental care. The RS function is given a sigmoidal shape for the following reasons. I assume that at low values of RE, RS increases only very gradually because some investment is necessary just to initiate reproduction (for example, enlarging the reproductive organs). RS then increases more rapidly as a function of RE but without achieving a very steep slope. RS finally levels off at high values of RE because of increased inefficiencies there (for example, inefficiencies in foraging; see Schoener 1971). I have graphed the value, f, at which net reproductive success for the female

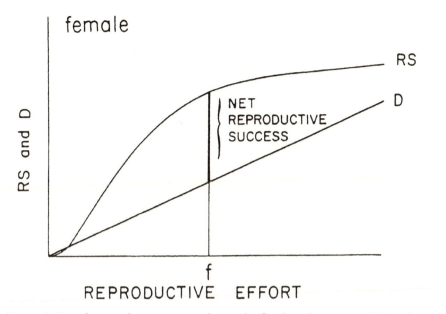

Figure 4. Female reproductive success during the first breeding season *(RS)* and diminution of future reproductive success *(D)* as functions of reproductive effort during first breeding. *D* is measured in units of first breeding (see text). At f the net reproductive success reaches a maximum. Species is one in which there is very little male parental investment.

reaches a maximum. Technically, due to competition, the shape of the RS function for any given female will depend partly on the reproductive effort devoted by other females; the graph therefore assumes that other females tend to invest near the optimal value, f, but an important feature of a female's RS is that it is *not* strongly dependent on the RE devoted by other females: the curve would not greatly differ if all other females invested much more or less. I have graphed D as a linear function of RE. So doing amounts to a definition of reproductive effort, that is, a given increment in reproductive effort during the first breeding season can be detected as a proportionately increased chance of dying between the first and second breeding seasons. Note that reproductive effort for the female is essentially synonymous with parental investment.

Male RS differs from female RS in two important ways, both of which stem from sexual selection. (1) A male's RS is highly dependent on the RE of other males. When other males invest heavily, an individual male will usually not outcompete them unless he invests as much or more. A considerable investment that is slightly below that of other males may result in zero RS. (2) A male's RS is potentially very high, much higher than that of a conspecific female, but only if he outcompetes other males. There should

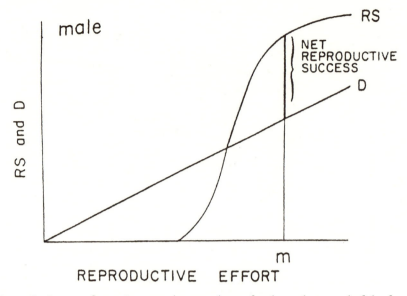

Figure 5. Same as figure 4 except that it is drawn for the male instead of the female. At m the net reproductive success reaches a maximum.

exist some factor or set of factors (such as size, aggressiveness, mobility) that correlates with high male RS. The effect of competition between males for females is selection for increased male RE, and this selection will continue until greater male than female RE is selected as long as the higher associated D is offset by the potentially very high RS. This argument is graphed in figure 5, where the steep slope of RS reflects the high interaction between one male's RS and the RE of the other males. Note that the argument here depends on the existence of a set of factors correlated with high male reproductive success. If these factors exists, natural selection will predispose the male to higher mortality rates than the female. Where a male can achieve very high RS in a breeding season (as in land-breeding seals, Bartholomew 1970), differential mortality will be correspondingly high.

Species with Appreciable Male Parental Investment

The analysis here applies to species in which males invest less parental care than, but probably more than one-half, what females invest. I assume that most monogamous birds are so characterized, and I have listed reasons and some data above supporting this assumption. The reasons can be summarized by saying that because of their initial large investment, females appear to be caught in a situation in which they are unable to force greater parental

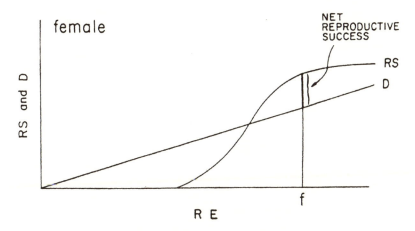

Figure 6. Female reproductive success and diminution in future reproductive success as functions of reproductive effort *(RE)* assuming male reproductive effort of m_1. Species is a hypothetical monogamous bird in which males invest somewhat less than females in parental care (see figures 7 and 8).

investment out of the males and would be strongly selected against if they unilaterally reduced their own parental investment.

Functions relating RS to parental investment are graphed for males and females in figures 6 and 7, assuming for each sex that the opposite sex shows the parental investment that results for it in a maximum net reproductive success. The female curve is given a sigmoidal shape for the reasons that apply to figure 4; in birds the female's initial investment in the eggs will go for nothing if more is not invested in brooding the eggs and feeding the young, while beyond a certain high RE further increments do not greatly affect RS. Assuming the female invests the value, f, male RS will vary as a function of male parental investment in a way similar to female RS, except the function will be displaced to the left (figure 7) and some RS will be lost due to the effects of the cuckoldry graphed in figure 8.

Because males invest in parental care more than one-half what females invest and because the offspring of a given female tend to be inseminated by a single male, selection does not favor males competing with each other to invest in the offspring of more than one female. Rather, sexual selection only operates on the male to inseminate females whose offspring he will not raise, especially if another male will raise them instead. Since selection presumably does not strongly favor female adultery and may oppose it (if, for example, detection leads to desertion by the mate), the opportunities for cuckoldry are limited: high investment in promiscuous activity will bring only limited RS. This argument is graphed in Figure 8. The predicted

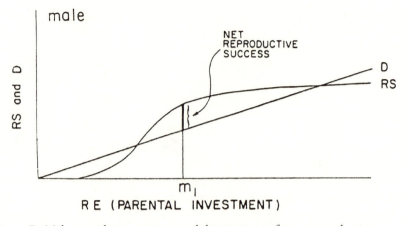

Figure 7. Male reproductive success and diminution in future reproductive success as functions of reproductive effort, assuming female reproductive effort of f. Species is same as in Figure 6. Reproductive effort of male is invested as parental care in one female's offspring. Net reproductive success is a maximum at m_1.

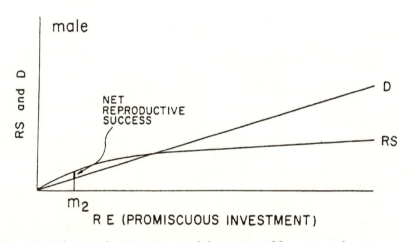

Figure 8. Male reproductive success and diminution of future reproductive success as a function of reproductive effort solely devoted to promiscuous behavior. Net reproductive success at m_2 is a maximum. Same species as in Figures 6 and 7.

differential mortality by sex can be had by comparing D (f) with D (m_1 + m_2).

It may seem ironic, but in moving from a promiscuous to a monogamous life, that is, in moving toward *greater* parental investment in his young, the male tends to *increase* his chances of surviving relative to the female. This tendency occurs because the increased parental investment disproportionately decreases the male's RE invested in male–male competition to inseminate females.

Note that in both cases above differential mortality tends to be self-limiting. By altering the ratio of possible sexual partners to sexual competitors differential mortality sets up forces that tend to keep the differential mortality low. In species showing little male parental investment differential male mortality increases the average number of females available for those males who survive. Other things being equal, this increase tends to make it more difficult for the most successful males to maintain their relative advantage. In monogamous birds differential female mortality induces competition among males to secure at least one mate, thereby tending to increase male mortality. Such competition presumably also increases the variance in male reproductive success above the sexual differential expected from cuckoldry.

Species with Greater Male than Female Parental Investment

Since the above arguments were made with reference to relative parental investment and not sex, they apply to species in which males invest more parental effort than females, except that there is never apt to be a female advantage to cuckolding other females, and this advantage is always alive with males. Where females invest more than one-half what males invest, one would predict differential male mortality. Where females invest less than one-half what males invest, one would predict competition, and a resulting differential female mortality.

Male–Male Competition

Competition between males does not necessarily end with the release of sperm. Even in species with internal fertilization, competition between sperm of different males can be an important component of male–male competition (see the excellent review by Parker 1970b). In rare cases, competition between males may continue after eggs are fertilized. For example, an adult male langur (*Presbytis entellus*) who ousts the adult male of a group may systematically kill the infants of that group (presumably fathered by the ousted male) thereby bringing most of the adult females quickly into estrus again (Sugiyama 1967). While clearly disadvantageous for the killed infants and their mothers, such behavior, benefiting the new male, may be an extreme product of sexual selection. Female mice spontaneously abort during the first four days of pregnancy when exposed to the smell of a strange male (Bruce 1960, reviewed in Sadleir 1967), a situation subject to several interpretations including one based on male-male competition.

Sperm competition may have important effects on competition between males prior to release of sperm. In those insects in which later-arriving sperm take precedence in fertilizing eggs, selection favors mating with a female just prior to release of eggs, thereby increasing competition at ovulation sites and intensifying selection for a postovulatory guarding phase by the male (see Parker 1970bcd, Jacobs 1955). I here concentrate on male-male competition prior to the release of sperm in species showing very little male parental investment.

The form of male-male competition should be strongly influenced by the distribution in space and time of the ultimate resource affecting male reproductive success, namely, conspecific breeding females. The distribution can be described in terms of three parameters: the extent to which females are clumped or dispersed in space, the extent to which they are clumped or dispersed in time, and the extent to which their exact position in space and time is predictable. I here treat females as if they are a passive resource for which males compete, but female choice may strongly influence the form of male–male competition, as, for example, when it favors males clumping together on display grounds (for example, S. T. Emlen 1968) which females then search out (see below under "Female Choice").

Distribution in Space

Cervids differ in the extent to which females are clumped in space or randomly dispersed (deVos, Broky & Geist 1967) as do antelopes (Eisenberg 1965), and these differences correlate in a predictable way with differences in male attributes. Generally male-male aggression will be the more severe the greater the number of females two males are fighting over at any given moment. Searching behavior should be more important in highly dispersed species especially if the dispersal is combined with unpredictability.

Distribution in Time

Clumped in time refers to highly seasonal breeders in which many females become sexually available for a short period at the same moment (for example, explosive breeding frogs; Bragg 1965, Rivero & Estevez 1969), while highly dispersed breeders (in time) are species (such as chimpanzees; Van Lawick-Goodall 1968) in which females breed more or less randomly throughout the year. One effect of extreme clumping is that it becomes more difficult for any one male to be extremely successful: while he is copulating with one female, hundreds of other females are simultaneously being inseminated. Dispersal in time, at least when combined with clumping in space, as in many primates, permits each male to compete for each newly available female and the same small number of males tend repeatedly to inseminate the receptive females (DeVore 1965).

Predictability

One reason males in some dragonflies (Jacobs 1955) may compete with each other for female oviposition sites is that those are highly predictable places at which to find receptive females. Indeed, males display several behaviors, such as testing the water with the tips of their abdomen, that apparently aid them in predicting especially good oviposition sites, and such sites can permit very high male reproductive success (Jacobs 1955). In the cicada killer wasp (*Sphecius spheciosus*) males establish mating territories around colony emergency holes, presumably because this is the most predictable place at which to find receptive females (Lin 1963).

The three parameters outlined interact strongly, of course, as when very strong clumping in time may strongly reduce the predicted effects of strong clumping in space. A much more detailed classification of species with non-obvious predictions would be welcome. In the absence of such models I present a partial list of factors that should affect male reproductive success and that may correlate with high male mortality.

Size

There are very few data showing the relationship between male size and reproductive success but abundant data showing the relationship between male dominance and reproductive success: for example, in elephant seals (LeBoeuf & Peterson 1969), black grouse (Koivisto 1965, Scott 1942), baboons (DeVore 1965) and rainbow lizards (Harris 1964). Since dominance is largely established through aggression and larger size is usually helpful in aggressive encounters, it is likely that these data partly reveal the relationship between size and reproductive success. (It is also likely that they reflect the relationship between experience and reproductive success.)

Circumstantial evidence for the importance of size in aggressive encounters can be found in the distribution of sexual size dimorphism and aggressive tendencies among tetrapods. In birds and mammals males are generally larger than females and much more aggressive. Where females are known to be more aggressive (that is, birds showing reversal in sex roles) they are also larger. In frogs and salamanders females are usually larger than males, and aggressive behavior has only very rarely been recorded. In snakes, females are usually larger than males (Kopstein 1941) and aggression is almost unreported. Aggression has frequently been observed between sexually active crocodiles and males tend to be larger (Allen Greer, personal communication). In lizards males are often larger than females, and aggression is common in some families (Carpenter 1967). Male aggressiveness is also common, however, in some species in which females are larger, for example, *Sceloporus*, (Blair 1960). There is a trivial reason for the lack of evidence of aggressiveness in most amphibians and reptiles: the species are difficult to observe and few behavioral data of any sort have been recorded. It is possible, however, that this correlation between human ignorance and species in which females are larger is not accidental. Humans tend to be more knowledgeable about those species that are also active diurnally and strongly dependent on vision, for example, birds and large mammals. It may be that male aggressiveness is more strongly selected in visually oriented animals because vision provides long-range information on the behavior of competitors. The male can, for example, easily observe another male beginning to copulate and can often quickly attempt to intervene (for example, baboons, DeVore 1965, and sage grouse, Scott 1942).

Mammals and birds also tend towards low, fixed clutch sizes and this may favor relatively smaller females, since large female size may be relatively unimportant in reproductive success. In many fish, lizards and salamanders female reproductive success as measured by clutch size is known to correlate strongly within species with size (Tinkle, Wilbur & Tilley 1970, Tilley 1968).

Measuring reproductive success by frequency of copulation, I have analyzed male and female reproductive success as a function of size in *Anolis garmani* (Figures 7.9 and 7.10). Both sexes show a significant positive correlation between size and reproductive success, but the trend in males is significantly stronger than the trend in females ($p < .01$). Consistent with this tendency, males grow faster at all sizes than females (Figure 7.11) and reach an adult weight two and one-half times that of adult females. The sex ratio of all animals is unbalanced in favor of females, which would seem to indicate differential mortality, but the factors that might produce the difference are not known. Males are highly aggressive and territorial, and large males defend correspondingly large territories with many resident females. No data are available on size and success in aggressive encounters, but in the closely related (and behaviorally very similar) *A. lineatopus*, 85 per cent of 182 disputes observed in the field were won by the larger animal (Rand

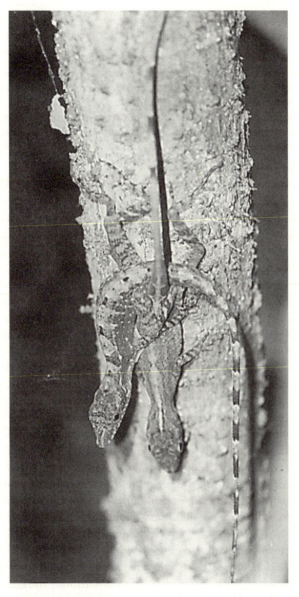

Figure 9. Male and female *Anolis* garmani copulating face down four feet up the trunk of a cocoanut tree. Photo by Joseph K. Long.

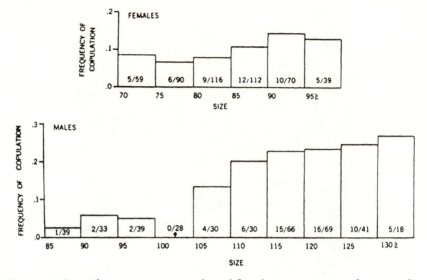

Figure 10. Reproductive success in male and female *A. garmani* as a function of size. Reproductive success is measured by the number of copulations observed per number of individuals (male or female) in each nonoverlapping 5 mm size category. Data combined from five separate visits to study area between summer 1969 and summer 1971.

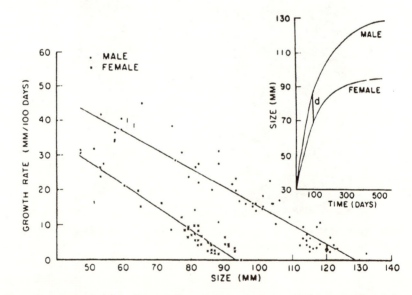

Figure 11. Male and female growth rates in *A. garmani* as a function of initial size based on summer 1970 recaptures of animals marked 3 to 4 months before. A line has been fitted to each set of data; d indicates how much larger a male is when a similar aged female reaches sexual maturity.

1967). Females lay only one egg at a time, but it is likely that larger adult females lay eggs slightly more often than smaller ones, and this may partly be due to advantages in feeding through size-dependent aggressiveness, since larger females wander significantly more widely than smaller adult ones. An alternate interpretation (based on ecological competition between the sexes) has been proposed for sexual dimorphism in size among animals (Selander 1966), and the interpretation may apply to *Anolis* (Schoener 1967).

Metabolic Rate

Certainly more is involved in differential male mortality than size, even in species in which males grow to a larger size than females. Although data show convincingly that nutritional factors strongly affect human male survival in *utero*, a sexual difference in size among humans is not detected until the twenty-fourth week after conception whereas differences in mortality appear as soon as the twelfth week. Sellers et al. (1950) have shown that male rats excrete four times the protein females do; the difference is removed by castration. Since males suffer more from protein-deficient diets than females (they gain less weight and survive less well) the sex-linked proteinuria, apparently unrelated to size, may be a factor in causing lower male survival in wild rats (Schein 1950). (The connection between proteinuria and male reproductive success is obscure.) Again, although human male survival is more adversely affected by poor nutritional conditions than female survival, Hamilton (1948) presents evidence that the higher metabolic rate of the male is an important factor increasing his vulnerability to many diseases which strike males more heavily than females. Likewise, Taber & Dasmann (1954) argue that greater male mortality in the deer, *Odocoileus hemionus*, results from a higher metabolic rate. High metabolic rate could relate to both aggressiveness and searching behavior.

Experience

If reproductive success increases more rapidly in one sex than the other as a function of age alone (for example, through age-dependent experience), then one would expect a postponement of sexual maturity in that sex and a greater chance of surviving through a unit of time than in the opposite sex. Thus, the adult sex ratio might be biased in favor of the earlier maturing sex but the sex ratio for all ages taken together should be biased in favor of the later maturing sex. Of course, if reproductive success for one sex increases strongly as a function of experience and experience only partly correlates with age, then the sex may be willing to suffer increased mortality if this mortality is sufficiently offset by increases in experience. Selander (1965) has suggested that the tendency of immature male blackbirds to exhibit some mature characteristics may be adaptive in that it increases the male's experience, although it also presumably increases his risk of mortality.

Mobility

Data from mammals (reviewed by Eisenberg 1965 and Brown 1966) and from some salamanders (Madison & Shoop 1970) and numerous lizards (Tinkle 1967 and Blair 1960) suggest that males often occupy larger home ranges and wander more widely than females *even when males are smaller* (Blair 1960). Parker (1970a) has quantified the importance of mobility and searching behavior in dung flies. If females are a dispersed resource, then male mobility may be crucial in exposing the male to a large number of available females. Again, males may be willing to incur greater mortality if this is sufficiently offset by increases in reproductive success. This factor should only affect the male during the breeding season (Kikkawa 1964) unless factors relevant to mobility (such as speed, agility or knowledge of the environment) need to be developed prior to the reproductive season. Lindburg (1969) has shown that macaque males, but not females, change troops more frequently during the reproductive season than otherwise and that this mobility increases male reproductive success as measured by frequency of copulation, suggesting that at least in this species, greater mobility can be confined to the reproductive season (see also Miller 1958). On the other hand, Taber and Dasmann (1954) present evidence that as early as six months of age male deer wander more widely from their mothers than females—a difference whose function, of course, is not known. Similar very early differences in mobility have been demonstrated for a lizard (Blair 1960) and for several primates, including man (Jensen, Bobbitt & Gordon 1968).

Female Choice

Although Darwin (1871) thought female choice an important evolutionary force, most writers since him have relegated it to a trivial role (Huxley 1938, Lack 1968; but see Fisher 1958 and Orians 1969). With notable exceptions the study of female choice has limited itself to showing that females are selected to decide whether a potential partner is of the right species, of the right sex and sexually mature. While the adaptive value of such choices is obvious, the adaptive value of subtler discriminations among broadly appropriate males is much more difficult to visualize or document. One needs both theoretical arguments for the adaptive value of such female choice and detailed data on how females choose. Neither of these criteria is met by those who casually ascribe to female (or male) choice the evolution of such traits as the relative hairlessness of both human sexes (Hershkovitz 1966) or the large size of human female breasts (Morris 1967). I review here theoretical considerations of how females might be expected to choose among the available males, along with some data on how females do choose.

Selection for Otherwise Neutral or Dysfunctional Male Attributes

The effects of female choice will depend on the way females choose. If some females exercise a preference for one type of male (genotype) while others mate at random, then other things being equal, selection will rapidly favor the preferred male type and the females with the preference (O'Donald 1962). If each female has a specific image of the male with whom she prefers to mate and if there is a decreasing probability of a female mating with a male as a function of his increasing deviation from her preferred image, then it is trivial to show that selection will favor distributions of female preferences and male attributes that coincide. Female choice can generate continuous male change only if females choose by a relative rather than an absolute criterion. That is, if there is a tendency for females to sample the male distribution and to prefer one extreme (for example, the more brightly colored males), then selection will move the male distribution toward the favored extreme. After a one generation lag, the distribution of female preferences will also move toward a greater percentage of females with extreme desires, because the granddaughters of females preferring the favored extreme will be more numerous than the granddaughters of females favoring other male attributes. Until countervailing selection intervenes, this female preference will, as first pointed out by Fisher (1958), move both male attributes and female preferences with increasing rapidity in the same direction. The female preference is capable of overcoming some countervailing selection on the male's ability to survive to reproduce, if the increased reproductive success of the favored males when mature offsets their chances of surviving to reproduce.

There are at least two conditions under which one might expect females to have been selected to prefer the extreme male of a sample. When two species, recently speciated, come together, selection rapidly favors females who can discriminate the two species of males. This selection may favor females who prefer the appropriate extreme of an available sample, since such a mechanism would minimize mating mistakes. The natural selection of females with such a mechanism of choice would then initiate sexual

selection in the same direction, which in the absence of countervailing selection would move the two male phenotypes further apart than necessary
to avoid mating error.

Natural selection will always favor female ability to discriminate male
sexual competence, and the safest way to do this is to take the extreme of
a sample, which would lead to runaway selection for male display. This case
is discussed in more detail below.

Selection for Otherwise Functional Male Attributes

As in other aspects of sexual selection, the degree of male investment in the
offspring is important and should affect the criteria of female choice. Where
the male invests little or nothing beyond his sex cells, the female has only
to decide which male offers the ideal genetic material for her offspring,
assuming that male is willing and capable of offering it. This question can
be broken down to that of which genes will promote the survival of her
offspring and which will lead to reproductive success, assuming the offspring
survive to adulthood. Implicit in these questions may be the relation between her genes and those of her mate: do they complement each other?

Where the male invests parental care, female choice may still involve the
above questions of the male's genetic contribution but should also involve,
perhaps primarily involve, questions of the male's willingness and ability to
be a good parent. Will he invest in the offspring? If willing, does he have
the ability to contribute much? Again, natural selection may favor female
attentiveness to complementarity: do the male's parental abilities complement her own? Can the two parents work together smoothly? Where males
invest considerable parental care, most of the same considerations that apply
to female choice also apply to male choice. The alternate criteria for female
choice are summarized in Table 1.

Sexual Competence

Even in males selected for rapid, repeated copulations the ability to do so
is not unlimited. After three or four successive ejaculations, for example,
the concentration of spermatozoa is very low in some male chickens (Parker,
McKenzie & Kempster 1940), yet males may copulate as often as 30 times
in an hour (Guhl 1951). Likewise, sperm is completely depleted in male
Drosophila melanogaster after the fifth consecutive mating on the same day
(Demerec & Kaufmann 1941, Kaufmann & Demerec 1942). Duration of
copulation is cut in half by the third copulation of a male dung fly on the
same day and duration of copulation probably correlates with sperm trans-
ferred (Parker 1970a). In some species females may be able to judge whether
additional sperm are needed (for example, house flies; Riemann, Moen &
Thorson 1967) or whether a copulation is at least behaviorally successful

Table 1. Theoretical criteria for female choice of males

I. All species, but especially those showing little or no male parental investment
   A. Ability to fertilize eggs
      (1) correct species
      (2) correct sex
      (3) mature
      (4) sexually competent
   B. Quality of genes
      (1) ability of genes to survive
      (2) reproductive ability of genes
      (3) complementarity of genes
II. Only those species showing male parental investment
   C. Quality of parental care
      (1) willingness of male to invest
      (2) ability of male to invest
      (3) complementarity of parental attributes

(for example, sea lions; Peterson & Bartholomew 1967), but in many species females may guarantee reproductive success by mating with those males who are most vigorous in courtship, since this vigor may correlate with an adequate supply of sperm and a willingness to transfer it.

When the male is completely depleted, there is no advantage in his copulating but selection against the male doing so should be much weaker than selection against the female who accepts him. At intermediate sperm levels, the male may gain something from copulation, but the female should again be selected to avoid him. Since there is little advantage to the male in concealing low reproductive powers, a correlation between vigor of courtship and sperm level would not be surprising. Females would then be selected to be aroused by vigorous courtship. If secondary structures used in display, such as bright feathers, heighten the appearance of vigorousness, then selection may rapidly accentuate such structures. Ironically the male who has been sexually most successful may not be ideal to mate with if this success has temporarily depleted his sperm supply. Males should not only be selected to recover rapidly from copulations but to give convincing evidence that they have recovered. It is not absurd to suppose that in some highly promiscuous species the most attractive males may be those who, having already been observed to mate with several females, are still capable of vigorous display toward a female in the process of choosing.

Good Genes

Maynard Smith (1956) has presented evidence that, given a choice, female *Drosophila subobscura* discriminate against inbred males of that species and that this behavior is adaptive: females who do not so discriminate leave about one-quarter as many viable offspring as those who do. Females may choose on the basis of courtship behavior: inbred males are apparently unable to perform a step of the typical courtship as rapidly as outbred males. The work is particularly interesting in revealing that details of courtship behavior may reveal a genetic trait, such as being inbred, but it suffers from an artificiality. If inbred males produce mostly inviable offspring, then, even in the absence of female discrimination, one would expect very few, if any, inbred males to be available in the adult population. Only because such males were artificially selected were there large numbers to expose to females in choice experiments. Had that selection continued one generation further, females who chose inbred males would have been the successful females.

Maynard Smith's study highlights the problem of analyzing the potential for survival of one's partner's genes: one knows of the adult males one meets that they have survived to adulthood; by what criterion does one decide who has survived better? If the female can judge age, then all other things being equal, she should choose older males, as they have demonstrated their capacity for long survival. All other things may not be equal, however, if old age correlates with lowered reproductive success, as it does in some ungulates (Fraser 1968) through reduced ability to impregnate. If the female can judge the physical condition of males she encounters, then she can discriminate against undernourished or sickly individuals, since they will be unlikely to survive long, but discrimination against such individuals may occur for other reasons, such as the presumed lowered ability of such males to impregnate successfully due to the weakened condition.

In some very restricted ways it may be possible to second-guess the future action of natural selection. For example, stabilizing selection has been demonstrated to be a common form of natural selection (see Mayr 1963) and under this form of selection females may be selected to exercise their own discrimination against extreme types, thereby augmenting the effects of any stabilizing selection that has occurred prior to reproduction. Mason (1969) has demonstrated that females of the California Oak Moth discriminate against males extreme in some traits, but no one has shown independent stabilizing selection for the same traits. Discrimination against extreme types may run counter to selection for diversity; the possible role of female choice in increasing or decreasing diversity is discussed below as a form of complementarity.

Reproductive success, independent of ability to survive, is easier for the female to gauge because she can directly observe differences in reproductive success before she chooses. A striking feature of data on lek behavior of birds is the tendency for females to choose males who, through competition with other males, have already increased their likelihood of mating. Female choice then greatly augments the effects of male-male competition. On the lek grounds there is an obvious reason why this may be adaptive. By mating with the most dominant male a female can usually mate more quickly, and hence more safely, than if she chooses a less dominant individual whose attempts at mating often result in interference from more dominant males. Scott (1942) has shown that many matings with less dominant individuals occur precisely when the more dominant individuals are unable, either because of sexual exhaustion or a long waiting line, to quickly service the female. Likewise, Robel (1970) has shown that a dominant female prevents less dominant individuals from mating until she has mated, presumably to shorten her stay and to copulate while the dominant male still can. A second reason why choosing to mate with more dominant males may be adaptive is that the female allies her genes with those of a male who, by his ability to dominate other males, has demonstrated his reproductive capacity. It is a common observation in cervids that females placidly await the outcome of male strife to go with the victor. DeVore (1965) has quantified the importance of dominance in male baboon sexual success, emphasizing the high frequency of interference by other males in copulation and the tendency for female choice, when it is apparent, to be exercised in favor of dominant males. The previous success may increase the skill with which males court females is suggested by work on the black grouse (Kruijt, Bossema & deVos, *in press*), and females may prefer males skillful at courting in part because their skill correlates with previous success.

In many species the ability of the male to find receptive females quickly may be more important than any ability to dominate other males. If this is so, then female choice may be considerably simplified: the first male to reach her establishes thereby a *prima facie* case for his reproductive abilities. In dung flies, in which females must mate quickly while the dung is fresh, male courtship behavior is virtually nonexistent (Parker 1970a). The male who first leaps on top of a newly arrived female copulates with her. This lack of female choice may also result from the *prima facie* case the first male establishes for his sound reproductive abilities. Such a mechanism of choice may of course conflict with other criteria requiring a sampling of the male population, but in some species this sampling could be carried out prior to becoming sexually receptive.

There are good data supporting the importance of complementarity of genes to female choice. Assortative mating in the wild has been demonstrated for several bird species (Cooch & Beardmore 1959, O'Donald 1959) and disassortative mating for a bird species and a moth species (Lowther 1961, Sheppard 1952). Petit and Ehrman (1969) have demonstrated the tendency in several *Drosophila* species for females to prefer mating with the rare type in choice experiments, a tendency which in the wild leads to a form of complementarity, since the female is presumably usually of the common type. These studies can all be explained plausibly in terms of selection for greater or lesser genetic diversity, the female choosing a male whose genes complement her own, producing an "optimal" diversity in the offspring.

Good Parent

Where male parental care is involved, females certainly sometimes choose males on the basis of their ability to contribute parental care. Orians (1969), for example, has recently reviewed arguments and data suggesting that polygyny evolves in birds when becoming the second mate of an already mated male provides a female with greater male parental contribution than becoming the first mate of an unmated male would. This will be so, for example, if the already mated male defends a territory considerably superior to the unmated male's. Variability in territory quality certainly occurs in most territorial species, even in those in which territories are not used for feeding. Tinbergen (1967), for example, has documented the tendency for central territories in the black-headed gull to be less vulnerable to predation. If females compete among themselves for males with good territories, or if males exercise choice as well, then female choice for parental abilities will again tend to augment intramale competition for the relevant resources (such as territories). The most obvious form of this selection is the inability of a non-territory-holding male to attract a female.

Female choice may play a role in selecting for increased male parental investment. In the roadrunner, for example, food caught by a male seems to act on him as an aphrodisiac: he runs to a female and courts her with the food, suggesting that the female would not usually mate without such a gift (Calder 1967). Male parental care invested after copulation is presumably not a result of female choice after copulation, since she no longer has anything to bargain with. In most birds, however, males defend territories which initially attract the females (Lack 1940). Since males without suitable territories are unable to attract a mate, female choice may play a role in maintaining male territorial behavior. Once a male has invested in a territory in order to attract a mate his options after copulating with her may be severely limited. Driving the female out of his territory would almost certainly result in the loss of his investment up until then. He could establish another territory, and in some species some males do this (von Haartman 1951), but in many species this may be difficult, leaving him with the option of aiding, more or less, the female he has already mated. Female choice, then, exercised *before* copulation, may indirectly force the male to increase his parental investment *after* copulation.

There is no reason to suppose that males do not compete with each other to pair with those females whose breeding potential appears to be high. Darwin (1871) argued that females within a species breeding early for nongenetic reasons (such as being in excellent physical condition) would produce more offspring than later breeders. Sexual selection, he argued, would favor males competing with each other to pair with such females. Fisher (1958) has nicely summarized this argument, but Lack (1968, p. 157) dismisses it as being "not very cogent," since "the date of breeding in birds has been evolved primarily in relation to two different factors, namely the food supply for the young and the capacity of the female to form eggs." These facts are, of course, fully consistent with Darwin's argument, since Darwin is merely supposing a developmental plasticity that allows females to breed earlier if they are capable of forming the eggs, and data presented elsewhere in Lack (1968) support the argument that females breeding earlier for nongenetic reasons (such as age or duration of pair bond) are more successful than those breeding later (see also, for example, Fisher 1969, Coulson 1966). Goforth and Baskett (1971) have recently shown that dominant males in a penned Mourning Dove population preferentially pair with dominant females; such pairs breed earlier and produce more surviving young than less dominant pairs. It would be interesting to have detailed data from other species on the extent to which males do compete for females with higher breeding potential. Males are certainly often initially aggressive to females intruding in their territories, and this aggressiveness may act as a sieve, admitting only those females whose high motivation correlates with early egg laying and high reproductive potential. There is good evidence that American women tend to marry up the socioeconomic scale, and physical attractiveness during adolescence facilitates such movement (Elder 1969). Until recently such a bias in female choice presumably correlated with increased reproductive success, but the value, if any, of female beauty for male reproductive success is obscure.

The importance of choice by both female and male for a mate who will not desert nor participate in sex outside the pair bond has been emphasized in an earlier section ("Desertion and cuckoldry"). The importance of complementarity is documented in a study by Coulson (1966).

Criteria Other than Male Characters

In many species male–male competition combined with the importance of some resource in theory unrelated to males, such as oviposition sites may mitigate against female choice for male characters. In the dragonfly *Parthemis tenera* males compete with each other to control territories containing good oviposition sites, probably because such sites are a predictable place at which to find receptive females and because sperm competition in insects usually favors the last male to copulate prior to oviposition (Parker 1970b). It is clear that the females choose the oviposition site and not the male (Jacobs 1955), and male courtship is geared to advertise good oviposition sites. A male maintaining a territory containing a good oviposition site is *not* thereby contributing parental investment unless that maintenance benefits the resulting young.

Female choice for oviposition sites may be an especially important determinant of male competition in those species, such as frogs and salamanders, showing external fertilization. Such female choice almost certainly predisposed these species to the evolution of male parental investment. Female choice for good oviposition sites would tend to favor any male investment in improving the site, and if attached to the site to attract other females the male would have the option of caring more or less for those eggs already laid. A similar argument was advanced above for birds. Internal fertilization and development mitigate against evolution of male parental care in mammals, since female choice can then usually only operate to favor male courtship feeding, which in herbivores would be nearly valueless. Female choice may also favor males who mate away from oviposition sites if so doing reduced the probability of predation.

Where females are clumped in space the effects of male competition may render female choice almost impossible. In a monkey troop a female preference for a less dominant male may never lead to sexual congress if the pair are quickly broken up and attacked by more dominant males. Apparent female acquiescence in the results of male–male competition may reflect this factor as much as the plausible female preference for the male victor outlined above.

Summary

The relative parental investment of the sexes in their young is the key variable controlling the operation of sexual selection. Where one sex invests considerably more than the other, members of the latter will compete among themselves to mate with members of the former. Where investment is equal, sexual selection should operate similarly on the two sexes. The pattern of relative parental investment in species today seems strongly influenced by the early evolutionary differention into mobile sex cells fertilizing immobile ones, and sexual selection acts to mold the pattern of relative parental investment. The time sequence of parental investment analyzed by sex is an important parameter affecting species in which both sexes invest considerable parental care: the individual initially investing more (usually the female) is vulnerable to desertion. On the other hand, in species with internal fertilization and strong male parental investment, the male is always vulnerable to cuckoldry. Each vulnerability has led to the evolution of adaptations to decrease the vulnerability and to counteradaptations.

Females usually suffer higher mortality rates than males in monogamous birds, but in nonmonogamous birds and all other groups, males usually suffer higher rates. The chromosomal hypothesis is unable to account for the data. Instead, an adaptive interpretation can be advanced based on the relative parental investment of the sexes. In species with little or no male parental investment, selection usually favors male adaptations that lead to high reproductive success in one or more breeding seasons at the cost of increased mortality. Male competition in such species can only be analyzed in detail when the distribution of females in space and time is properly described. Data from field studies suggest that in some species, size, mobility, experience and metabolic rate are important to male reproductive success.

Female choice can augment or oppose mortality selection. Female choice can only lead to runaway change in male morphology when females choose by a relative rather than absolute standard, and it is probably sometimes adaptive for females to so choose. The relative parental investment of the sexes affects the criteria of female choice (and of male choice). Throughout, I emphasize that sexual selection favors different male and females reproductive strategies and that even when ostensibly cooperating in a joint task male and female interests are rarely identical.

NOTES

Reprinted from Bernard Campbell, Editor, *Sexual Selection and the Descent of Man 1871–1971* (Chicago: Aldine Publishing Company, 1972).

I thank E. Mayr for providing me at an early date with the key reference. I thank J. Cohen, I. DeVore, W. H. Drury, M. Gadgil, W. D. Hamilton, J. Roughgarden, and T. Schoener for comment and discussion. I thank M. Sutherland (Harvard Statistics Department) for statistical work on my *A. garmani* data, H. Hare for help with references, and V. Hogan for expert typing of drafts of the paper. I thank especially E. E. Williams for comment, discussion and unfailing support throughout. The work was completed under a National Science Foundation predoctoral fellowship and partly supported by NSF Grant B019801 to E. E. Williams.

1. Selection should favor males producing such an abundance of sperm that then fertilize all a female's available eggs with a single copulation. Furthermore, to decrease competition among offspring, natural selection may favor females who prefer single paternity for each batch of eggs (see Hamilton 1964). The tendency for females to copulate only once or twice per batch of eggs is supported by data for many species (see, for example, Bateman 1948, Savage 1961, Burns 1968 but see also Parker 1970).

2. In particular, I assume an approximately 50/50 sex ratio at conception (Fisher 1958) and no differential mortality by sex, because I later derive differential mortality as a function of reproductive strategies determined by sexual selection. (Differential maturation, which affects the adult sex ratio, can also be treated as a function of sexual selection.) For most species the disparity in parental investment between the sexes is so great that the assumptions here can be greatly relaxed.

REFERENCES

Bartholomew, G. A. 1970. A model for the evolution of pinniped polygyny. *Evolution* 24: 546–559.

Bastock, M. 1967. *Courtship: An ethological study.* Chicago: Aldine.

Bateman, A. J. 1948. Intrasexual selection in *Drosophila. Heredity* 2: 349–368.

Beebe, W. 1925. The variegated Tinamou *Crypturus variegatus variegatus* (Gmelin). *Zoologica* 6: 195–227.

Beer, J. R., L. D. Frenzel, & C. F. MacLeod. 1958. Sex ratios of some Minnesota rodents. *American Midland Naturalist* 59: 518–524.

Beverton, J. M., & S. J. Holt. 1959. A review of the lifespan and mortality rates of fish in nature and their relation to growth and other physiological characteristics. In *The lifespan of animals,* ed. G. Wolstenhome & M. O'Connor, pp. 142–177. London: J. & A. Churchill.

Blair, W. F. 1960. *The Rusty Lizard.* Austin: University of Texas.

Bouliere, Z. F., & Verschuren, J. 1960. *Introduction a l'ecologie des ongules du Parc National Albert.* Bruxelles: Institut des Parcs Nationaux du Congo Belge.

Bragg, A. N. 1965. *Gnomes of the night.* Philadelphia: University of Pennsylvania Press.

Brown, L. E. 1966. Home range and movement of small mammals. *Symposium of the Zoological Society of London* 18: 111–142.

Bruce, H. 1960. A block to pregnancy in the mouse caused by the proximity of strange males. *Journal of Reproduction and Fertility* 1: 96–103.

Burns, J. M. 1968. Mating frequency in natural populations of skippers and butterflies as determined by spermatophore counts. *Proceedings of the National Academy of Sciences* 61: 852–859.

Calder, W. A. 1967. Breeding behavior of the Roadrunner, *Geococcyx californianus. Auk* 84: 597–598.

Carpenter, C. 1967. Aggression and social structure in Iguanid lizards. In *Lizard ecology,* ed. W. Milstead, Columbia, Mo.: University of Missouri.

Chapman, A. B., L. E. Casida, & A. Cote. 1938. Sex ratios of fetal calves. *Proceedings of the American Society of Animal Production* 1938, pp. 303–304.

Cooch, F. G., & M. A. Beardmore. 1959. Assortative mating and reciprocal difference in the Blue-Snow Goose complex. *Nature* 183: 1833–1834.

Corbet, P. C. Longfield, & W. Moore. 1960. *Dragonflies.* London: Collins.

Coulson, J. C. 1960. A study of the mortality of the starling based on ringing recoveries. *Journal of Animal Ecology* 29: 251–271.

———. 1966. The influence of the pair-bond and age on the breeding biology of the kittiwake gull *Rissa tridactyla. Journal of Animal Ecology* 35: 269–279.

Cowan, I. M. 1950. Some vital statistics of big game on overstocked mountain range. *Transactions of North American Wildlife Conference* 15: 581–588.

Darley, J. 1971. Sex ratio and mortality in the brown-headed cowbird. *Auk* 88: 560–566.

Darwin, C. 1871. *The descent of man, and selection in relation to sex.* London: John Murray.

Demerec, M., & Kaufmann, B. P. 1941. Time required for *Drosophila* males to exhaust the supply of mature sperm. *American Naturalist* 75: 366–379.

DeVore, I. 1965. Male dominance and mating behavior in baboons. In *Sex and behavior,* ed. Frank Beach. New York: John Wiley and Sons.

deVos, A. P. Broky, & V. Geist. 1967. A review of social behavior of the North American cervids during the reproductive period. *American Midland Naturalist* 77: 390–417.

Dunn, E. R. 1941. Notes on *Dendrobates auratus*. *Copeia* 1941, pp. 88–93.

Eaton, T. H. 1941. Notes on the life history of *Dendrobates auratus*. *Copeia* 1941, pp. 93–95.

Eisenberg, J. F. 1965. The social organizations of mammals. *Handbuch der Zoologie* 10 (7): 1–92.

Elder, G. 1969. Appearance and education in marriage mobility. *American Sociological Review* 34: 519–533.

Emlen, J. M. 1968. A note on natural selection and the sex-ratio. *American Naturalist* 102: 94–95.

Emlen, J. T. 1940. Sex and age ratios in the survival of the California Quail. *Journal of Wildlife Management* 4: 91–99.

Emlen, S. T. 1968. Territoriality in the bullfrog, *Rana catesbeiana*. *Copeia* 1968, pp. 240–243.

Engelmann, F. 1970. *The physiology of insect reproduction.* Oxford: Pergamon Press.

Fiedler, K. 1954. Vergleichende Verhaltensstudien an Seenadeln, Schlangennadeln und Seepferdchen (Syngnathidae). *Zeitsch. Tierpsych.* 11: 358–416. 358–416.

Fisher, H. 1969. Eggs and egg-laying in the Laysan Albatross, *Diomedea immutabilis*. *Condor* 71: 102–112.

Fisher, R. A. 1958. *The genetical theory of natural selection.* New York: Dover Publications.

Fraser, A. F. 1968. *Reproductive behavior in ungulates.* London and New York: Academic Press.

Gadgil, M., & W. H. Bossert. 1970. Life historical consequences of natural selection. *American Naturalist* 104: 1–24.

Goethals, G. W. 1971. Factors affecting permissive and nonpermissive rules regarding premarital sex. In *Studies in the sociology of sex: a book of readings,* ed. J. M. Henslin. New York: Appleton-Century-Croft.

Goforth, W., & T. Baskett. 1971. Social organization of penned Mourning Doves. *Auk* 88: 528–542.

Guhl, A. M. 1951. Measurable differences in mating behavior of cocks. *Poultry Science* 30: 687.

Haartman, L. von. 1951. Successive polygamy. *Behavior* 3: 256–274.

———. 1969. Nest-site and evolution of polygamy in European passerine birds. *Ornis Fennica* 46: 1–12.

Hamilton, J. B. 1948. The role of testicular secretions as indicated by the effects of castration in man and by studies of pathological conditions and the short life-span associated with maleness. *Recent Progress in Hormone Research* 3: 257–322.

Hamilton, J. B., & M. Johanson. 1965. Influence of sex chromosomes and castration upon lifespan: studies of meal moths, a species in which sex chromosomes are homogenous in males and heterogenous in females. *Anatomical Record* 24: 565–578.

Hamilton, J. B., & G. E. Mestler. 1969. Mortality and survival: comparison of eunuchs with intact men and women in a mentally retarded population. *Journal of Gerontology* 24: 395–411.

Hamilton, J. B., R. S. Hamilton, & G. E. Mestler. 1969. Duration of life and causes of death in domestic cats: influence of sex, gonadectomy and inbreeding. *Journal of Gerontology* 24: 427–437.

Hamilton, W. D. 1964. The genetical evolution of social behavior. *Journal of Theoretical Biology* 7: 1–52.

———. 1967. Extraordinary sex ratios. *Science* 156: 477–488.

Harris, V. A. 1964. *The life of the Rainbow Lizard*. London Hutchinson Tropical Monographs.

Hays, F. A. 1947. Mortality studies in Rhode Island Reds II. *Massachusetts Agricultural Experiment Station Bulletin* 442: 1–8.

Hershkovitz, P. 1966. Letter to *Science* 153: 362.

Hirth, H. F. 1963. The ecology of two lizards on a tropical beach. *Ecological Monographs* 33: 83–112.

Höhn, E. O. 1967. Observations on the breeding biology of Wilson's Phalarope (*Steganopus tricolor*) in Central Alberta. *Auk* 84: 220–244.

Huxley, J. S. 1938. The present standing of the theory of sexual selection. In *Evolution*, ed. G. DeBeer. New York: Oxford University Press.

Jacobs, M. 1955. Studies in territorialism and sexual selection in dragonflies. *Ecology* 36: 566–586.

Jensen, G. D., Bobbitt, R. A., & Gordon, B. N. 1968. Sex differences in the development of independence of infant monkeys. *Behavior* 30: 1–14.

Johns, J. E. 1969. Field studies of Wilson's Phalarope. *Auk* 86: 660–670.

Kaufmann, B. P., & Demerec, M. 1942. Utilization of sperm by the female *Drosophila melanogaster*. *American Naturalist* 76: 445–469.

Kessel, E. L. 1955. The mating activities of baloon flies. *Systematic Zoology* 4: 97–104.

Kikkawa, J. 1964. Movement, activity and distribution of small rodents *Clethrionomys glareolus* and *Apodemus sylvaticus* in woodland. *Journal of Animal Ecology* 33: 259–299.

Kluijver, H. N. 1933. Bijrage tot de biologie en de ecologie van den spreeuw (*Sturnus vulgaris* L.) gedurende zijn voortplantingstijd. *Versl. Plantenziekten-kundigen dienst, Wageningen* 69: 1–145.

Koivisto, I. 1965. Behavior of the black grouse during the spring display. *Finnish Game Research* 26: 1–60.

Kolman, W. 1960. The mechanism of natural selection for the sex ratio. *American Naturalist* 94: 373–377.

Kopstein, F. 1941. Uber Sexualdimorphismus bei Malaiischen Schlangen. *Temminckia*, 6: 109–185.

Kruijt, J. P, I. Bossema, & G. J. deVos. *In Press*. Factors underlying choice of mate in Black Grouse. *15th Congr. Intern. Ornith.*, The Hague, 1970.

Kruijt, J. P., & J. A. Hogan. 1967. Social behavior on the lek in Black Grouse, *Lyrurus tetrix tetrix* (L.) *Ardea* 55: 203–239.

Lack, D. 1940. Pair-formation in birds. *Condor* 42: 269–286.

———. 1954. *The natural regulation of animal numbers*. New York: Oxford University Press.

———. 1968. *Ecological adaptations for breeding in birds*. London: Methuen.

LeBoeuf, B. J., & R. S. Peterson. 1969. Social status and mating activity in elephant seals. *Science* 163: 91–93.

Lee, R., 1969, !Kung Bushman violence. Paper presented at meeting of American Anthropological Association, November, 1969.

Leigh, E. G. 1970. Sex ratio and differential mortality between the sexes. *American Naturalist* 104: 205–210.

Lin, N. 1963. Territorial behavior in the Cicada killer wasp *Sphecius spheciosus* (Drury) (Hymenoptera: Sphecidae.) I. *Behaviour* 20: 115–133.

Lindburg, D. G. 1969. Rhesus monkeys: mating season mobility of adult males. *Science* 166: 1176–1178.

Lowther, J. K. 1961. Polymorphism in the white-throated sparrow, *Zonotrichia albicollis* (Gmelin). *Canadian Journal of Zoology* 39: 281–292.

Ludwig, W., & C. Boost. 1951. Über Beziehungen zwischen Elternalter, Wurfgrösse und Geschlechtsverhältnis bei Hunden. *Zeitschrift fur indukt. Abstammungs und Vererbungslehre* 83: 383–391.

Madison, D. M., & Shoop, C. R. 1970. Homing behavior, orientation, and home range of salamanders tagged with tantalum-182. *Science* 168: 1484–1487.

Mason, L. G. 1969. Mating selection in the California Oak Moth (Lepidoptera, Droptidae). *Evolution* 23: 55–58.

Maynard Smith, J. 1956. Fertility, mating behaviour and sexual selection in *Drosophila subobscura*. *Journal of Genetics* 54: 261–279.

Mayr, E. 1939. The sex ratio in wild birds. *American Naturalist* 73: 156–179.

———. 1963. *Animal species and evolution*. Cambridge: Harvard University Press.

Miller, R. S. 1958. A study of a wood mouse population in Wytham Woods, Berkshire. *Journal of Mammalogy* 39: 477–493.

Morris, D. 1952. Homosexuality in the Ten-spined Stickleback (*Pygosteus pungitius*). *Behaviour* 4: 233–261.

———. 1967. *The naked ape*. New York: McGraw Hill.

Myers, J., & C. Krebs. 1971. Sex ratios in open and closed vole populations: demographic implications. *American Naturalist* 105: 325–344.

Nevo, R. W. 1956. A behavior study of the red-winged blackbird. 1. Mating and nesting activities. *Wilson Bulletin* 68: 5–37.

O'Donald, P. 1959. Possibility of assortative mating in the Arctic Skua. *Nature* 183: 1210.

———. 1962. The theory of sexual selection. *Heredity* 17: 541–552.

Orians, G. H. 1969. On the evolution of mating systems in birds and mammals. *American Naturalist* 103: 589–604.

Parker, G. A. 1970a. The reproductive behavior and the nature of sexual selection in *Scatophaga stercoraria* L. (Diptera: Scatophagidae) 2. The fertilization rate and the spatial and temporal relationships of each sex around the site of mating and oviposition. *Journal of Animal Ecology* 39: 205–228.

———. 1970b. Sperm competition and its evolutionary consequences in the insects. *Biological Reviews* 45: 525–568.

———. 1970c. The reproductive behaviour and the nature of sexual selection in *Scatophaga stercoraria* L. (Diptera: Scatophagidae). VI. The adaptive significance of emigration from the oviposition site during the phase of genital contact. *Journal of Animal Ecology* 40: 215–233.

———. 1970d. The reproductive behaviour and the nature of sexual selection in *Scatophaga stercoraria* L. (Diptera: Scatophagidae). VI. The adaptive sig-evolution of the passive phase. *Evolution* 24: 774–788.

Parker, J. E., F. F. McKenzie, & H. L. Kempster. 1940. Observations on the sexual behavior of New Hampshire males. *Poultry Science* 19: 191–197.

Peterson, R. S., & G. A. Bartholomew. 1967. *The natural history and behavior of the*

California Sea Lion. Special Publications #1, American Society of Mammalogists.

Petit, C., & L. Ehrman. 1969. Sexual selection in *Drosophila*. In *Evolutionary biology*, vol. 5, ed. T. Dobzhansky, M. K. Hecht, W. C. Steere. New York: Appleton-Century-Crofts.

Potts, G. R. 1969. The influence of eruptive movements, age, population size and other factors on the survival of the Shag (*Phalacrocorax aristotelis* L.) *Journal of Animal Ecology* 38: 53–102.

Rand, A. S. 1967. Ecology and social organization in the iguanid lizard *Anolis lineatopus*. *Proc. U.S. Nat. Mus.* 122: 1–79.

Riemann, J. G., D. J. Moen, & B. J. Thorson. 1967. Female monogamy and its control in house flies. *Insect Physiology* 13: 407–418.

Rivero, J. A., & A. E. Estevez. 1969. Observations on the agonistic and breeding behavior of *Leptodactylus pentadactylus* and other amphibian species in Venezuela. *Breviora No.* 321:1–14.

Robel, R. J. 1966. Booming territory size and mating success of the Greater Prairie Chicken (*Tympanuchus cupido pinnatus*). *Animal Behaviour* 14: 328–331.

———. 1970. Possible role of behavior in regulating greater prairie chicken populations. *Journal of Wildlife Management* 34: 306–312.

Robinette, W. L., J. S. Gashwiler, J. B. Low, & D. A. Jones. 1957. Differential mortality by sex and age among mule deer. *Journal of Wildlife Management* 21: 1–16.

Rockstein, M. 1959. The biology of ageing insects. In *The lifespan of animals*, ed. G. Wolstenhome & M. O'Connor, pp. 247–264. London: J. A. Churchill.

Rowley, I. 1965. The life history of the Superb Blue Wren *Malarus cyaneus*. *Emu* 64: 251–297.

Royama, T. 1966. A re-interpretation of courtship feeding. *Bird Study* 13: 116–129.

Sadleir, R. 1967. *The ecology of reproduction in wild and domestic mammals*. London: Methuen.

Savage, R. M. 1961. *The ecology and life history of the common frog*. London: Sir Isaac Pitman and Sons.

Schein, M. W. 1950. The relation of sex ratio to physiological age in the wild brown rat. *American Naturalist* 84: 489–496.

Schoener, T. W. 1967. The ecological significance of sexual dimorphism in size in the lizard *Anolis conspersus*. *Science* 155: 474–477.

———. 1971. Theory of feeding strategies. *Annual Review of Ecology and Systematics*, 2: 369–404.

Scott, J. W. 1942. Mating behavior of the Sage Grouse. *Auk* 59: 477–498.

Selander, R. K. 1965. On mating systems and sexual selection. *American Naturalist* 99:129–141.

———. 1966. Sexual dimorphism and differential niche utilization in birds. *Condor* 68: 113–151.

Sellers, A., H. Goodman, J. Marmorston, & M. Smith. 1950. Sex differences in proteinuria in the rat. *American Journal of Physiology* 163: 662–667.

Sheppard, P. M. 1952. A note on non-random mating in the moth *Panaxia dominula* (L.). *Heredity* 6: 239–241.

Stephens, M. N. 1952. Seasonal observations on the Wild Rabbit (*Oryctolagus cuniculus cuniculus* L.) in West Wales. *Proceedings of the Zoological Society of London* 122: 417–434.

Stokes, A., & H. Williams. 1971. Courtship feeding in gallinaceous birds. *Auk* 88: 543–559.

Sugiyama, U. 1967. Social organization of Hanuman langurs. In *Social communication among primates*, ed. S. Altmann. Chicago: University of Chicago Press.

Taber, R. D., & R. F. Dasmann. 1954. A sex difference in mortality in young Columbian Black-tailed Deer. *Journal of Wildlife Management* 18: 309–315.

Tilley, S. 1968. Size-fecundity relationships and their evolutionary implications in five desmognathine salamanders. *Evolution* 22: 806–816.

Tinbergen, N. 1935. Field observations of East Greenland birds. 1. The behavior of the Red-necked Phalarope (*Phalaropus lobatus* L.) in Spring. *Ardea* 24: 1–42.

———. 1967. Adaptive features of the Black-headed Gull *Larus ridibundus* L. *Proceedings of the International Ornithological Congress* 14: 43–59.

Tinkle, D. W. 1967. The life and demography of the Side-blotched Lizard, *Uta stansburiana*. *Miscellaneous Publications of the Museum of Zoology, University of Michigan* 132: 1–182.

Tinkle, D., H. Wilbur, & S. Tilley. 1970. Evolutionary strategies in lizard reproduction. *Evolution* 24: 55–74.

Tyndale-Biscoe, C. H., & R.F.C. Smith. 1969. Studies on the marsupial glider, *Schoinobates volans* (Kerr). 2. Population structure and regulatory mechanisms. *Journal of Animal Ecology* 38: 637–650.

Van Lawick-Goodall, J. 1968. The behavior of free-living chimpanzees in the Gombe Stream Reserve. *Animal Behaviour Monographs* 1: 181–311.

Verner, J. 1964. Evolution of polygamy in the long-billed marsh wren. *Evolution* 18: 252–261.

———. 1965. Selection for sex ratio. *American Naturalist* 99: 419–421.

Verner, J., & M. Willson. 1969. Mating systems, sexual dimorphism, and the role of male North American passerine birds in the nesting cycle. *Ornithological Monographs* 9: 1–76.

Williams, G. C. 1966. *Adaptation and natural selection*. Princeton: Princeton University Press.

Willson, M., & E. Pianka. 1963. Sexual selection, sex ratio, and mating system. *American Naturalist* 97: 405–406.

Wood, D. H. 1970. An ecological study of *Antechinus stuartii* (Marsupialia) in a Southeast Queensland rain forest. *Australian Journal of Zoology* 18: 185–207.

Postscript

There is an enormous literature on the evolution of sex differences. As of June 2001 there were more than 3,000 science citations to my parental investment paper alone. This is in part because I defined a term—parental investment—and then stuck the term on the discipline. It was then useful for many other people when using the term to say "as defined by Trivers 1972." I say this because, of the 3,000 citations, perhaps fully one-third are to the particular page on which the term was defined. There are, of course, many more thousands of papers that deal with the two sexes in one form or another without referring to my paper. Perhaps the best the reader can do who wishes to be updated on the subject is to go to Andersson's (1994) general review of sexual selection, an excellent book.

The most important new work on the subject was the realization that sexual reproduction without male parental investment is a very inefficient mode of reproduction (compared to asexual reproduction). This, in turn, led to a burst of theoretical activity on the meaning of sex and the realization that parasites were likely to play an especially prominent role in generating sex in the host—the better to permit the host to recombine new combinations against the co-evolving parasites. This, in turn, led to the supposition that selection for parasite-resistant genes is expected to be an important part of mate choice and that bright color in male and elaborate song may both serve to reveal males relatively free of parasite infestation and damage (Hamilton and Zuk 1982).

3

THE TRIVERS-WILLARD EFFECT

Dan Willard was a graduate student in mathematics at Harvard University, and he wanted to meet some women. There were very few women in mathematics at Harvard in those days (1970), so Dan started to prowl the halls in search of courses with more agreeable sex ratios. He soon found one in Professor DeVore's renowned primate behavior course. This course attracted almost three hundred students each year, about two-thirds of them women. Dan developed an interest in primate behavior. I was a graduate teaching assistant in Professor DeVore's course, and my duties included giving a few lectures to the class. One lecture was on the sex ratio at birth, and another one was on the logic of mate choice. Dan Willard put pieces of the two lectures together and came up with a brand new argument for adaptive variation in the sex ratio at birth.

In the lecture on the sex ratio at birth, I explained to the class Fisher's sex ratio theory, which, under outbreeding, resulted in 50/50 sex ratios. This is because any deviations from a 1 : 1 sex ratio are unstable: whichever sex is produced in greater numbers becomes less valuable to its parents than offspring of the opposite sex, because it has a lower chance of reproducing. For example, an excess of males means that each male has less than one female on average to fertilize. I emphasized that as long as sex ratios canceled out across a population to give an overall sex ratio of 1 : 1, each was equally adaptive. Thus, a 5 : 1 sex ratio might cancel out with a 1 : 5 sex ratio in a population of individuals otherwise tending to produce 1 : 1 sex ratios, and 1 : 5, 5 : 1, and 1 : 1 would be equally adaptive. This turned out to be a key point in Willard's argument.

In the lecture on mate choice I had noted that in the human species, a more or less monogamous one, there was evidence that women tended on average to marry up the socioeconomic scale. A consequence of this bias was that some women on the upper end of the scale would be left without men to marry, while the same thing would be true for some men at the lower end of the scale. Since there is a strong tendency for the socioeconomic status of the parents to determine the socioeconomic status of the children, Willard realized that it made sense for parents at the upper end of the scale to produce a male-biased sex ratio so as to take advantage of females eager to marry up the scale, while parents at the lower end of the scale would show the opposite bias, preferring to produce daughters, who could always marry up, rather than sons who might have no one with whom to pair. This kind of sex ratio adjustment met the Fisherian test for sex ratios that suffered no overall bias; that is, the two kinds of sex ratios tend to cancel out to give an overall sex ratio of 1 : 1. (They need not cancel out exactly, as David Haig recently pointed out to me: if females had perfect information and could predict which sons would be successful, they would be selected to cut down on the number produced.)

I think Willard put together this theory during my lecture on mate choice because he came up to me after the lecture and presented his theory in the form of a question: Wouldn't it make sense, he asked, for humans to adjust the sex ratios they produce according to their socioeconomic status so as to take advantage of women's tendency to marry up the scale? I liked the argument the moment I heard it. For one thing, the sex ratio adjustment he described was stable, and a change in the key underlying parameter (i.e., chance of marrying up) would produce compensatory changes in the sex ratio adjustment. For another thing, I knew of a fact that immediately lent support to Willard's theory, namely, that the sex ratio at birth, in both England and the United States, is associated with socioeconomic class in exactly the right direction. It is not often that a student generates a new, true theory that correctly predicts a fact of which the student is unaware!

I encouraged Dan Willard to develop his theory. I gave him, as I remember it, a couple of sex ratio references in mammals and urged him to generalize the argument as much as he could, that is, to figure out the form of it that would apply to species more broadly. There was no response. The course soon ended and Dan presumably went on to other courses in further search of women, but in any case he neglected his little sex ratio theory.

One day I sat down to develop the argument myself, and I did it in the usual style. That is, I took an argument that applied to the human species and tried to generalize it as far as I could. Parental investment seemed like an immediately attractive variable to substitute for socioeconomic status, since this at once extended the argument to all other mammals, among other species, and produced several independent predictions, for example, based on maternal litter size or maternal condition. The logic of the Trivers-

Willard effect, as it came to be known, was very simple. If a son's reproductive success increased more rapidly as a function of parental investment
than did a daughter's, then parents would be selected, when having less
than average resources to invest per offspring (e.g., because of a larger litter
size), to invest in an excess of females, and vice versa when there was more
investment available per offspring. Independently, with litter size held constant, one expected a positive correlation between the sex ratio and maternal
condition or, as I discuss below, maternal dominance.

Over about a five-week period of time the manuscript was written more
or less as you see it here, except that I was the lone author and there was a
warm acknowledgment to Dan Willard for first suggesting the idea. I sent
the manuscript to Dan, care of the mathematics department, while at the
same time I submitted the paper to *Science*. Dan contacted me shortly after
the paper was accepted by *Science* and said that, in all fairness, we ought to
be co-authors. I readily agreed (happy, I suppose, that he was not asking for
senior authorship!) but now I also felt somewhat ambivalent about all the
work I had done, so I wrote a new acknowledgment, which embarrasses me
to this day and reads . . . "R.L.T. and D.E.W. independently conceived the
basic theory. The collection of data and writing of the paper were performed
by R.L.T. alone." What nonsense! If it were really true that Dan and I had
conceived the idea independently I can assure you that no one today would
have heard of Dan Willard—at least not where sex ratio theory is concerned!
If I had wanted to divide credit between the two of us I should have said
"The idea was Dan's; the exposition, Bob's."

Incidentally, I like to pat myself on the back for always trying to generalize outward in my thinking, but the title of this paper is evidence of such
a tendency taken too far. I actually talked myself into believing that this was
the first argument for adaptive variation in the sex ratio, when I already
knew Hamilton's famous 1967 paper "Extraordinary Sex Ratios," which
among other things explored adaptive variation in sex ratio in response to
variation in inbreeding. Also, as it turns out, students of solitary wasps had
already shown in one or two species that females would lay female eggs on
larger hosts and male eggs on smaller ones, so adjustment of the sex ratio
to resources available per offspring had itself already been shown.

I remember sharing the manuscript with Ernst Mayr. He liked the paper
but thought I gave too much emphasis to differential mortality by sex during
the uterine period, as a mechanism to bring about the expected correlations.
It was well known in humans and other mammals that males suffered higher
mortality than females, even in utero, but he thought there was evidence
that the sex ratio might be adjusted at conception as well, and he certainly
saw no reason, mechanistically, to deny such a possibility. Since under some
circumstances it would be the most efficient way of making the adjustment,

it should not be left out. I was only too glad to follow his advice. I found in those days that he was usually the best person to see concerning a paper because his instincts were so good, and his knowledge very extensive.

About three weeks before the paper appeared, I flew to Washington, D.C., in December 1972 to deliver a talk on parent–offspring conflict at the AAAS meetings (see chap. 4). There I met for the first time Richard Alexander of the University of Michigan and his graduate students. Alexander was to deliver a talk at the same symposium on the subject of parental manipulation. We were having drinks at a bar when Paul Sherman, one of his graduate students (later Professor of Biology at Cornell), told me that Dr. Alexander had an interesting idea regarding sex ratio. He imagined that in societies that practice infanticide it might show a bias by socioeconomic status due to the tendency of men at the top end of the scale to have multiple wives. This was remarkably similar to Dan Willard's first idea, though perhaps more specifically limited to the phenomenon of infanticide. I said, "Curious that you should say so," and told him about our sex ratio paper, then only days from publication. This was one of a remarkable series of parallels in thinking between myself and Richard Alexander. When I published my paper on reciprocal altruism he was, I understand, lecturing on "reciprocal beneficence," and while I had a theory of parent–offspring relations, he had a theory of parental manipulation. Alexander had achieved something rare for scientists. He had retooled himself in middle age so as to master an emerging discipline, elements of which he had known only weakly or not at all. He was an expert in crickets, grasshoppers, and their allies and had done evolutionary biology of the classic sort that included taxonomy, ecology, behavior, zoogeography, and so on. But like the rest of us, he was a good bit weaker on his genetics and his formal evolutionary theory, so I was conscious then that his decision to retool was initially costly in that a younger scientist such as myself was able to beat him to the punch, in effect, on some important topics.

The same fate was to befall me later in life when I tried to master genetics in order to contribute to our understanding of the evolution of selfish genetic elements (see chap. 9). Here I have been mostly reduced to a bystander, appreciating the work of younger people, better trained in both genetics and formal modeling, and more energetic at putting together the new literature. People such as David Haig, Laurence Hurst, and my genetics coauthor Austin Burt.

When the paper was published I confidently predicted that this was the perfect paper "because it would take them twenty years to prove us wrong." *I* was to be proven wrong myself in a most agreeable fashion only thirteen years later. But first, the paper itself.

⚬⚬ ◯ ⚬⚬

Natural Selection of Parental Ability to Vary the Sex Ratio of Offspring

ROBERT L. TRIVERS AND DAN E. WILLARD

Abstract.   Theory and data suggest that a male in good condition at the end of the period of parental investment is expected to outreproduce a sister in similar condition, while she is expected to outreproduce him if both are in poor condition. Accordingly, natural selection should favor parental ability to adjust the sex ratio of offspring produced according to parental ability to invest. Data from mammals support the model: As maternal condition declines, the adult female tends to produce a lower ratio of males to females.

Fisher (*1*) showed, and others (*2*) reformulated, that natural selection favors those parents who invest equally in both their sons and their daughters. When the parents invest the same in an average son as in an average daughter, natural selection favors a 50/50 sex ratio (ratio of males to females) at conception (*3,4*). (For simplicity, we assume here that parents are investing equally in average offspring of either sex.) Individuals producing offspring in sex ratios that deviate from 50/50 are not selected against as long as these deviations exactly cancel out and result in a sex ratio at conception of 50/50 for the local breeding population. Such a situation is highly unstable, since random deviations from the 50/50 ratio in local populations rapidly favor those individuals producing their young in ratios of 50/50. We show here that under certain well-defined conditions, natural selection favors systematic deviations from a 50/50 sex ratio at conception, and that these deviations tend to cancel out in the local breeding population.

Imagine a population of animals (for instance, caribou) in which the condition of adult females varies from good to poor (as measured, for example, by weight). Assume that a female in good condition is better able to bear and nurse her calf than is a female in poor condition, so that at the end of the period of parental investment (PI), the healthiest, strongest, and heaviest calves will tend to be the offspring of the adult females who were in the best condition during the period of PI. Assume that there is some tendency for differences in the condition of calves at the end of the period of PI to be maintained into adulthood. Finally, assume that such adult differences in condition affect male reproductive success (RS) more strongly than they affect female RS. That is, assume that male caribou in good condition tend to exclude other males from breeding, thereby inseminating many more females themselves, while females in good condition, through their greater ability to invest in their young, show only a moderate increase in RS. Under these assumptions, an adult female in good condition who produces a son will leave more surviving grandchildren than a similar female who produces a daughter, while an adult female in poor condition who produces a daughter will leave more surviving grandchildren than a similar female who produces a son.

In short, natural selection favors the following reproductive strategy. As females deviate from the mean adult female condition they should show an increasing tendency to bias the production of their young toward one sex or the other. Whenever variance around some mean condition is a predictable attribute of adults in a species, natural selection will arrange the deviations away from a 50/50 sex ratio at conception so that the deviations will tend to cancel out. Other things being equal, species showing especially high variance in male RS (compared to variance in female RS) should show, as a function of differences in maternal condition, especially high variance in sex ratios produced.

The model we are advancing depends on three assumptions, for which there are both supporting data and theoretical arguments.

1. The condition of the young at the end of PI will tend to be correlated with the condition of the mother during PI. This has been shown for many species (5–7) and is probably true of almost all animals with small brood sizes. It is sometimes true of species with large, highly variable brood sizes but need not be (7).

2. Differences in the condition of young at the end of the period of PI will tend to endure into adulthood. Although animals show some capacity for compensatory growth, we would be surprised if this claim were not often true. It has been demonstrated experimentally for laboratory and farm animals (8). In rats, for example, differences in weanling size due to differences in litter size are maintained into adulthood (9). Throughout life, human twins lag behind their singleton counterparts in height and weight (10) and in RS (11).

3. Adult males will be differentially helped in RS (compared to adult females) by slight advantages in condition. In all species showing negligible PI by males, male RS, is expected to vary more than female RS, and considerable evidence supports this claim (4, 12). In theory, slight advantages in condition should (because of male competition to inseminate females) have disproportionate effects on male RS compared to the effects on female RS.

We assume that sex ratio at birth in mammals is a measure of tendency to invest in one sex more than in the other. With this assumption, available data from several species support the prediction that females in better condition tend to invest in males (13–15). Adverse environmental conditions for the mother during pregnancy are correlated with a reduced sex ratio at birth in deer and humans. Experimentally induced stress of rabbits in utero apparently reduces the sex ratio at birth. In dogs, deer, and humans, two variables that correlate with decreased maternal investment per offspring (maternal parity and litter size) correlate with reduced sex ratios at birth. Likewise, increasing litter size in mink and sheep correlates inversely with sex ratio at birth. Naturally occurring variations in sex ratio at birth can be large; in two seal species (16), females pupping early in the season produce sex ratios larger than 120/100, while females pupping late produce a complementary ratio (less than 80/100).

Since females in good condition are assumed to outreproduce females in poor condition, it is not possible for genes producing one sex ratio to accumulate among females in poor condition and genes for the complementary sex ratio to accumulate among females in good condition. Instead, natural selection must favor one or more genes that adjust the sex ratio produced by an adult female to her own condition at the time of PI. In species such as mammals, in which males determine sex of offspring, female control of the sex ratio must involve differential mortality by sex, either of sperm cells (17) or of the growing young during PI. If, as in caribou, maternal PI extends over a period of time in which maternal condition may unpredictably deteriorate, then the female who can make adjustments during that period should out-compete the female who adjusts the sex ratio only at the very beginning of PI. In general, of course, the earlier the adjustment, the better. Differential male mortality during the period of parental investment ought to be part of the mechanism by which a female adjusts the sex ratio of her young in such a way as to maximize her eventual reproductive success. Differential male mortality in utero has been demonstrated for deer, cows, and humans; and most of the differential mortality takes place early in pregnancy (18).

As it applies to mammals, the model amounts in part to an adaptive interpretation of early differential male mortality. We know of no alternate functional model for such differential mortality. The usual nonfunctional argument (that the unguarded X chromosome of the male predisposes him to differential mortality) not only fails as a general explanation of differential mortality by sex (19), it also fails to account for the influence of maternal condition on differential mortality in utero and for species differences in the degree of early differential mortality. Careful attempts to measure the contribution of the unguarded X chromosome of the human male to his differential mortality in utero have concluded that the contribution must be negligible (20). That variations in sex ratio as large as those observed in nature should be a matter of indifference to the individuals producing them seems most unlikely.

The application of the model to humans is complicated by the tendency for males to invest parental effort in their young (which reduces variance in male RS), and by the importance of kin interactions among adults (21). Despite these complications, the model can be applied to humans differentiated on a socioeconomic scale, as long as the RS of a male at the upper end of the scale exceeds his sister's, while that of a female at the lower end of the scale exceeds her brother's. A tendency for the female to marry a male whose socioeconomic status is higher than hers will, other things being equal, tend to bring about such a correlation, and there is evidence of such a bias in female choice in the United States (22). The corresponding prediction is satisfied: Sex ratio at birth correlates with socioeconomic status (14).

If the model is correct, natural selection favors deviations away from 50/50 investment in the sexes, rather than deviations in sex ratios per se. In species with a long period of PI after birth of young, one might expect biases in parental behavior toward offspring of different sex, according to parental condition; parents in better condition would be expected to show a bias toward male offspring.

REFERENCES AND NOTES

1. R. A. Fisher, *The Genetical Theory of Natural Selection* (Clarendon, Oxford, 1930).

2. Discussed by E. Leigh, *Amer. Natur.* 104, 205 (1970).

3. As pointed out by Fisher (*1*), male differential mortality during the period of parental investment will mean (other things being equal) that *parents Invest less, on the average*, in each male conceived than in each female conceived. The sex ratio in such species should be higher than 50/50 at conception and lower than 50/50 at the end of the period of parental investment. For a definition of parental investment, see (*4*).

4. R. L. Trivers, in *Sexual Selection and the Descent of Man, 1871–1971*, B. Campbell, Ed. (Aldine-Atherton, Chicago, 1972), pp. 136–179.

5. Sheep: L. R. Wallace, *J. Physiol. London* 104, 34 (1945); mice: N. Bateman, *Physiol. Zool.* 27, 163 (1954); humans: J. McClung, *Effects of High Altitude on Human Birth* (Harvard Univ. Press, Cambridge, Mass., 1969).

6. Deer: W. L. Robinette, J. S. Geshwiler, J. B. Low, D. A. Jones, *J. Wildl. Manage.* 21, 1 (1957).

7. T. W. Schoener, *Annu. Rev. Ecol. Syst.* 2. 369 (1971): R. M. F. S. Sadleir, *The Ecology of Reproduction in Wild and Domestic Mammals* (Methuen, London, 1969).

8. J. Moustgaard, in *Reproduction in Domestic Animals*, H. H. Cole and P. T. Cupps, Eds. (Academic Press, New York, ed. 2, 1969); sheep: R. W. Phillips and W. M. Dawson, *Proc. Amer. Soc. Anim. Prod.* 30, 296 (1938); salmon: N. Ryman, *Hereditas* 70, 119 (1972).

9. G. C. Kennedy, *Ann. N.Y. Acad. Sci.* 157, 1049 (1969).

10. M. G. Bulmer, *The Biology of Twinning in Man* (Clarendon, Oxford, 1970), p. 64.

11. G. Wyshak and C. White, *Hum. Biol.* 41, 66 (1969). As our theory would predict, there is a slight but consistent tendency for a male twin to show a greater reduction in RS (compared to his singleton counterpart) than a female twin shows (compared to her singleton counterpart).

12. A. J. Bateman, *Heredity* 2, 349 (1948). In species showing greater PI by males than by females, reviewed by Trivers (*4*), female RS is expected to vary more strongly than male RS. In such species, parents in poor condition should prefer to produce males.

13. Deer: Robinette *et al.* (*6*); humans: Shapiro *et al.* (*14*); pigs: R. R. Maurer and R. H. Foote, *J. Reprod. Fert.* 25. 329 (1971); dogs: (*13a*); mink: R. Apelgren [Vâra Palsdjur 12, 349 (1941)], cited in R. K. Enders. *Proc. Amer. Phil. Soc.* 96, 691 (1952); sheep: K. Rasmussen, *Sci. Agr.* 21, 759 (1941).

13a. W. Ludwig and C. Boost, *Z. Induct. Abstamm. Verebungsl.* 83, 383 (1951).

14. S. Shapiro, E. R. Schlesinger, R.E.L. Nesbitt, Jr., *Infant. Perinatal, Maternal, and Childhood Mortality in the United States* (Harvard Univ. Press, Cambridge, Mass., 1968).

15. After this paper was accepted for publication, R. Klester (Department of Biology, Harvard) kindly brought to our attention independent data tending to confirm the theory. Adult female red deer who fail to breed the preceding year (and are therefore presumably in better than normal condition during the present year) appear to produce a much higher sex ratio than do adult females who bred the preceding year [F. F. Darling, *A Herd of Red Deer* (Oxford Univ. Press, London, 1937), pp. 46–48].

16. Grey seal: J. C. Coulson and G. Hickling, *Nature* 190, 281 (1961); Weddell seal: I. Stirling, *J. Mammal.* 52, 842 (1971).

17. It is a common observation of animal breeders that the later one mates a female mammal in her estrous cycle the greater the chance of producing males [(*13a*); also W. H. James, *Lancet* 1971-I, 112 (1971)]. Late matings minimize the time between copulation and fertilization and would therefore minimize differential mortality by sex of the sperm cell, presumed to operate against male-producing sperm.

18. Deer: Robinette *et al.* (*6*); cows: A. B. Chapman, L. E. Cassida, A. Cote, *Proc. Amer. Soc. Anim. Prod.* 30, 303 (1938); humans: S. Shapiro, E. W. Jones, P. M. Densen, *Milbank Mem. Fund Quart.* 40, 7 (1962).

19. J. B. Hamilton, R. S. Hamilton, G. E. Mestler, *J. Gerontol.* 24, 427 (1969). Arguments and data were reviewed by Trivers (4, p. 152).

20. A. C. Stevenson and M. Bobrow, *J. Med. Genet.* 4, 190 (1967).

21. If members of one sex perform more altruistic acts toward kin of the opposite sex than the other way around, then one can show that parents will be selected to invest, on the average more than 50 percent of their resources in producing offspring of the more altruistic sex. This factor may be important in explaining the apparent human overproduction of sons (R. L. Trivers, in preparation).

22. G. Elder, *Amer. Sociol. Rev.* 34, 519 (1969); H. Carter and P. C. Glick, *Marriage and Divorce: A Social and Economic Study* (Harvard Univ. Press, Cambridge, Mass., 1970).

23. R.L.T. and D.E.W. independently conceived the basic theory. The collection of data and writing of the paper were performed by R.L.T. alone. We thank I. DeVore, B. J. LeBouef, and T. Schoener for detailed comments. We thank H. Hare

for help finding references. R.L.T. thanks I. DeVore for advice and unfailing support throughout. The work was completed under an NIH postdoctoral fellowship to R.L.T. and partly supported by NIMH grant 13156 to I. DeVore.

Postscript

In the first few years after this paper was published, one or two studies appeared providing limited support for the theory and two or three papers appeared failing to confirm a Trivers-Willard effect. The low point for the paper came in 1979, when G. C. Williams, in a paper entitled "The Question of Adaptive Sex Ratio in Outcrossed Vertebrates," questioned whether there could be any Trivers-Willard effect in vertebrates and purported to show evidence all but ruling out such an effect.

George's argument took the following form. For evidence, he claimed that a Trivers-Willard effect ought to cause a deviation from binomial sex ratios in different litter sizes, and since in a few data sets he failed to find such an effect, he claimed that evidence argued against the existence of such an effect. I had a great deal of respect for George Williams, especially in demarcating the line between a real adaptation and a false one, and I thought that if I were going to make a mistake in this regard I would tend to make the mistake of overinterpretation, of seeing an effect of natural selection where there was none. So, George's paper gave me some uneasy moments. On the other hand, I thought his argument regarding the evidence was a weak position to take. It is easy to imagine patterns of sex ratio correlation that result in overall binomial distribution but still show significant sex ratio correlations with relevant underlying variables, such as maternal condition. Furthermore, the test Williams used, deviation from binomial distribution, was a very weak test and required large samples to spot weak deviations. In principle, it also seemed unwise to rope off a phenomenon like this in advance, as Williams's paper appeared to do.

Nevertheless, Williams (1979) gave me some pause, and I sometimes wondered whether the whole thing had merely been a mirage. Such thoughts were shattered for good in 1984 when Timothy Clutton-Brock and colleagues working on the red deer showed a Trivers-Willard effect and every component of the logic thereof. Relatively dominant mothers produced a high proportion of sons (about three out of five) while females low in dominance produced more daughters than sons. At the same time, they showed that dominance had effects on the parental investment per offspring in the expected direction and that these effects lasted into adulthood. Finally, and the most extraordinary to me, they actually had lifetime reproductive success of males and females as a function of the dominant status of their mothers. These data clearly showed that a son's reproductive success was a rising function of maternal dominance but a daughter's reproductive success was unassociated with maternal dominance.

Clutton-Brock, Albon, and Guinness (1984) was also a vivid reminder to me of how I had underestimated the power of evolutionary biology to change while overestimating the ability of the social sciences to do so. In 1970 I was fond of saying, "Twenty years from now you will not be able to walk down the hall of any social science department without hearing people say, 'I wonder why natural selection favors *that*.' " It is now thirty years later and you can walk down the halls of most social science departments without fear that you will ever hear the words "natural selection." But biology, being a unified discipline, has a tremendous capacity to respond to new theory by altering empirical work and producing data of a very high precision. In evolutionary biology you also have the advantage that many people are wedded to a particular groups of organisms—*Anolis* lizards, say—and are only too happy to give up less interesting work on their creatures for more promising (and fundable) studies, while social scientists are stuck with one species and divide themselves along lines of methodology, which are then more difficult to change.

Perhaps the most striking advance since my original paper is the appreciation in humans of how behavior toward the very young may be adjusted according to Trivers-Willard logic. That is, the upper classes tend to invest disproportionately in sons, and the lower classes, in daughters. Certainly some such bias is introduced by the sex ratio bias with socioeconomic status itself, but Voland (1988) showed something very striking from German historical data of the seventeenth century. Rank-ordering people from farmers to day-laborers (i.e., high to low socioeconomic status), it could be shown that a female wishing to maximize her chances of surviving to one year of age would, if she could choose, prefer to be born into the lowest classes instead of the higher. Later work has shown similar kinds of effects in a wide range of societies, from the Ifaluk of the South Pacific (Betsig and Turke 1986), to the Mukugodo of Kenya (Cronk 1991, 2000), North American societies (Gaulin and Robbins 1991) and European and the United States ones (Mueller 1993). A greater age difference between husband and wife is associated with more sons in the U.K. (Manning, Anderton and Shutt 1997). A recent excellent paper showing only modest effects is Koziel and Ulijaszek 2001. For stress and sex ratio effects in Denmark, see Hansen, Moeller and Olsen 1999.

Meanwhile, progress has been achieved in understanding which primate species show what kind of sex ratio pattern (van Schaik and Hrdy 1991) as well as very recent work showing how density may affect sex ratio correlations in red deer (Kruuk et al 1999). For evidence of sex ratio effects in horses and in chimpanzees, see Cameron et al 1999 and Boesch 1997. For a novel application to plants, see Freeman et al 1994.

4

PARENT–OFFSPRING CONFLICT

In early 1971 I decided to work on what I thought of as a biological theory of the family, a theory that would have variables such as sex and age and other relevant parameters and would derive how natural selection was acting on members of the family. There was only one problem: individuals in a family were related to each other, and somehow you had to take this relatedness into account when describing natural selection acting on the participants. I puzzled about the matter for some time and, in my usual style, consulted with advanced graduate students and relevant faculty, but without getting any help. What surprises me so much about this, in retrospect, was that I already knew Hamilton's kinship theory—he had, in fact, solved the very problem that was bedeviling me, and I had by then lectured on his work. But for some reason I still narrowly conceived his work as explaining altruistic traits per se, not as covering all interactions between kin. It was shortly *after* lecturing on the topic that it occurred to me that the key issue had already been solved, but I know that I greeted this discovery with relief. Hamilton had solved the difficult problem; the rest was easy. The key parameter turned out to be r, or degree of relatedness, the chance that one individual shares an identical copy of any given gene with another individual by direct descent (typically, 1/2 in both directions for parent–offspring). I narrowed the paper to mother–offspring conflict, as I had first called it, because I was thinking about mammals, but then later, in the usual syle, made the argument more general, changing it to parent–offspring conflict. (When my father saw the change in the title, he muttered to himself "I was afraid of that.")

Thinking Like an Economist

One day I sat down to write the paper. I was in Jamaica doing work on lizards. I sat out in the sun to get my tropical tan and simply wrote out the first couple of sections of the paper. Graphing parent–offspring conflict over the time of weaning was easy to do. You just used benefits (B) and costs (C), stuck in degree of relatedness (r), and considered both individuals, mother and offspring, in the same graph. Following this, I drew the graph for conflict at any time during the period of parental investment, again relying only on B, C, and r. Later, when I would lecture on this work to economists, they would come up afterward and say, "We like the way you think—just like an economist!" Some wondered whether I had learned graphing techniques by studying economics. Fortunately, I had never had a course in economics, and the similarity in graphing techniques, I think, occurred just because of similarity in logic. Economists thought in terms of something they called "utility," not reproductive success, but still something that could be conceptualized in terms of benefits and costs. Of course, they lacked degrees of relatedness in their system of logic.

Initially, I thought only in terms of parent–offspring conflict over parental investment. It was gratifying to see that such well-known phenomena as weaning conflict in mammals and fledging conflict in birds might be seen as natural consequences of selection acting on imperfect degrees of relatedness between parent and offspring. At Harvard, I often watched fledging conflict in pigeons. Both parents acted very solicitously to newly hatched chicks, stroking the neck of a chick, for example, to induce gaping and then feeding it, but toward the end of parental care, parents were harassed almost continuously by their fully grown offspring and often flew onto very narrow ledges to escape the incessant begging. I do not remember when it occurred to me that the same argument could be applied to the behavioral tendencies of the offspring, so as to show a bias in socialization or manipulation—parents trying to make the offspring act more altruistic and less selfish than it would otherwise act on its own. It may have occurred to me on a very memorable trip to India and East Africa with Irven DeVore, but I know that when it did occur to me, I realized that the argument had much deeper implications than I had been aware of. At least in our own species, with its very long period of parental investment, with language and other means of manipulation, the character and personality of the offspring would be molded in an arena of conflict.

I certainly knew about adult male discipline of juvenile baboons from the work of Professor DeVore. Since the males doing the disciplining were centrally dominant males, heavily involved in breeding the females, the adult discipline may indeed have been a form of parental discipline. I saw an example of this in 1972 at the Gombe Stream Reserve, where Jane Goodall studied her chimpanzees. There were a couple of baboon troops that roamed through her study area, and these were being studied as well. One day I was wandering along a path by myself when I saw an adult male baboon, an adult female, and what looked to be her two youngsters, one about six months old and one about two and one-half. It looked like a little family unit. The adults were, as I remember, grooming each other, and the two youngsters were playing with each other nearby. The six-month-old reared up alone on its hind legs, but could not maintain balance and fell on its back, where it continued to flail its arms in order to right itself. This caught the attention of the adult male, and he immediately strode over to the juvenile, who knew at once what was coming and ran up the nearest tree. Unfortunately, this tree only reached about five feet off the ground, and the adult male grabbed the juvenile, shook and released it, and returned to the adult female. One wanted to cry out, "But it wasn't his fault!" and the principle is, of course, a familiar one in human life: never mind the details, the older of the two is to blame. Professor DeVore had noted this principle years earlier as it applied to baboon discipline.

It was also on that trip that I noticed how severe weaning conflict could be in nature. First with langur monkeys in India, and then with baboons in Kenya and Tanzania, we saw many instances of individuals being weaned, who spent days on end harassing their mothers, making loud ick-ick-icking sounds. Indeed, a common way of finding baboon troops in the morning was to go to the kinds of trees they prefer to sleep in, and listen for weaning conflict. Baboons are otherwise mostly silent, and these observations suggested that the conflict would be sufficiently intense to put both mother and offspring (and other individuals) at risk.

Incidentally, while I was at Gombe, I was invited to lecture to the assembled scientists, just as were Professors DeVore and Robert Hinde, a well-known animal behaviorist from Cambridge University. Hinde had done some very nice experiments on mother–offspring separation in rhesus monkeys that supported the logic I was developing. Dr. Hinde lectured to the Gombe group on his findings, and two days later when I spoke on parent–offspring conflict I naturally seized the opportunity to present my reinterpretation of his results. At the end of the lecture, Hinde came up to speak to me, and I braced myself for what was coming. I had by then some experience with the tender egos of elderly academics. He said to me that he had been working on this subject for fifteen years and that, in one night, I had completely changed the way he thought about it. It almost brings tears to my eyes now (literally!) to remember his behavior at the time. The far more common reaction would have been to attack my work from every angle, and it was especially rare to have such an open-minded response so generously worded. I have liked the man ever since! When I showed him my manuscript, he liked it but had a verbal disagreement, which he took time to explain to me. I had repeatedly said "the parents should do this," "the offspring should do that," using "should," as a theoretician might, to refer to expected behavior on the theory that was being advanced. Hinde did not like this usage. At first I tried to argue against his position, but I soon saw the wisdom of the point he was making. "Should" had other connotations in everyday speech, in particular the sound of a moral imperative. It was a bad idea to mix the two uses. The language of expectation–this is what you *expect* from theory—was the preferable language. I think I took the lesson to heart. Theory can be quite assertive enough as it is—to the point of arrogance sometimes—without overloading the language. Quite the opposite, in fact, is desirable, namely, to take the humbler stance that theory leads only to *expectations* regarding reality, and not direct knowledge, so that empirical confirmation is always necessary.

One Little Mistake

When the paper was nearly completed and I was making final, very minor changes, I came across a sentence saying that Hamilton did not "develop a theory of parent–offspring relations" and, of course, by implication I had. I decided that this was pompous, an overassertion, so I changed the sentence in what I imagined to be a more humble direction. I said, "Hamilton did not *apply* his theory to parent–offspring relations." But in so doing, I immediately made an error. He had, perhaps, not applied his theory to parent–offspring relationships in *vertebrates*, but in his famous 1964 paper he had certainly applied his theory to haplodiploid families and, furthermore, had applied it to parent–offspring relations in plants! He had pointed out that when a plant was fertilized by multiple fathers, you might expect greater variance in seed size, due to competitive interactions between less related seeds.

This little mistake always illustrated to me something I noticed while writing papers: you do not necessarily improve things by rewriting. Sometimes you could take a paper downhill in your efforts to get it just right. In this case, I certainly had, because I had gone from what was an arguably true statement to one that was manifestly false. All in the name of humility! Incidentally, after the paper was published, I wrote Bill once and, fully conscious of how derivative my own paper was, asked him when the things in my paper had first occurred to him. He wrote back a very nice letter saying that he thought much of what I said had passed through his mind at one time or another, but perhaps not always left an imprint sufficient to its importance.

Attributing Too Much to Parent–Offspring Conflict

After my paper was published, I think during a visit to the University of Michigan in 1979, I remember walking with Bill Hamilton after lunch when I advanced yet some other twist on the parent–offspring conflict theme and Bill said softly, "I think you attribute too much to parent–offspring conflict." I readily agreed with him, but I may not have known then how true his statement was. The logic coming out of parent–offspring conflict had a beguiling quality, much as did Freudian doctrine. Namely, twists and turns in the form of the conflict could be used to generate almost any personality characteristic in the offspring that you desired. And so it promised to provide a coherent system of logic that gave meaning and context to a wide range of human psychology. In retrospect, the main problem with pushing the argument too far was the importance of genetic variability itself. If there is strong selection for the production of genetic variability, then there may be significant behavioral differences among individuals that have no special meaning beyond the fact that the genetic system is continually generating new genetic combinations and new genetic traits. This large genetic variability is being selected in a variety of directions and contexts, not only, of course, that of parent–offspring relations. A deeper analysis of social life also suggests that it would not be *logical* for the relationship to have such a large influence on the personality of the child.

There is no doubt that I felt something of an aggressive edge toward my parents in the context of parent–offspring conflict. I had forgotten, until recently reminded by my mother, that I had delivered a talk at Ball State University, on the evolution of parent–offspring conflict in about 1976, while my father was on the faculty as a diplomat-in-residence. Apparently, I insisted that both parents be at the lecture. My mother, who is an accomplished mimic, recently mimicked for me her own pose during the talk, not quite the Nancy Reagan my-man-can-do-no-wrong look, but rather her head perched back at an angle, her lips pursed, and her eyes fixed on me as if to say that this is a very interesting talk that certainly has nothing directly to do with her! I know that I had a tendency back then to blame my mental breakdowns on my parents, and although I later came to believe that only the *first* breakdown, if any, should be assigned to their column, at the time I was biased toward seeing underinvestment and parental denial as having helped fuel offspring mental illness. I am much more inclined now to emphasize genetic proclivities in the offspring (or even an infectious agent such as a virus). I do very much believe in seeing, whenever possible, the downside and upside to traits or situations when they are logically linked. Thus, if my intellectual creativity and mental instability were logically linked as reflecting the upside and downside of the same genetic and developmental trajectory, then I best see them together.

Kids

There was a period of two or three years around the time the paper was published and for the next couple of years where I was unusually sensitive to parent–offspring interactions (especially in others!). Of course, I championed the offspring's viewpoint. One strong sensitivity I developed then— you might even call it an acute allergic response—was to the word "kid." I thought it absurd (and psychologically revealing) that an entire society of people named their children after the young of another species that they neither kept, ate, nor had any special affection for. I thought it no accident that this occurred in the most dominant society in the world, since I imagined that *within* societies greater parental dominance would correlate with less attention to the offspring's viewpoint. I saw adults tempted into phenotypic indulgence as their resources increased and disinclined to invest much directly in their children. In any case, I often challenged my students to come forward if they could with another society in which most people most of the time referred to their children by the name of another animal. An occasional student would appear and tell me of a usage, for example gavron, or goat, in Chile or Argentina, but invariably these cases referred to occasional uses of the name of another species, usually with a negative connotation. When I moved to California in 1978 I encountered in Santa Cruz a more odious term used with surprising frequency, "rug rats," and it seemed in some cases accurately to capture the parents' true conceptualization of their children, namely, as disagreeable objects under foot.

My sensitivity to "kids" was so great that there was a time when I tried to correct each and every person who used the term, but I soon saw that trying to convert Americans one by one was a hopeless task, and after that I restricted myself to correcting them whenever they used the term about my own children. So, if they asked, "How many kids do you have?" I would answer, "None. I don't have enough property to keep any goats," or some such response. Many people argued with me, of course, that their own use of "kids" was always positive, and while I conceded that in many individual cases this might be true, I still argued that, at the very least, unconsciously the child was being diminished in status, "kid" itself being a diminutive term. Does "stop kid abuse" sound as emotionally effective as "stop child abuse?" "Child" at times seemed to evoke immaturity too much and was not as suitable for older children as for younger ones, but the term I preferred— though I rarely used it myself—was the Jamaican Rastafarian term "youth." That choice of language was deliberate in Rasta culture, emerging from a downtrodden section of downtrodden people, which naturally emphasized individual respect.

Parent–Offspring Conflict

ROBERT L. TRIVERS

Abstract.   When parent–offspring relations in sexually reproducing species are viewed from the standpoint of the offspring as well as the parent, conflict is seen to be an expected feature of such relations. In particular, parent and offspring are expected to disagree over how long the period of parental investment should last, over the amount of parental investment that should be given, and over the altruistic and egoistic tendencies of the offspring as these tendencies affect other relatives. In addition, under certain conditions parents and offspring are expected to disagree over the preferred sex of the potential offspring. In general, parent–offspring conflict is expected to increase during the period of parental care, and offspring are expected to employ psychological weapons in order to compete with their parents. Detailed data on mother–offspring relations in mammals are consistent with the arguments presented. Conflict in some species, including the human species, is expected to extend to the adult reproductive role of the offspring: under certain conditions parents are expected to attempt to mold an offspring, against its better interests, into a permanent nonreproductive.

In classical evolutionary theory parent–offspring relations are viewed from the standpoint of the parent. If parental investment (PI) in an offspring is defined as anything done by the parent for the offspring that increases the offspring's chance of surviving while decreasing the parent's ability to invest in other offspring (Trivers, 1972), then parents are classically assumed to allocate investment in their young in such a way as to maximize the number surviving, while offspring are implicitly assumed to be passive vessels into which parents pour the appropriate care. Once one imagines offspring as *actors* in this interaction, then conflict must be assumed to lie at the heart of sexual reproduction itself—an offspring attempting from the very beginning to maximize its reproductive success (RS) would presumably want more investment than the parent is selected to give. But unlike conflict between unrelated individuals, parent–offspring conflict is expected to be circumscribed by the close genetic relationship between parent and offspring. For example, if the offspring garners more investment than the parent has been selected to give, the offspring thereby decreases the number of its surviving siblings, so that any gene in an offspring that leads to an additional investment decreases (to some extent) the number of surviving copies of itself located in siblings. Clearly, if the gene in the offspring exacts too great a cost from the parent, that gene will be selected against even though it confers some benefit on the offspring. To specify precisely how much cost an offspring should be willing to inflict on its parent in order to gain a given benefit, one must specify how the offspring is expected to weigh the survival of siblings against its own survival.

The problem of specifying how an individual is expected to weigh siblings against itself (or any relative against any other) has been solved in outline by Hamilton (1964), in the context of explaining the evolution of altruistic behavior. An altruistic act can be defined as one that harms the organism performing the act while benefiting some other individual, harm and benefit being defined in terms of reproductive success. Since any gene that helps itself spread in a population is, by definition, being selected for, altruistic behavior in the above sense can be selected only if there is a sufficiently large probability that the recipient of the act also has the gene. More precisely, the benefit/cost ratio of the act, times the chance that the recipient has the gene, must be greater than one. If the recipient of the act is a relative of the altruist, then the probability that the recipient has the gene by descent from a common ancestor can be specified. This conditional probability is called the *degree of relatedness*, r_0. For an altruistic act directed at a relative to have survival value its benefit/cost ratio must be larger than the inverse of the altruist's r_0 to the relative. Likewise an individual is expected to forego a selfish act if its cost to a relative, times the r_0 to that relative, is greater than the benefit to the actor.

The rules for calculating degrees of relatedness are straightforward for both diploid and haplodiploid organisms, even when inbreeding complicates the relevant genealogy (see the addendum in Hamilton, 1971). For example, in a diploid species (in the absence of inbreeding) an individual's r_0 to his or her full-siblings is $\frac{1}{2}$; to half-siblings, $\frac{1}{4}$; to children, $\frac{1}{2}$; to cousins, $\frac{1}{8}$. If in calculating the selective value of a gene one not only computes its effect on the reproductive success of the individual bearing it, but adds to this its effects on the reproductive success of related individuals, appropriately' devalued by the relevant degrees of relatedness, then one has computed what Hamilton (1964) calls *inclusive fitness*. While Hamilton pointed out that the parent–offspring relationship is merely a special case of relations between any set of genetically related individuals, he did not apply his theory to such relations. I present here a theory of parent–offspring relations which follows directly from the key concept of inclusive fitness and from the assumption that the offspring is at all times capable of an active role in its relationship to its parents. The form of the argument applies equally well to haplodiploid species, but for simplicity the discussion is mostly limited to diploid species. Likewise, although many of the arguments apply to any sexually reproducing species showing parental investment (including many plant species), the arguments presented here are particularly relevant to understanding a species such as the human species in which parental investment is critical to the offspring throughout its entire prereproductive life (and often later as well) and in which an individual normally spends life embedded in a network of near and distant kin.

Parent–Offspring Conflict over the Continuation of Parental Investment

Consider a newborn (male) caribou calf nursing from his mother. The benefit to him of nursing (measured in terms of his chance of surviving) is large, the cost to his mother (measured in terms of her ability to produce additional offspring) presumably small. As time goes on and the calf becomes increasingly capable of feeding on his own, the benefit to him of nursing decreases while the cost to his mother may increase (as a function, for example, of the calf's size). If cost and benefit are measured in the same units, then at some point the cost to the mother will exceed the benefit to her young and the net reproductive success of the mother decreases if she continues to nurse. (Note that later-born offspring may contribute less to the mother's eventual RS than early-born, because their reproductive value may be lower [Fisher, 1930], but this is automatically taken into account in the cost function.)

The calf is not expected, so to speak, to view this situation as does his mother, for the calf is completely related to himself but only partially related to his future siblings, so that he is expected to devalue the cost of nursing (as measured in terms of future sibs) by his r_0 to his future sibs, when comparing the cost of nursing with its benefit to himself. For example, if future sibs are expected to be full-sibs, then the calf should nurse until the cost to the mother is more than twice the benefit to himself. Once the cost to the mother is more than twice the benefit to the calf, continued nursing is opposed by natural selection acting on both the mother and the calf. As long as one imagines that the benefit/cost ratio of a parental act changes continuously from some large number to some very small number near zero, then there must occur a period of time during which $\frac{1}{2} < B/C < 1$. This period is one of expected conflict between mother and offspring, in the sense that natural selection working on the mother favors her halting parental investment while natural selection acting on the offspring favors his eliciting the parental investment. The argument presented here is graphed in Figure

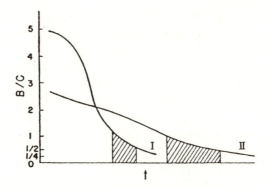

Figure 1. The benefit/cost ratio (B/G) of a parental act (such as nursing) toward an offspring as a function of time. Benefit is measured in units of reproductive success of the offspring and cost in comparable units of reproductive success of the mother's future offspring. Two species are plotted. In species I the benefit/cost ratio decays quickly; in species II, slowly. Shaded areas indicate times during which parent and offspring are in conflict over whether the parental care should continue. Future sibs are assumed to be full-sibs. If future sibs were half-sibs, the shaded areas would have to be extended until B/C = ¼.

1. (Note, as argued below, that there are specialized situations in which the offspring may be selected to consume *less* PI than the parent is selected to give.)

This argument applies to all sexually reproducing species that are not completely inbred, that is, in which siblings are not identical copies of each other. Conflict near the end of the period of PI over the continuation of PI is expected in all such species. The argument applies to PI in general or to any subcomponent of PI (such as feeding the young, guarding the young, carrying the young) that can be assigned a more or less independent cost-benefit function. Weaning conflict in mammals is an example of parent-offspring conflict explained by the argument given here. Such conflict is known to occur in a variety of mammals, in the field and in the laboratory, for example, baboons (DeVore, 1963), langurs (Jay, 1963), rhesus macaques (Hinde and Spencer-Booth, 1971), other macaques (Rosenblum, 1971), vervets (Struhsaker, 1971), cats (Schneirla et al., 1963), dogs (Rheingold, 1963), and rats (Rosenblatt and Lehrman, 1963). Likewise, I interpret conflict over parental feeding at the time of fledging in bird species as conflict explained by the present argument, for example, Herring Gulls (Drury and Smith, 1968), Red Warblers (Elliott, 1969), Verreaux's Eagles (Rowe, 1947), and White Pelicans (Schaller, 1964).

Weaning conflict is usually assumed to occur either because transitions in nature are assumed always to be imperfect or because such conflict is assumed to serve the interests of both parent and offspring by informing each of the needs of the other. In either case, the marked inefficiency of weaning conflict seems the clearest argument in favor of the view that such conflict results from an underlying conflict in the way in which the inclusive fitness of mother and offspring are maximized. Weaning conflict in baboons, for example, may last for weeks or months, involving daily competitive interactions and loud cries from the infant in a species otherwise strongly selected for silence (DeVore, 1963). Interactions that inefficient *within* a multicellular organism would be cause for some surprise, since, unlike mother and offspring, the somatic cells within an organism are identically related.

One parameter affecting the expected length (and intensity) of weaning conflict is the offspring's expected r_0 to its future siblings. The lower the offspring's r_0 to its future siblings, the longer and more intense the expected weaning conflict. This suggests a simple prediction. Other things being equal, species in which different, unrelated males commonly father a female's successive offspring are expected to show stronger weaning conflict than species in which a female's successive offspring are usually fathered by the same male. As shown below, however, weaning conflict is merely a special case of conflict expected throughout the period of parental investment, so that this prediction applies to the intensity of conflict prior to weaning as well.

Conflict Throughout the Period of PI over the Amount of PI

In Figure 1 it was assumed that the amount of investment for each day (or moment in time) had already been established, and that mother and young were only selected to disagree over when such investment should be ended. But it can be shown that, in theory, conflict over the amount of investment that should at each moment be given, is expected throughout the period of PI.

At any moment in the period of PI the female is selected to invest that amount which maximizes the difference between the associated cost and benefit, where these terms are defined as above. The infant is selected to induce that investment which maximizes the difference between the benefit and a cost devalued by the relevant r_0. The different optima for a moment in time in a hypothetical species are graphed in Figure 2. With reasonable assumptions about the shape of the benefit and cost curves, it is clear that the infant will, at each instant in time, tend to favor greater parental investment than the parent is selected to give. The period of transition discussed in the previous section is a special case of this continuing competition, namely, the case in which parent and offspring compete over whether *any* investment should be given, as opposed to their earlier competition over

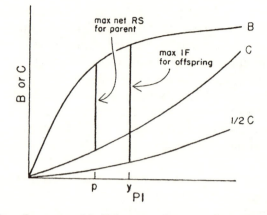

Figure 2. The benefit, cost, and half the cost of a parental act toward an offspring at one moment in time as a function of the amount the parent invests in the act (PI). Amount of milk given during one day of nursing in a mammal would be an example of PI. At p the parent's inclusive fitness $(B - C)$ is maximized; at y the offspring's inclusive fitness $(B - C/2)$ is maximized. Parent and offspring disagree over whether p or y should be invested. The offspring's future siblings are assumed to be full-siblings. IF = inclusive fitness.

how much should be given. Since parental investment begins before eggs are laid or young are born, and since there appears to be no essential distinction between parent–offspring conflict outside the mother (mediated primarily by behavioral acts) and parent–offspring conflict inside the mother (mediated primarily by chemical acts), I assume that parent–offspring conflict may in theory begin as early as meiosis.

It must be emphasized that the cost of parental investment referred to above (see Fig. 2) is measured *only* in terms of decreased ability to produce *future* offspring (or, when the brood size is larger than one, decreased ability to produce *other* offspring). To appreciate the significance of this definition, imagine that early in the period of PI the offspring garners more investment than the parent has been selected to give. This added investment may decrease the parent's later investment in the offspring at hand, either through an increased chance of parental mortality during the period of PI, or through a depletion in parental resources, or because parents have been selected to make the appropriate adjustment (that is, to reduce later investment below what otherwise would have been given). In short, the offspring may gain a temporary benefit but suffer a later cost. This self-inflicted cost is subsumed in the benefit function (B) of Figure 2, because it decreases the benefit the infant receives. It is not subsumed in the cost function (C) because this function refers only to the mother's future offspring.

The Time Course of Parent Offspring Conflict

If one could specify a series of cost-benefit curves (such as Fig. 2) for each day of the period of PI, then the expected time course of parent–offspring conflict could be specified. Where the difference in the offspring's inclusive fitness at the parent's optimum PI (p in Fig. 2) and at the offspring's optimum PI (y) is large, conflict is expected to be intense. Where the difference is slight, conflict is expected to be slight or nonexistent. In general, where there is a strong difference in the offspring's inclusive fitness at the two different optima (p and y), there will also be a strong difference in the parent's inclusive fitness, so that both parent and offspring will simultaneously be strongly motivated to achieve their respective optimal values of PI. (This technique of comparing cost-benefit graphs can be used to make other predictions about parent–offspring conflict, for example, that such conflict should decrease in intensity with increasing age, and hence decreasing reproductive value, of the parent; see Figure 3.) In the absence of such day-by-day graphs three factors can be identified, all of which will usually predispose parent and offspring to show greater conflict as the period of PI progresses.

1. Decreased chance of self-inflicted cost.   As the period of PI progresses, the offspring faces a decreased chance of suffering a later self-inflicted cost for garnering additional investment at the moment. At the end of the period of PI any additional investment forced on the parent will only affect later offspring, so that at that time the interests of parent and offspring are maximally divergent. This time-dependent change in the offspring's chance of suffering a self-inflicted cost will, other things being equal, predispose parent and offspring to increasing conflict during the period of PI.

2. Imperfect replenishment of parental resources.   If the parent is unable on a daily basis to replenish resources invested in the offspring, the parent will suffer increasing depletion of its resources, and, as time goes on, the cost of such depletion should rise disproportionately, even if the amount of resources invested per day declines. For example, a female may give less milk per day in the first half of the nursing period than in the second half (as in pigs: Gill and Thomson, 1956), but if she is failing throughout to replenish her energy losses, then she is constantly increasing her deficit (although at a diminishing rate) and greater deficits may be associated with disproportionate costs. In some species a parent does not feed itself during much of the period of PI and at least during such periods the parent must be depleting its resources (for example, female elephant seals during the nursing period: Le Boeuf et al., 1972). But the extent to which parents who feed during the period of PI fail to replenish their resources is usually not known. For

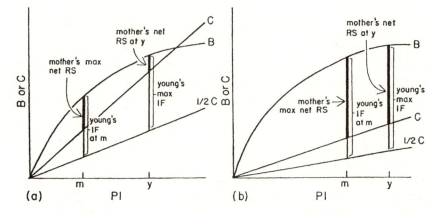

Figure 3. The benefit and cost of a parental act (as in Fig. 2) toward (*a*) an off-spring born to a young female and (*b*) an offspring born to an old female. One assumes that the benefit to the offspring of a given amount of PI does not change with birth order but that the cost declines as a function of the declining reproductive value (Fisher, 1930) of the mother: she will produce fewer future offspring anyway. The difference between the mother's inclusive fitness at m and y is greater for (*a*) than for (*b*). The same is true for the offspring. Conflict should be correspondingly more intense between early born young and their mothers than between late born young and their mothers.

some species it is clear that females typically show increasing levels of depletion during the period of PI (e.g., sheep: Wallace, 1948).

3. *Increasing size of the offspring.* During that portion of the period of PI in which the offspring receives all its food from its parents, the tendency for the offspring to begin very small and steadily increase in size will, other things being equal, increase the cost to the parent of maintaining and enlarging it. (Whether this is always true will depend, of course, on the way in which the offspring's growth rate changes as a function of increasing size.) In addition, as the offspring increases in size the relative energetic expense to it of competing with its parents should decline.

The argument advanced here is only meant to suggest a general tendency for conflict to increase during the period of PI, since it is easy to imagine circumstances in which conflict might peak several times during the period of PI. It is possible, for example, that weight at birth in a mammal such as humans is strongly associated with the offspring's survival in subsequent weeks, but that the cost to the mother of bearing a large offspring is considerably greater than some of her ensuing investment. In such circumstances, conflict *prior* to birth over the offspring's weight at birth may be more intense than conflict over nursing in the weeks after birth.

Data from studies of dogs, cats, rhesus macaques, and sheep appear to support the arguments of this and the previous section. In these species, parent–offspring conflict begins well before the period of weaning and tends to increase during the period of PI. In dogs (Rheingold, 1963) and cats (Schneirla et al., 1963) postnatal maternal care can be divided into three periods according to increasing age of the offspring. During the first, the mother approaches the infant to initiate parental investment. No avoidance behavior or aggression toward the infant is shown by the mother. In the second, the offspring and the mother approach each other about equally, and the mother shows some avoidance behavior and some aggression in response to the infant's demands. The third period can be characterized as the period of weaning. Most contacts are initiated by the offspring. Open avoidance and aggression characterize the mother.

Detailed quantitative data on the rhesus macaque (Hinde and Spencer-Booth, 1967, 1971), and some parallel data on other macaques (Rosenblum, 1971), demonstrate that the behavior of both mother and offspring change during the period of postnatal parental care in way consistent with theory. During the first weeks after she has given birth, the rhesus mother's initiative in setting up nipple contacts is high but it soon declines rapidly. Concurrently she begins to reject some of the infant's advances, and after her own initiatives toward nipple contact have ceased, she rejects her infant's advances with steadily increasing frequency until at the end of the period of investment all of the offspring's advances are rejected. Shortly after birth, the offspring leaves the mother more often than it approaches her, but as time goes on the initiative in maintaining mother–offspring proximity shifts to the offspring. This leads to the superficially paradoxical result that as the offspring becomes increasingly active and independent, spending more and more time away from its mother, its initiative in maintaining mother–offspring proximity *increases* (that is, it tends to approach the mother more often then it leaves her). According to the theory presented here, this result reflects the underlying tendency for parent–offspring conflict to increase during the period of PI. As the interests of mother and offspring diverge, the offspring must assume a greater role in inducing whatever parental investment is forthcoming.

Data on the production and consumption of milk in sheep (Wallace, 1948) indicate that during the first weeks of the lamb's life the mother typically produces more milk than the lamb can drink. The lamb's appetite determines how much milk is consumed. But after the fourth week, the mother begins to produce less than the lamb can drink, and from that time on it is the mother who is the limiting factor in determining how much milk is consumed. Parallel behavioral data indicate that the mother initially permits free access by her lamb(s) but after a couple of weeks begins to prevent some suckling attempts (Munro, 1956; Ewbank, 1967). Mothers who are in poor condition become the limiting factor in nursing earlier than do mothers in good condition, and this is presumably because the cost of a given amount of milk is considerably higher when the mother is in poor condition, while the benefit to the offspring remains more or less unchanged. Females who produce twins permit either twin to suckle on demand during the first three weeks after birth, but in the ensuing weeks they do not permit one twin to suckle unless the other is ready also (Ewbank, 1964; Alexander, 1960).

Disagreement over the Sex of the Offspring

Under certain conditions a potential offspring is expected to disagree with its parents over whether it should become a male or a female. Since one can not assume that potential offspring are powerless to affect their sex, sex ratios observed in nature should to some extent reflect the offspring's preferred value as well as the parents'.

Fisher (1930) showed that (in the absence of inbreeding) parents are selected to invest as much in the total of their daughters as in the total of their sons. When each son produced costs on average the same as each daughter, parents are selected to produce a sex ratio of 50/50. In such species, the expected reproductive success (RS) of a son is the same as that of a daughter, so that an offspring should be indifferent as to its sex. But if (for example) parents are selected to invest twice as much in a typical male as in a typical female, then they will be selected to produce twice as many females as males, and the expected RS of each son will be twice that of each daughter. In such a species a potential offspring would prefer to be a male, for it would then achieve twice the RS it would as a female, without suffering a comparable decrease in inclusive fitness through the cost forced on its parents, because the offspring is selected to devalue that cost by the offspring's expected r_0 to the displaced sibling. For the example chosen, the exact gain in the offspring's inclusive fitness can be specified as follows. If the expected RS of a female offspring is defined as one unit of RS, then, in being made male, the offspring gains one unit of RS, but it deprives its mother of an additional daughter (or half a son). This displaced sibling (whether a female or half of a male) would have achieved one unit of RS, but this unit is devalued from the offspring's standpoint by the relevant r_0. If the displaced sibling would have been a full-sibling, then this unit of RS is devalued by ½, and the offspring, in being made a male, achieves a ½ unit net increase in inclusive fitness. If the displaced sibling would have been a half-sibling, the offspring, in being made a male, achieves a ¾ unit net increase in inclusive fitness. The parent, on the other hand, experiences initially only a trivial decrease in RS, so that *initially* any gene in the offspring tending to make it a male against its parents' efforts would spread rapidly.

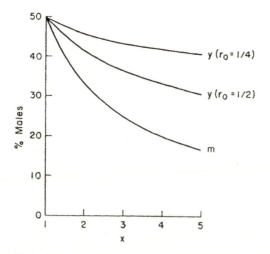

Figure 4. The optimal sex ratio (per cent males) for the mother (m) and the young (y) where the mother invests more in a son than in a daughter by a factor of x (and assuming no paternal investment in either sex). Two functions are given for the offspring, depending on whether the siblings it displaces are full-siblings ($r_0 = \frac{1}{2}$) or half-siblings ($r_0 = \frac{1}{4}$). Note the initial rapid divergence between the mother's and the offspring's preferred sex ratio as the mother moves from equal investment in a typical individual of either sex (x = 1) to twice as much investment in a typical male (x = 2).

As a hypothetical gene for offspring control of sex begins to spread, the number of males produced increases, thereby lowering the expected RS of each male. This decreases the gain (in inclusive fitness) to the offspring of being made a male. If the offspring's equilibrial sex ratio is defined as that sex ratio at which an offspring is indifferent as to whether it becomes a male or a female, then this sex ratio can be calculated by determining the sex ratio at which the offspring's gain in RS in being made a male is exactly offset by its loss in inclusive fitness in depriving itself of a sister (or half a brother). The offspring's equilibrial sex ratio will depend on both the offspring's expected r_0 to the displaced siblings and on the extent to which parents invest more in males than in females (or vice versa). The general solution is given in the Appendix. Parent and offspring equilibrial sex ratios for different values of r_0 and different values of x (PI in a typical son/PI in a typical daughter) are plotted in Figure 4. For example, where the r_0 between siblings is $\frac{1}{2}$ and where parents invest twice as much in a son as in a daughter (x = 2), the parents' equilibrial sex ratio is 1 : 2 (males : females) while that of the offspring is 1 : 1.414.

As long as all offspring are fathered by the same male, he will prefer the same sex ratio among the offspring that the mother does. But consider a species such as caribou in which the female produces only one offspring a year and assume that a female's successive offspring are fathered by different, unrelated males. If the female invests more in a son than in a daughter, then she will be selected to produce more daughters than sons. The greater cost of the son is not borne by the father, however, who invests nothing beyond his sperm, and who will not father the female's later offspring, so the father's equilibrial sex ratio is an equal number of sons and daughters. The offspring will prefer some probability of being a male that is intermediate between its parents' preferred probabilities, because (unlike the father) the offspring is related to the mother's future offspring but (unlike the mother) it is less related to them than to itself.

In a species such as just described (in which the male is heterogametic) the following sort of competitive interaction is possible. The prospective father produces more Y-bearing sperm than the female would prefer and she subjects the Y-bearing sperm to differential mortality. If the ratio of the sperm reaching the egg has been reduced to near the mother's optimal value, then the egg preferentially admits the Y-bearing sperm. If the mother ovulated more eggs than she intends to rear, she could then choose which to invest in, according to the sex of the fertilized egg, unless a male egg is able to deceive the mother about its sex until the mother has committed herself to investing in him. Whether such interactions actually occur in nature is at present unknown.

One consequence of the argument advanced here is that there is an automatic selective agent tending to keep maternal investment in a son similar to that in a daughter, for the greater the disparity between the investment in typical individuals of the two sexes, the greater the loss suffered by the mother in competitive interactions with her offspring over their preferred sex and in producing a sex ratio further skewed away from her preferred ratio (see Fig. 4). This automatic selection pressure may partly account for the apparent absence of strongly size-dimorphic young (at the end of P1) in species showing striking adult sexual dimorphism in size.

The argument presented here applies to any tendency of the parent to invest differentially in the young, whether according to sex or some other variable, except that in many species sex is irreversibly determined early in ontogeny and the offspring is expected at the very beginning to be able to discern its own sex and hence the predicted pattern of investment it will receive, so that, unlike other forms of differential investment, conflict is expected very early, namely, at the time of sex determination.

The Offspring as Psychological Manipulator

How is the offspring to compete effectively with its parent? An offspring can not fling its mother to the ground at will and nurse. Throughout the period of parental investment the offspring competes at a disadvantage. The offspring is smaller and less experienced than its parent, and its parent controls the resources at issue. Given this competitive disadvantage the offspring is expected to employ psychological rather than physical tactics. (Inside the mother the offspring is expected to employ chemical tactics, but some of the analysis presented below should also apply to such competition.) It should attempt to *induce* more investment than the parent wishes to give.

Since an offspring will often have better knowledge of its real needs than will its parent, selection should favor parental attentiveness to signals from its offspring that apprize the parent of the offspring's condition. In short, the offspring cries when hungry or in danger and the parent responds appropriately. Conversely, the offspring signals its parent (by smiling or wagging its tail) when its needs have been well met. Both parent and offspring benefit from this system of communication. But once such a system has evolved, the offspring can begin to employ it out of context. The offspring can cry not only when it is famished but also when it merely wants more food than the parent is selected to give. Likewise, it can begin to withhold its smile until it has gotten its way. Selection will then of course favor parental ability to discriminate the two uses of the signals, but still subtler mimicry and deception by the offspring are always possible. Parental experience with preceding offspring is expected to improve the parent's ability to make the appropriate discrimination. Unless succeeding offspring can employ more confusing tactics than earlier ones, parent–offspring interactions are expected to be increasingly biased in favor of the parent as a function of parental age.

In those species in which the offspring is more helpless and vulnerable the younger it is, its parents will have been more strongly selected to respond positively to signals of need emitted by the offspring, the younger that offspring is. This suggests that at any stage of ontogeny in which the offspring is in conflict with its parents, one appropriate tactic may be to revert to the gestures and actions of an earlier stage of development in order to induce the investment that would then have been forthcoming. Psychologists have long recognized such a tendency in humans and have given it the name of regression. A detailed functional analysis of regression could be based on the theory presented here.

The normal course of parent–offspring relations must be subject to considerable unpredictable variation in both the condition of the parent and (sometimes independently) the condition of the offspring. Both parents must be sensitive to such variation and must adjust their behavior appropriately. Low investment coming from a parent in poor condition has a different meaning than low investment coming from a parent in good condition. This suggests that from an early age the offspring is expected to be a psychologically sophisticated organism. The offspring should be able to evaluate the cost of a given parental act (which depends in part on the condition of the parent at that moment) and its benefit (which depends in part on the condition of the offspring). When the offspring's interests diverge from those of its parent, the offspring must be able to employ a series of psychological maneuvers, including the mimicry and regression mentioned above. Although it would be expected to learn appropriate information (such as whether its psychological maneuvers were having the desired effects), an important feature of the argument presented here is that the offspring cannot rely on its parents for disinterested guidance. One expects the offspring to be preprogrammed to resist some parental teaching while being open to other forms. This is particularly true, as argued below, for parental teaching that affects the altruistic and egoistic tendencies of the offspring.

If one event in a social relationship predicts to some degree future events in that relationship, the organism should be selected to alter its behavior in response to an initial event, in order to change the probability that the predicted events will occur. For example, if a mother's lack of love for her offspring early in its life predicts deficient future investment, then the offspring will be selected to be sensitive to such early lack of love, whether investment at that time is deficient or not, in order to increase her future investment. The best data relevant to these possibilities come from the work of Hinde and his associates on groups of caged rhesus macaques. In a series of experiments, a mother was removed from her 6-month-old infant, leaving the infant in the home cage with other group members. After 6 days, the mother was returned to the home cage. Behavioral data were gathered before, during, and after the separation (see points 1 and 2 below). In a parallel series of experiments, the infant was removed for 6 days from its mother, leaving her in the home cage, and the same behavioral data were gathered (see point 3 below). The main findings can be summarized as follows:

1. *Separation of mother from her offspring affects their relationship upon reunion.*   After reunion with its mother, the infant spends more time on the mother than it did before separation—although, had the separation not occurred, the infant would have reduced its time on the mother. This increase is caused by the infant, and occurs despite an increase in the frequency of maternal rejection (Hinde and Spencer-Booth, 1971). These effects can be detected at least as long as 5 weeks after reunion. These data are consistent with the assumption that the infant has been selected to interpret its mother's disappearance as an event whose recurrence the infant can help prevent by devoting more of its energies to staying close to its mother.

2. *The mother–offspring relationship prior to separation affects the offspring's behavior on reunion.*   Upon reunion with its mother, an infant typically shows distress, as measured by callings and immobility. The more frequently an infant was rejected *prior* to separation, the more distress it shows upon

reunion. This correlation holds for at least 4 weeks after reunion. In addition, the more distressed the infant is, the greater is its role in maintaining proximity to its mother (Hinde and Spencer-Booth, 1971). These data support the assumption that the infant interprets its mother's disappearance in relation to her predeparture behavior in a logical way: the offspring should assume that a rejecting mother who temporarily disappears needs more offspring surveilance and intervention than does a nonrejecting mother who temporarily disappears.

3. *An offspring removed from its mother shows, upon reunion, different effects than an offspring whose mother has been removed.* Compared to an infant whose mother had been removed, an infant removed from its mother shows, upon reunion, and for up to 6 weeks after reunion, less distress and more time off the mother. In addition, the offspring tends to play a smaller role in maintaining proximity to its mother, and its experiences less frequent maternal rejections (Hinde and Davies, 1972a,b). These data are consistent with the expectation that the offspring should be sensitive to the *meaning* of events affecting its relationship to its mother. The offspring can differentiate between a separation from its mother caused by its own behavior or some accident (infant removed from group) and a separation which may have been caused by maternal negligence (mother removed from group). In the former kind of separation, the infant shows less effects when reunited, because, from its point of view, such a separation does not reflect on its mother and no remedial action is indicated. A similar explanation can be given for differences in the mother's behavior.

Parent Offspring Conflict over the Behavioral Tendencies of the Offspring

Parents and offspring are expected to disagree over the behavioral tendencies of the offspring insofar as these tendencies affect related individuals. Consider first interactions among siblings. An individual is only expected to perform an altruistic act toward its full-sibling whenever the benefit to the sibling is greater than twice the cost to the altruist. Likewise, it is only expected to forego selfish acts when $C > 2B$ (where a selfish act is defined as one that gives the actor a benefit, B, while inflicting a cost, C, on some other individual, in this case, on a full-sibling). But parents, who are equally related to all of their offspring, are expected to encourage all altruistic acts among their offspring in which $B > C$, and to discourage all selfish acts in which $C > B$. Since there ought to exist altruistic situations in which $C < B < 2C$, parents and offspring are expected to disagree over the tendency of the offspring to act altruistically toward its siblings. Likewise, whenever for any selfish act harming a full-sibling $B < C < 2B$, parents are expected to discourage such behavior and offspring are expected to be relatively re-fractory to such discouragement.

This parent–offspring disagreement is expected over behavior directed toward other relatives as well. For example, the offspring is only selected to perform altruistic acts toward a cousin (related through the mother) when $B > 8C$. But the offspring's mother is related to her own nephews and nieces by $r_0 = \frac{1}{4}$ and to her offspring by $r_0 = \frac{1}{2}$, so that she would like to see any altruistic acts performed by her offspring toward their maternal cousins whenever $B > 2C$. The same argument applies to selfish acts, and both arguments can be made for more distant relatives as well. (The father is unrelated to his mate's kin and, other things being equal, should not be distressed to see his offspring treat such individuals as if they were unre-lated.)

The general argument extends to interactions with unrelated individuals, as long as these interactions have some effect, however remote and indirect, on kin. Assume, for example, that an individual gains some immediate ben-efit, B, by acting nastily toward some unrelated individual. Assume that the unrelated individual reciprocates in kind (Trivers, 1971), but assume that the reciprocity is directed toward both the original actor and some relative, e.g., his sibling. Assuming no other effects of the initial act, the original actor will be selected to perform the nasty act as long as $B > C_1 + \frac{1}{2}(C_2)$, where C_1 is the cost to the original actor of the reciprocal nastiness he receives and C_2 is the cost to his sibling of the nastiness the sibling receives. The actor's parents viewing the interaction would be expected to condone the initial act only if $B > C_1 + C_2$. Since there ought to exist situations in which $C_1 + \frac{1}{2}(C_2) < B < C_1 + C_2$, one expects conflict between offspring and parents over the offspring's tendency to perform the initial nasty act in the situation described. A similar argument can be made for altruistic be-havior directed toward an unrelated individual if this behavior induces al-truism in return, part of which benefits the original altruist's sibling. Parents are expected to encourage such altruism more often than the offspring is expected to undertake on his own. The argument can obviously be extended to behavior which has indirect effects on kin other than one's sibling.

As it applies to human beings, the above argument can be summarized by saying that a fundamental conflict is expected during socialization over the altruistic and egoistic impulses of the offspring. Parents are expected to socialize their offspring to act more altruistically and less egoistically than the offspring would naturally act, and the offspring are expected to resist such socialization. If this argument is valid, then it is clearly a mistake to view socialization in humans (or in any sexually reproducing species) as only or even primarily a process of "enculturation," a process by which parents teach offspring their culture (e.g., Mussen et al., 1969, p. 259). For example, one is not permitted to assume that parents who attempt to impart such virtues as responsibility, decency, honesty, trustworthiness, generosity, and self-denial are merely providing the offspring with useful information on appropriate behavior in the local culture, for all such virtues are likely to affect the amount of altruistic and egoistic behavior impinging on the parent's kin, and parent and offspring are expected to view such behavior differently. That some teaching beneficial to the offspring transpires during human socialization can be taken for granted, and one would expect no conflict if socialization involved *only* teaching beneficial to the offspring. According to the theory presented here, socialization is a process by which parents attempt to mold each offspring in order to increase their own inclusive fitness, while each offspring is selected to resist some of the molding and to attempt to mold the behavior of its parents (and siblings) in order to increase its inclusive fitness. Conflict during socialization need not be viewed solely as conflict between the culture of the parent and the biology of the child; it can also be viewed as conflict between the biology of the parent and the biology of the child. Since teaching (as opposed to molding) is expected to be recognized by offspring as being in their own self-interest, parents would be expected to overemphasize their role as teachers in order to minimize resistance in their young. According to this view, then, the prevailing concept of socialization is to some extent a view one would expect adults to entertain and disseminate.

Parent-offspring conflict may extend to behavior that is not on the surface either altruistic or selfish but which has consequences that can be so classified. The amount of energy a child consumes during the day, and the way in which the child consumes this energy, are not matters of indifference to the parent when the parent is supplying that energy, and when the way in which the child consumes the energy affects its ability to act altruistically in the future. For example, when parent and child disagree over when the child should go to sleep, one expects in general the parent to favor early bedtime, since the parent anticipates that this will decrease the offspring's demands on parental resources the following day. Likewise, one expects the parent to favor serious and useful expenditures of energy by the child (such as tending the family chickens, or studying) over frivolous and unnecessary expenditures (such as playing cards)—the former are either altruistic in themselves, or they prepare the offspring for future altruism. In short, we expect the offspring to perceive some behavior, that the parent favors, as being dull, unpleasant, moral, or any combination of these. One must at least entertain the assumption that the child would find such behavior more enjoyable if in fact the behavior maximized the offspring's inclusive fitness.

Conflict Over the Adult Reproductive Role of the Offspring

As a special case of the preceding argument, it is clear that under certain conditions conflict is expected between parent and offspring over the adult reproductive role of the offspring. To take the extreme case, it follows at once from Hamilton's (1964) work that individuals who choose not to reproduce (such as celibate priests) are not necessarily acting counter to their genetic self-interest. One need merely assume that the nonreproducer thereby increases the reproductive success of relatives by an amount which, when devalued by the relevant degrees of relatedness, is greater than the nonreproducer would have achieved on his own. This kind of explanation has been developed in some detail to explain nonreproductives in the haplodiploid Hymenoptera (Hamilton, 1972). What is clear from the present argument, however, is that it is even more likely that the nonreproducer will thereby increase his *parents'* inclusive fitness than that he will increase his own. This follows because his parents are expected to value the increased reproductive success of kin relatively more than he is.

If the benefits of nonreproducing are assumed, for simplicity, to accrue only to full-siblings and if the costs of nonreproducing are defined as the surviving offspring the nonreproducer would have produced had he or she chosen to reproduce, then parent–offspring conflict over whether the offspring should reproduce is expected whenever $C<B<2C$. Assuming it is sometimes possible for parents to predict while an offspring is still young what the cost and benefit of its not reproducing will be, the parents would be selected to mold the offspring toward not reproducing whenever $B>C$. Two kinds of nonreproductives are expected: those who are thereby increasing their own inclusive fitness ($B>2C$) and those who are thereby lowering their own inclusive fitness but increasing that of their parents ($C<B<2C$). The first kind is expected to be as happy and content as living creatures ever are, but the second is expected to show internal conflict over its adult role and to express ambivalence over the past, particularly over the behavior and influence of its parents. I emphasize that it is not necessary for parents to be conscious of molding an offspring toward nonreproduction in order for such molding to occur and to increase the parent's inclusive fitness. It remains to be explored to what extent the etiology of sexual preferences (such as homosexuality) which tend to interfere with reproduction can be explained in terms of the present argument.

Assuming that parent and offspring agree that the offspring should reproduce, disagreement is still possible over the form of that reproduction. Whether an individual attempts to produce few offspring or many is a decision that affects that individual's opportunities for kin-directed altruism, so that parent and offspring may disagree over the optimal reproductive effort of the offspring. Since in humans an individual's choice of mate may affect his or her ability to render altruistic behavior toward relatives, mate choice is not expected to be a matter of indifference to the parents. Parents are expected to encourage their offspring to choose a mate that will enlarge the offspring's altruism toward kin. For example, when a man marries his cousin, he increases (other things being equal) his contacts with relatives, since the immediate kin of his wife will also be related to him, and marriage will normally lead to greater contact with her immediate kin. One therefore might expect human parents to show a tendency to encourage their offspring to marry more closely related individuals (e.g., cousins) than the offspring would prefer. Parents may also use an offspring's marriage to cement an alliance with an unrelated family or group, and insofar as such an alliance is beneficial to kin of the parent in addition to the offspring itself, parents are expected to encourage such marriages more often than the offspring would prefer. Finally, parents will more strongly discourage marriage by their offspring to individuals the local society defines as pariahs, because such unions are likely to besmirch the reputation of close kin as well.

Because parents may be selected to employ parental investment itself as an incentive to induce greater offspring altruism, parent–offspring conflict may include situations in which the offspring attempts to terminate the period of PI *before* the parent wishes to. For example, where the parent is selected to retain one or more offspring as permanent "helpers at the nest" (Skutch, 1961), that is, permanent nonreproductives who help their parents raise additional offspring (or help those offspring to reproduce), the parent may be selected to give additional investment in order to tie the offspring to the parent. In this situation, selection on the offspring may favor any urge toward independence which overcomes the offspring's impulse toward additional investment (with its hidden cost of additional dependency). In short, in species in which kin-directed altruism is important, parent–offspring conflict may include situations in which the offspring wants *less* than the parent is selected to give as well as the more common situation in which the offspring attempts to garner *more* PI than the parent is selected to give.

Parent-offspring relations early in ontogeny can affect the later adult reproductive role of the offspring. A parent can influence the altruistic and egoistic tendencies of its offspring whenever it has influence over any variable that affects the costs and benefits associated with altruistic and egoistic behavior. For example, if becoming a permanent nonreproductive, helping one's siblings, is more likely to increase one's inclusive fitness when one is small in size relative to one's siblings (as appears to be true in some polistine wasps: Eberhard, 1969), then parents can influence the proportion of their offspring who become helpers by altering the size distribution of their offspring. Parent–offspring conflict over early PI may itself involve parent–offspring conflict over the eventual reproductive role of the offspring. This theoretical possibility may be relevant to human psychology if parental decision to mold an offspring into being a nonreproductive involves differential investment as well as psychological manipulation.

The Role of Parental Experience in Parent-Offspring Conflict

It cannot be supposed that all parent–offspring conflict results from the conflict in the way in which the parent's and the offspring's inclusive fitnesses are maximized. Some conflict also results, ironically, because of an overlap in the interests of parent and young. When circumstances change, altering the benefits and costs associated with some offspring behavior, both the parent and the offspring are selected to alter the offspring's behavior appropriately. That is, the parent is selected to mold the appropriate change in the offspring's behavior, and if parental molding is successful, it will strongly reduce the selection pressure on the offspring to change its behavior spontaneously. Since the parent is likely to discover the changing circumstances as a result of its own experience, one expects tendencies toward parental molding to appear, and spread, before the parellel tendencies appear in the offspring. Once parents commonly mold the appropriate offspring behavior, selection still favors genes leading toward voluntary offspring behavior, since such a developmental avenue is presumably more efficient and more certain than that involving parental manipulation. But the selection pressure for the appropriate offspring genes should be weak, and if circumstances change at a faster rate than this selection operates, there is the possibility of continued parent–offspring conflict resulting from the greater experience of the parent.

If the conflict described above actually occurs, then (as mentioned in an earlier section) selection will favor a tendency for parents to overemphasize their experience in all situations, and for the offspring to differentiate between those situations in which greater parental experience is real and those situations in which such experience is merely claimed in order to manipulate the offspring.

Appendix: The Offspring's Equilibrial Sex Ratio

Let the cost of producing a female be one unit of investment, and let the cost of producing a male be x units, where x is larger than one. Let the expected reproductive success of a female be one unit of RS. Let the sex ratio produced be 1 : y (males : females), where y is larger than one. At this sex ratio the expected RS of a male is y units of RS, so that, in being made a male instead of a female, an offspring gains y − 1 units of RS. But the offspring also thereby deprives its mother of x − 1 units of investment. The offspring's equilibrial sex ratio is that sex ratio at which the offspring's gain in RS in being made a male (y − 1) is exactly offset by its loss in inclusive fitness which results because it thereby deprives its mother of x − 1 units of investment. The mother would have allocated these units in such a way as to achieve a 1 : y sex ratio; that is, she would have allocated $x/(x + y)$ of the units to males and $y/(x + y)$ of the units to females. In short, she would have produced $(x - 1)/(x + y)$ sons, which would have achieved RS of $y(x - 1)/(x + y)$, and she would have produced $y(x - 1)/(x + y)$ daughters, which would have achieved RS of $y(x - 1)/(x + y)$. The offspring is expected to devalue this loss by the offspring's r_0 to its displaced siblings. Hence, the offspring's equilibrial sex ratio results when

$$y - 1 = \frac{r_0\, y(x - 1)}{x + y} + \frac{r_0\, y(x - 1)}{x + y}$$

$$= (2r_0 y)\, \frac{x - 1}{x + y}$$

Rearranging gives

$$y^2 + y(x - 2r_0\, x + 2r_0 - 1) - x = 0$$
$$y^2 + (x - 1)\,(1 - 2r_0)\, y - x = 0$$

The general solution for this quadratic equation is

$$y = \frac{-\,(x - 1)\,(1 - 2r_0)}{2} +$$

$$\frac{\sqrt{(x - 1)^2\,(1 - 2r_0)^2 + 4x}}{2}$$

Where $r_0 = \frac{1}{2}$, the equation reduces to $y = \sqrt{x}$. In other words, when the offspring displaces full–siblings (as is probably often the case), the offspring's equilibrial sex ratio is $1 : \sqrt{x}$, while the parent's equilibrial sex ratio is $1 : x$. These values, as well as the offspring's equilibrial sex ratio where $r_0 = \frac{1}{4}$, are plotted in Figure 4. The same general solution holds if parents invest more in females by a factor of x, except that the resulting sex ratios are then reversed (e.g., $\sqrt{x} : 1$ instead of $1 : \sqrt{x}$).

NOTES

I thank I. DeVore for numerous conversations and for detailed comments on the manuscript. For additional comments I thank W. D. Hamilton, J. Roughgarden, T. W. Schoener, J. Seger, and G. C. Williams. For help with the appendix I thank J. D. Weinrich. Finally, for help with the references I thank my research assistant, H. Hare, and the Harry Frank Guggenheim Foundation, which provides her salary. Part of this work was completed under an N.I.H. postdoctoral fellowship and partly supported by N.I.M.H. grant MH-13611 to I. DeVore.

REFERENCES

Alexander, G. 1960. Maternal behaviour in the Merino ewe. Anim. Prod. 3:105–114.

DeVore, I. 1963. Mother-infant relations in free-ranging baboons, p. 305–335. *In* H. Rheingold [ed.], Maternal behavior in mammals. Wiley, N.Y.

Drury, W. H., and W. J. Smith. 1968. Defense of feeding areas by adult Herring Gulls and intrusion by young. Evolution 22: 193–201.

Eberhard, M.J.W. 1969. The social biology of polistine wasps. Misc. Publ. Mus, Zool. Univ. Mich. 140:1–101.

Elliott, B. 1969. Life history of the Red Warbler. Wilson Bull. 81: 184–195.

Ewbank, R. 1964. Observations on the suckling habits of twin lambs. Anim. Behav. 12: 34–37.

Ewbank, R. 1967. Nursing and suckling behaviour amongst Clun Forest ewes and lambs. Anim. Behav. 15: 251–258.

Fisher, R. A. 1930. The genetical theory of natural selection. Clarendon, Oxford.

Gill, J. C., and W. Thomson. 1956. Observations on the behavior of suckling pigs. Anim. Behav. 4: 46–51.

Hamilton, W. D. 1964. The genetical evolution of social behavior. J. Theoret. Biol. 7:1–52.

Hamilton, W. D. 1971. The genetical evolution of social behavior, p. 23–39. Reprinted, with addendum. *In* G. C. Williams [ed.], Group selection. Aldine-Atherton, Chicago.

Hamilton, W. D. 1972. Altruism and related phenomena, mainly in social insects. Annu. Rev. Ecol. Syst. 3: 193–232.

Hinde, R. A., and Y. Spencer-Booth. 1967. The behaviour of socially living rhesus monkeys in their first two and a half years. Anim. Behav. 15: 169–196.

Hinde, R. A., and Y. Spencer-Booth. 1971. Effects of brief separation from mother on rhesus monkeys. Science 173:111–118.

Hinde, R. A., and L. M. Davies. 1972*a*. Changes in mother-infant relationship after separation in rhesus monkeys. Nature 239: 41–42.

Hinde, R. A., and L. M. Davies 1972*b*. Removing infant rhesus from mother for 13 days compared with removing mother from infant. J. Child Psychol. Psychiat. 13: 227–237.

Jay, P. 1963. Mother-infant relations in langurs, p. 282–304. *In* H. Rheingold [ed.], Maternal behaviour in mammals. Wiley, N.Y.

Le Boeuf, B. J., R. J. Whiting, and R. F. Gantt. 1972. Perinatal behavior of northern elephant seal females and their young. Behaviour 43: 121–156.

Munro, J. 1956. Observations on the suckling behaviour of young lambs. Anim. Behav. 4: 34–36.

Mussen, P. H., J. J. Conger, and J. Kagan. 1969. Child development and personality. 3rd ed. Harper and Row, N.Y.

Rheingold, H. 1963. Maternal behavior in the dog, p. 169–202. *In* H. Rheingold [ed.], Maternal behavior in mammals. Wiley, N.Y.

Rosenblatt, J. S., and D. S. Lehrman. 1963. Maternal behavior of the laboratory rat, p. 8–57. *In* H. Rheingold [ed.], Maternal behavior in mammals. Wiley, N.Y.

Rosenblum, L. A. 1971. The ontogeny of mother-infant relations in macaques, p. 315–367. *In* H. Moltz [ed.], The ontogeny of vertebrate behavior. Academic Press, N.Y.

Rowe, E. G. 1947. The breeding biology of *Aquila verreauxi* Lesson. Ibis 89: 576–606.

Schaller, G. B. 1964. Breeding behavior of the White Pelican at Yellowstone Lake, Wyoming. Condor 66: 3–23.

Schneirla, T. C., J. S. Rosenblatt, and E. Tobach. 1963. Maternal behavior in the cat, p. 122–168. *In* H. Rheingold [ed.], Maternal behavior in mammals. Wiley, N.Y.

Skutch, A. F. 1961. Helpers among birds. Condor 63: 198–226.

Struhsaker, T. T. 1971. Social behaviour of mother and infant vervet monkeys (*Cercopithecus aethiops*). Anim. Behav. 19:233–250.

Trivers, R. L. 1971. The evolution of reciprocal altruism. Quart. Rev. Biol. 46:35–57.

Trivers, R. L. 1972. Parental investment and sexual selection, p. 136–179. *In* B. Campbell [ed.], Sexual selection and the descent of man, 1871–1971. Aldine-Atherton, Chicago.

Wallace, L. R. 1948. The growth of lambs before and after birth in relation to the level of nutrition. J. Agri. Sci. 38: 93–153.

Postscript

One initial response to the publication of "Parent–Offspring Conflict" was a spate of models, explicitly genetic, verifying that, indeed, selection on parent and offspring could be divergent and was best described by Hamilton's rules (nicely reviewed in Godfray 1995). These papers were necessitated, in part, by an influential paper by Richard Alexander (1974), which argued that parents would inevitably "win" in conflicts with their offspring.

The most important work on the subject has been that of David Haig, on genomic imprinting (e.g., Haig 1997) and on genetic conflicts in human pregnancy (Haig 1993, 1996). Haig's 1993 paper, was originally entitled "Mother–Fetal Conflict," but the paper aroused such strong passions in some—the metaphor of conflict and even warfare was being erroneously extended into the blissful mother–fetal symbiosis—that he changed it to the more neutral title "Genetic Conflicts in Human Pregnancy."

I had known that mother–fetal conflict was an expected consequence of Hamilton's kinship theory. I used to teach that birth itself was an arbitrary moment in the relationship, from the standpoint of relatedness. I believed that the conflict would be largely biochemical and hormonal, although I used to amuse myself and my classes at UC Santa Cruz by telling pregnancy stories from my own marriage that suggested behavioral conflict as well. For sibling conflict, the matter was obvious on inspection. When my first wife Lorna carried our twins, Natasha and Natalia, we used to amuse ourselves watching twin fights break out late in pregnancy in which you could see one elbow extended and then another or a fist or foot, as they pummeled each other. My wife also pointed out to me when pregnant with our first child, Jonathan, how he used to kick her upright when she leaned forward while sitting. She believed that although this was more comfortable for her, it increased the pressure on him and that he defended himself by dramatically increasing the pressure on her, that is, kicking her upright!

I was also deeply affected by an incident that suggested new vistas for parent–parent conflict affecting offspring behavior. While my first wife was perhaps seven months pregnant with Jonathan, we got into a short and verbally violent argument. Four or five hours later when tempers had cooled and we were back to normal, Lorna noticed that Jonathan was unusually active within her, which she attributed to our recent argument. I will never forget the sensation I experienced, a kind of internal shudder. I thought we were arguing in private—no witnesses! My crime loomed up larger in my mind. I knew the child could hear perfectly well in the womb, and I imagined that he could easily associate maternal stress hormones experienced via the placenta with my loud and ugly voice, in effect forming a hypothesis about me before meeting me!

Incidentally, in my experience women differ considerably in their consciousness of the baby inside them. My wife, Debra, seems especially sensitive. Driving down the New Jersey Turnpike one night at about 80 miles an hour, she asked me to slow down because the baby inside her was bracing himself in response to my driving. Suddenly I could imagine the full horror of the situation from his standpoint: hurtling through space in a dark room, unable to control its movement or see who was attempting to control it, experiencing sudden deacceleration when I braked, rapid sideways movements when I changed lanes, all the while attempting to brace himself should the room come to a sudden halt.

What I did not know then was that there was an extensive literature in medicine and anatomy on early placental development, as well as complex hormonal changes throughout pregnancy, that could be reorganized along the theme of mother–offspring conflict. What Haig (1993) did was to provide a beautifully detailed and impressive argument for the importance of mother–fetal conflict in the evolution of the placenta and its early behavior as well as an understanding of fetal hormonal effects on the mother. Recently it has even been shown that fetal cells in humans may persist in a mother's blood for more than 25 years after birth (Bianchi et al. 1996).

An example of direct tests in birds of the classic theory can be found in Pugesek (1990). A nice application of the logic to a species with helpers at the nest is found in Emlen and Wrege (1992). Evidence that mothers manipulate testosterone levels in their eggs (with effects on sibling conflict) is found in Schwabl (1993). For effects of inbreeding on maternal investment in mice, see Margulis (1997). For a nice review of prenatal influences in rodents on later reproductive behavior, see Clark and Galef (1995). For sacrifice of offspring survival to maximize maternal reproduction in a fish, see Einum and Fleming (2000). For evidence of a similar trade-off in humans, see Jones (1978). For an excellent treatment of the notion that peer influences outweigh parental ones in humans, see Harris (1995). For sibling rivalry, see Mock and Parker (1997). For a rich, multi-layered approach to mother–offspring relations, see Hrdy (1999). For evidence that begging behavior may have a substantial cost in a bird, see Rodriguez-Girones, Zuniga and Redondo (2001). For novel parent–offspring conflict over the sex ratio, with expected effects on sex chromosome evolution, see Beukeboom, de Jong and Pen (2001).

A most interesting paper has recently analyzed the genetics underlying parental investment in a bug, showing that parents genetically inclined to invest more have offspring which are genetically inclined not to be particularly demanding of investment (Agrawal et al. 2001). It would be very useful to see this kind of work pursued in a variety of species—including our own—and with an attempt to measure ever more sharply the precise negative genetic correlation that evolves between offspring demands and parental investment.

5

HAPLODIPLOIDY AND THE SOCIAL INSECTS

Richard Feynman, the great physicist, once warned of the dangers of think tanks—of trying to do intellectual work, especially theoretical work, without any teaching responsibilities. At first it sounds like a wonderful idea, to have that much more time to devote to what you really care about. But in actual fact it was often a death trap. Inevitably, there will be times when you have no ideas. What do you do then? If you have a teaching position you concentrate on teaching, and the teaching is likely to rejuvenate you mentally. The act of preparing materials for other minds requires you to think things through more carefully than if you are merely explaining them to yourself. In addition, of course, you may get valuable feedback from your students, truly naive questions sometimes being the best at generating new possibilities for further thought. Lacking all of this, your "off" times would grow longer, depression more likely, fear of failure greater, and the greater would be the chance of falling into permanent torpor. I believe my social insect paper is a humble illustration of what Feynman had in mind. The paper really emerged from two different lectures that I gave on the subject to students, neither lecture planned with any additional work in mind.

I Lecture on Haplodiploidy to Anthropology Undergraduates

My friend Irven DeVore arranged a teaching position for me, as an Instructor in Anthropology at Harvard, in the spring semester of 1971, before I had finished my Ph.D. There were 51 students in the course, of whom the best turned out to be Barbara Smuts, later a professor of psychology and anthropology at the University of Michigan, who has done valuable work on friendship in baboons, sexual coercion in primates, and a host of related topics.

Jeffrey Kurland, later a professor of anthropology at Penn State, was the teaching assistant. In any case, I taught social theory based on natural selection. I gave one lecture on Hamilton's kinship theory and decided to give a follow-up lecture on its application to the haplodiploid social insects. My reason for this was that there was, at that time, still very little direct evidence from vertebrates that Hamilton's logic applied, while there appeared to be striking evidence for it in the haplodiploid Hymenoptera (ants, bees, and wasps). Because males arise from unfertilized eggs and are haploid—that is, have only one set of chromosomes—they put that full set of chromosomes in each sperm cell. All sperm cells produced by a male are genetically identical to each other and to the male producing them. The result is that full sisters, related through both parents, are unusually closely related. Half of their genotype is already identical by virtue of having the same father, and half of the remaining half is identical through decent from the common mother: $\frac{1}{2} + \frac{1}{4} = \frac{3}{4}$. Females are diploid, that is, have two sets of chromosomes, only one of which they place in any given egg. Thus, females are related to their own offspring by $\frac{1}{2}$, while being related to their sisters by $\frac{3}{4}$. It is very unusual, in nature, for individuals to be more closely related to someone other than their own offspring (identical twins is another example). Thus, we might expect to see a tendency in ants, bees, and wasps for females to help their mothers to raise fertile sisters in preference to their own reproduction.

About a week before my lecture, I sat down to write the lecture and, in particular, to derive the full set of haplodiploid degrees of relatedness. I did so easily by simply computing the chance that an identical copy of a gene was located in another individual if it was known to be located in one individual. Being the cautious scientist, I naturally checked to make sure that my degrees of relatedness were right by looking at Hamilton's 1964 paper. I was astonished to see that my own degrees of relatedness differed from Hamilton's in four separate instances. I studied and restudied Hamilton's derivation but could not make sense of what he was saying. He seemed to be inventing ghost chromosomes or "cipher" chromosomes, in order to make males diploid and I could not understand the justification for this reasoning. I went around to those who should be in the know, both professors and fellow graduate students, to no avail. I then sat down and wrote Hamilton himself, explaining to him my dilemma and asking him for a quick reply. "I had no choice," I said, "but to teach the degrees of relatedness that I myself understood," but I implored him to send me corrections as soon as possible so that the class would not be misled.

I then received a very nice and most encouraging letter from Hamilton. He wrote from England to say that, as far as he knew, besides himself there was only one other person who had caught the error in his original paper, Ross Crozier, an Australian geneticist. He said that he had noticed the error a few years before but was only now correcting it in the appendix to a paper still in press. I do not remember what else he wrote in the letter, but I certainly remember the warm glow that followed. In effect, I was being told that by deriving things myself and paying attention to elementary logic as I understood it, I could correct an error even of the master himself. I cannot tell you the feeling of self-confidence that this gave me in an otherwise alien domain. I suspect I would never have done my subsequent work on social insects had I not early on learned the pleasures of thinking through haplo-diploid degrees of relatedness on my own. At that time it did not occur to me that using the correct degrees of relatedness still resulted in some prob-lems that could only be solved by considering the sex ratio in addition to degrees of relatedness. I think Hamilton was slowed himself in that direction by starting with erroneous rs.

I Lecture to Graduate Students on Hamilton's System

This next lecture was to graduate students in biology at Harvard, a consid-erably more formidable audience. Preparing my lecture, now only the night before, I soon discovered that Hamilton's system was insufficient as it stood. Although full sisters were related to each other by $3/4$, they were related to their brothers by only $1/4$, so if they invested the same amount in raising sisters as in brothers, they would by Fisher's sex ratio theory gain the same inclusive fitness as if they reproduced on their own. In other words, the high r of $3/4$ to sisters is exactly offset by an r of $1/4$ to brothers: there is no *average* bias toward raising siblings in haplodiploid species. Working late into the night—and, I feel certain, with a full set of stimulants at hand—I gave a disjointed and poor performance the next day. Caught between two sys-tems, I left the students confused as to the form the logic should take. But lecturing poorly in front of fifteen or twenty students is, to me, intensely embarrassing, an embarrassment that would encourage me to write up a handout for the class. The handout would clean up the system and serve to show the students that, though we may have gotten off to a shaky start, we would soon clarify matters. So I worked all week long stating the logic that now seemed required. This was simply that natural selection would still show a bias toward females helping their mothers reproduce, as long as initially they biased their sib investment toward sisters over brothers or, alternatively, raised sisters related by $3/4$ and the sons of laying workers (their full sisters) related by $3/8$, to give an average of $9/16$, greater than the $1/2$ to her progeny. This turned out to be the introduction to the paper that you are about to read. If this enlarged argument was correct, then one expected ratios of investment in social Hymenoptera such as ants to vary between $1:1$ and $1:3$. Under complete worker control they might reach $1:3$, while with no worker effects the ratio of investment should reflect the queen's best interests, 1 : 1. Dressed up later, this became the beginning sections of the paper that was published.

Ed Wilson Helps Me Test the Theory

I had the good fortune then of teaching in a department with two of the greatest ant scientists of all time right down the hall. These were Ed Wilson, the famous evolutionist, and Bert Holldobler (later at University of Würtburg), less well known than Dr. Wilson but a superb experimentalist and a wonderful student of ants. "The world's largest ant collection," as Ed Wilson was pleased to point out, was housed right next door to my office. Tray after tray of ants from all over the world, each glued to the edge of a little cardboard piece, were pierced by pins, along with several tiny cards giving information on locality and such. In short, I was in the ideal situation to test my theory, at least as it applied to ants. Ed Wilson soon put me in touch with the ant literature on sex ratios of reproductive (winged forms or alates). And the collections later permitted me to measure the relative size of a male and a female reproductive without having to send all over the world for the relevant specimens.

The first thing you noticed, when you looked at sex ratios of ants from the literature, was the frequency of male-biased sex ratios! This was discouraging for the theory since we were expecting the ratio of investment to be biased toward females. Wilson quickly assured me that the female was always larger than the male and that this might help resolve the conflict. The first species we tested was *Prenolepis imparis*, an ant with a well-measured sex ratio of reproductives (i.e., excluding workers—from here on I will just refer to the sex ratio) of about 8 : 1 (males : females). One day in late spring Ed Wilson brought in some alates from his backyard in Lexington, Massachusetts. The winged forms were just leaving *Prenolepis* nests in great numbers, and he was able to catch a nice sample of males and females. We had agreed ahead of time, Holldobler, Wilson, and myself, that the dry weight of reproductives would be the best measure, at this stage, of their relative cost to the colony. It was not an exact measure, of course, but a better measure than wet weight since we assumed that the material going into the rest of the ant was more expensive, on average, than the water. In any case, I half expected the new theory to be blown out of the water on its first test, but lo and behold, when we weighed females compared to males, we got a ratio of 25 : 1. Eight males, but each female twenty five times as expensive, is almost exactly a 1 : 3 ratio of investment!

Wilson then, as I remember it, went on a collecting expedition to Florida and brought back for my purposes freshly produced winged forms of the fire ant *Solenopsis invicta*. This fire ant had a well-established sex ratio of almost exactly 1 : 1. An entomologist had placed traps over a great number of fire ant nests and captured all winged forms emerging from these nests. He had then counted the entire sample by hand, refusing to subsample it, as many of us would have been tempted to do. The actual sex ratio was slightly more than 100,000 : 100,000. I now expected the theory to suffer from the opposite problem. If sexual dimorphism was the same as in *Prenolepis*, I would now have a very badly female-biased ratio of investment, way beyond the 1 : 3 that my theory could explain. But when we measured the winged forms of *Solenopsis*, we found that a female weighed exactly 3.5 times as much as a male did: ratio of investment, 1 : 3.5. Thus, over almost one order of magnitude the theory was confirmed. Incidentally, I had been expecting something between 1 : 1 and 1 : 3, perhaps between 1 : 2 and 1 : 3, in deference to the workers' greater relative numbers, but the discovery of 1 : 3 in the first two species convinced me that this was, in fact, the expected rule for the entire group. As I joked to myself at the time, This theory is so true it will actually sustain an honest scientific test.

I set about to do exactly this, to measure the ratio of investment in every relevant context. In termites, since they were eusocial but diploid, in solitary bees and wasps since they were haplodiploid but nonsocial, and so on. In doing so, I employed Mrs. Hope Hare as my assistant, to measure the relative weight of the two sexes. We decided on a sample of five for each sex. So, except for the fortunate few cases where Professor Wilson or others would bring us alates in sufficient number to measure them at once, we usually had to remove each pinned ant from its glued corner, dissolve the glue with water, then redry the ant before weighing. And then we would have to reglue the ant and make sure that all the right pieces of paper and the right ants were on the right pin before it went into the specimen tray in the right place. In the case of the solitary bees and wasps, the chore was more onerous still because the pin went through the thorax of the bee and the glue had to be dissolved, the bee dried and weighed, and then the pin put through the same hole in the thorax, reglued, and so on. Mrs. Hare was essential for this work. I am convinced that had I been doing this work myself I would have destroyed, through my clumsiness, the first two or three specimens, perhaps thrown the specimen tray through the window, and declared the matter over with.

Ants Enslaved by Slave-makers Raise a 1 : 1 Ratio of Investment of Slave-makers

One "prediction" that I am especially proud of in this paper is the prediction that slave-making ants will tend to produce a 1 : 1 ratio of investment because all of the work done raising the young is performed by unrelated slaves, members of different ant species enslaved by the slave-making workers, who themselves do little or nothing in the nest and live largely outside it. Since the slaves are unrelated to the reproductives that they are rearing, they have no self-interest in the ratio of investment itself. The queen is thus expected to evolve new ploys for use in her own nest to which the enslaved workers, having no self-interest at stake, are not expected to evolve counterploys. This prediction came about in the following fashion. Dr. Wilson had put me onto an ant field biologist, Mary Talbot, who had devoted the better part of her adult life to digging up ant nests and scrupulously counting all individuals by category. Thus, the most reliable sex ratios of reproductives had been published by her, and she had an equal wealth of unpublished data that she was only too happy to share with me. Indeed, I was astonished to receive original field data in her handwriting in the mail, and could only wonder whether she had taken the precaution of making a copy for her own records.

In any case, one evening I received unpublished data from her in the mail for two ant species, *Leptothorax curvispinosus* and *L. duloticus*. *Leptothorax curvispinosus*, in turn, was the species being enslaved by *duloticus*. So you had sex ratios for nests of *curvispinosus* working for their own mother versus nests where *curvispinosus* were working for an unrelated slave-maker queen. As I sat looking over her tables of sex ratios for the two species, I noticed that there seemed to be many more males, relative to females, being produced in the *duloticus* nests than in the *curvispinosus* nests, and undoubtedly thought to myself, I wonder why that might be? Then I went off to bed. I remember waking early the next morning, six or six-thirty—a full two hours ahead of schedule—and almost literally jumping out of bed, saying "But that's exactly what you would expect!" I then spent a few moments clearly formulating the logic given above. Since I assumed that sexual dimorphism in size had probably changed little, if at all, in the two species of *Leptothorax*, the sex ratios ought to give the relative ratios of investment. I ran to Talbot's data and found, sure enough, that *duloticus* nests were producing almost three times as many males, compared to females, as were *curvispinosus* nests. Now in the actual paper, the matter is rendered as blind prediction followed by confirming test. As you can see from this account, of course, the notion itself was stimulated by a glance at some relevant data. But in those days I was still very much in the mold of presenting papers in the form prediction-test, prediction-test and was willing, unfortunately, to misrepresent the actual order of things to suit this design. (Data for a second slave-maker showed the same 1:1 ratio and so have later data.)

Ernst Mayr Appears to Me in a Dream

After I had gathered the empirical evidence that was the heart of the paper, I was presented with an interesting new fact that required interpretation. It appeared, in ant nests, as if the queen typically dominated male production, winning that contest with her daughters, but that nests typically produced the ratio of investment preferred by the daughters and not the mothers. Why should this be? Why should the mother dominate one variable and the daughters the other? I thought about this problem for a few weeks without any progress, when I fell asleep one night and had a dream about Ernst Mayr. Professor Mayr and I were in an ant nest, reduced to the size of the ants, workers trundling by while a large physogastric queen spewed out eggs in a corner. Ernst Mayr was pointing at the queen and saying, "Bob, it's the chance of the queen dying, it's the chance of the queen dying!" I woke up around six-thirty in the morning, mumbling the words, "It's the chance of the queen dying." Soon enough I thought to myself, Well, Ernst Mayr has never been wrong in real life, as why assume that he would be any less accurate in my dreams?

I immediately set about trying to figure out how the chance of the queen dying might be the key parameter. As I remember, I spent a few weeks trying various alternatives but eliminating each. Then it occurred to me one day that in a dispute between the queen and a worker, there was a striking asymmetry regarding the effects of damage to either party. The worker could be killed and you would barely be able to measure the loss to *either* party's inclusive fitness, while death of the queen would have a large negative effect on both parties' inclusive fitness. From the worker's standpoint it removed the one individual capable of producing highly related sisters but it would also, in many species, kill off the nest itself, since in those only the fertilized queen is capable of producing more workers. In any case, it seemed obvious that in an aggressive encounter between worker and queen, the worker would be selected to give way to the queen and to avoid, at all costs, inflicting any injury on her. In short, the chance of the queen dying dictates that the queen should win. In aggressive contests with laying workers, the queen would be expected easily to dominate or even eliminate them. There are relatively few such individuals operating in a typical colony, and it should probably be easy for the queen to evolve the ability to detect them, since in order to become egg layers, they have to undergo a series of internal neuroendocrinological changes. By contrast, it was very difficult to see how a queen could aggressively dominate the ratio of investment in the reproductives. It is inevitably made up of thousands, if not hundreds of thousands, of separate acts by thousands and sometimes hundreds of thousands of workers, and I could envision no way for a queen to aggressively assert her preferred ratio. Thus, my dream of Ernst Mayr had helped put me onto an explanation that gave the paper a certain completeness, since it could now explain the novel facts regarding ratios of investment revealed by the empirical work undertaken for other reasons.

Incidentally, after the paper was accepted in *Science*, but before it was published, I ran into Professor Mayr on the staircase of the Museum of Comparative Biology (having told him about the dream the morning after it occurred), and as is, alas, a weakness of mine, I pushed the matter one step too far and asked jokingly whether because of the dream he wanted me to cite "Ernst Mayr, personal communication" in the paper. Mayr did not laugh or, I believe, even smile but said, "Use your own judgment, Bob" and then bounded up the stairs—in other words, "Don't bother me with trivia like this." Mayr, incidentally, liked the paper. I remember he wrote on an earlier draft: "Very impressive. What's next?" That was utterly characteristic of Dr. Mayr, to be on to the future as soon as possible. Being the class act that he was, when he entered my office that day and found only Mrs. Hare working, he said, "But wait, you're a coauthor here, I can give you my comments just as well as to him" and then proceeded to give her his critical comments on the manuscript.

Wilson and His Enemies

Ed Wilson had a wonderful sense of humor, and this was often his saving grace, especially when the great sociobiology debate rolled forth in the mid 1970s and he was pilloried as a biological determinist providing some of the theoretical underpinnings for fascism. He kept up a steady stream of jokes, parodies, and so on to keep himself and others amused. For example, he wrote the abstract of an imaginary paper published in China describing an experiment in which 5,000 pairs of identical and 5,000 pairs of fraternal twins were separated at birth and randomly assigned to different households for rearing. Everyone was administered a "socialist consciousness" scale, and this trait was shown to have a significant heritability. Finally, cross-cultural work on the *average* level of socialist consciousness across several societies showed a higher level in countries that had recently undergone a socialist revolution, apparently as a result of the bloodletting that occurred after each revolution. We sometimes amused ourselves imagining what would happen when the revolution finally came to the United States. I think both of us agreed that Richard Lewontin would be shot shortly after the revolution and that Richard Levins would be pulling the trigger, but that Stephen Jay Gould would certainly survive. His phenomenal ability to rationalize almost any position and to spew forth supporting verbiage by the gallon would serve him well in almost any society as a midlevel apparatchik.

Of course, Wilson's enemies found even his humor distasteful if not more darkly menacing. I remember when the story swept the museum that Wilson had suggested dropping *Macrotermes* on the North Vietnamese in order to blockade the Ho Chi Minh Trail (*Macrotermes* is a termite that builds ten-foot-tall nests that are almost as solid as concrete), and that his political enemies suggested this was yet more evidence of Wilson's impulse to use biology in the service of fascism. But I thought it was a wonderful joke, and I can imagine Ed's face twisted in good humor as he propounds his secret weapon. After all, the alternative was napalm.

Wilson had other jokes on the Vietnam War that I also did not find hostile. For example, I can remember him showing me some ants, a species named *Pheidole dentata*, which had a very large soldier form and a regular-sized worker. The soldier form had massive mandibles and a large head with the appropriate muscles to move the mandibles and was used by the colony primarily in defense against other ants and enemies. These soldiers often hovered near the entrance to the nest and, in fact, were only drawn elsewhere by solicitations from the other workers. We were watching some of these reluctant soldiers in response to an artificially induced conflict with a neighboring colony when Wilson leaned over and said to me, "You know, Bob, you can almost hear those soldiers stridulating 'Hell no, we won't go. Hell no, we won't go.' " This was vintage Wilson. He tricked you into bending over, thinking you were about to be shown some arcane detail of ant biology, when all you were going to get is a little human joke.

Incidentally, Wilson kept a great number of ant species right in his office. He had perfected or adapted a technique for painting the inside upper walls of a Plexiglas container with an ant repellant so that ants remained within the container. And often a test tube with a moistened piece of cotton and whatnot would be provided for the queen and her eggs and so on. He lived, literally, in a world of ants and was always in a position to make new observations. While I was there he was helped in this by Bert Hölldobler, so these men were often arranging encounters between colonies of, say, weaver ants. His most spectacular display was a colony of the leaf-cutting ant *Atta*, which filled nearly an entire room. Here he had a vast array of interconnected Plexiglas cubes with tubes between them in which the ants grew a particular species of fungus on the vegetation that Wilson kindly provided. These colonies in nature can reach at least two million individuals at a time, last for decades, and occupy a hundred cubic meters of space. It was a wonderful achievement to see a fragment of this world captured all around you, so that you almost had the experience of being inside the ant colony when you were in that room.

Getting the Paper Published

Ed Wilson arranged for the paper to be an invited paper from *Science*. An invited paper had the benefit that the journal was already behind your article. It would still be sent out for review, however, and could still be rejected, but there was an initial bias in your favor. I wrote up the paper in the format of *Science* and, after due diligence in rewriting, submitted the paper for consideration. *Science* sent the paper out to two reviewers, and their opinions came back badly split. Both were written by entomologists, I believe; one was short and positive and the other somewhat longer and negative. *Science* then sent the paper out for another round of reviews and once again, the responses came back evenly divided. There was, on the one hand a beautiful evaluation written by Bill Hamilton, which said, among other things, that this was an important paper and, even if it occasionally glided over difficulties or failed to discuss alternatives in greater detail, deserved to be published as it stood. Most of the rest of Hamilton's comments were taken up with his disagreement with our assertion that he had created an "inconsistent system." He suggested, instead, that we state that overemphasis on pairwise comparisons of degrees of relatedness had obscured some of the deeper features of the system, language we incorporated into the next rewrite of our paper.

The other review, by Richard Alexander (presumably with the help of Paul Sherman), was itself already a manuscript. It had an abstract, subheadings, a bibliography, and so on. And, in fact, after various revisions, the manuscript was published as a lead article itself in *Science* (Alexander and Sherman 1977). Alexander called into question every aspect of the data he could, but he mostly emphasized an alternative hypothesis, namely, local mate competition between males over access to close relatives, such as sisters. This was an alternative that I never took seriously.

For local mate competition to work, as it has been discovered numerous times in insects, brothers compete over access to sisters and selection favors mothers, or parents more broadly, who bias the sex ratio of progeny toward females. By one simple model of this effect, fully half of the matings every generation would have to be between siblings to generate a 1 : 3 ratio of investment from this force alone. I regarded this as negligibly likely for most ant species, in which nests over a wide area produce winged forms in great numbers that meet together in the sky and in which siblings reared in the same nest were known to avoid each other sexually, certainly until they flew. Why on earth would you evolve these mass synchronous flights of winged forms if you intended half the matings to be between siblings? In addition, my familiarity with ant sex ratios suggested that all-female and all-male nests were unusually common. In these, of course, sibling matings were excluded at the outset. In any case, *Science* finally decided to settle the matter, one way or the other, and sent it out for a third round of review with only *one* reviewer, George Williams. He wrote a positive review, and the paper was accepted for publication.

The paper had an immediate and strong effect. Fully fifteen pages long, it was the longest lead article I had seen in *Science* and it claimed to provide strong evidence in support of major theories and suppositions then emerging in the field—Hamilton's kinship theory itself, Fisher's sex ratio theory, the supposition that haplodiploidy played a unique role in the evolution of the ants, bees, and wasps, and the supposition that offspring are capable of acting in their own interests, counter to those of their parents. The paper was warmly praised in *Nature*, I attracted large audiences almost everywhere I spoke on the subject, upwards toward a thousand, for example, at Stanford University in 1977 and even S.J. Gould left seventeenth-century France long enough to notice the paper in his *Natural History* column. The paper suggested that we evolutionary biologists, or as we came to be called, sociobiologists, were not just talking a good game (or telling "just so stories") but actually making an occasional advance as well. It was a high time in my personal life as well. I married my first wife, Lorna, a beautiful Jamaican country woman of great strength and intelligence. We shared in the excitement of the ant project, even collecting allates on a brief honeymoon to Cape Cod, perhaps more fondly remembered now by me than by her. The future seemed bright indeed. Intellectually we were revolutionizing the subject of animal behavior, and I still naively imagined that our proper influence within the social sciences would soon be widely felt.

Haplodiploidy and the Evolution of the Social Insects

The unusual traits of the social insects are uniquely explained by Hamilton's kinship theory

ROBERT L. TRIVERS AND HOPE HARE

In 1964 Hamilton (*1*) proposed a general theory for the way in which kinship is expected to affect social behavior. An important modification of Darwin's theory of natural selection, it specified the conditions under which an organism is selected to perform an altruistic act toward a related individual. It likewise specified the conditions under which an individual is selected to forego a selfish act because of the act's negative consequences on the reproductive success of relatives. Broad in scope, the theory provided an explanation for most instances of altruistic behavior, and it promised to provide the basis for a biological theory of the family.

Although many facts from diploid organisms (and some quantitative data) are explained by Hamilton's theory (2–4), the theory has received its main support from the study of the social insects, in particular the social Hymenoptera (ants, bees, wasps). Because species of the Hymenoptera, are haplodiploid (males, haploid; females, diploid), there exist asymmetries in the way in which individuals are related to each other, so that predictions based on these asymmetries can be tested in the absence of quantitative measures of reproductive success. A set of such predictions has been advanced (1, 5, 6), but heavy reliance on pairwise comparisons of degrees of relatedness has obscured some of the more striking implications of haplodiploidy. These emerge when kinship theory is combined with Fisher's sex ratio theory (7–9) in such a way as to predict, under a variety of conditions, the ratio of investment in the two sexes, a social parameter which can be measured with sufficient precision to test the proposed theory. Such a test leads us, in turn, to a new theory concerning the evolution of worker-queen relations in the social Hymenoptera. If the work we describe is approximately valid, it lends support to the view that social behavior has evolved not only in response to a large array of ecologically defined selection pressures (4, 5, 10) but also according to some simple, underlying social and genetical principles.

Hamilton's Kinship Theory

An altruistic act is defined as one that harms the organism performing the act while benefiting some other individual, harm and benefit being measured in terms of reproductive success (RS). Genes inducing such behavior in their bearers will be positively selected if the recipient of the altruism is sufficiently closely related so that the genes themselves enjoy a net benefit. The conditional probability that a second individual has a given gene if a related individual is known to have the gene is called the degree of relatedness, or r (11). For natural selection to favor an altruistic act directed at a relative, the benefit of the act times the altruist's r to the relative must be greater than the cost of the act. Likewise, an individual is selected to forego a selfish act if its cost to a relative times the relevant r is greater than the benefit to the actor. The rules for calculating r's are straightforward in both diploid and haplodiploid species, even under inbreeding (12). If in calculating the selective value of a gene one computes its effect on the RS of the individual bearing it and adds to this its effects on the RS of related individuals, devalued by the relevant r's, then one has computed what Hamilton (1) calls inclusive fitness. Kinship theory asserts that each living creature is selected to attempt throughout its lifetime to maximize its own inclusive fitness.

In sexually reproducing species the offspring's inclusive fitness and the parent's are maximized in similar, but not identical, ways (1, 6, 8). This has the obvious consequence that parent and offspring are expected to show conflict over each other's altruistic and egoistic tendencies. Neither party is expected to see its interests fully realized, and data on both the existence and form of parent-offspring conflict appear to support this view (6, 8, 13). Since a human being typically grows to maturity dependent on a family whose members divide among themselves many resources critical to reproduction, the processes of human psychological development are expected to be strongly affected by kin interactions and designed strongly to affect such interactions. For this reason, kinship theory appears to be a necessary component of any functional theory of human psychological development.

Degrees of Relatedness in Haplodiploid Species

The social Hymenoptera account for nearly 2 percent of all described animal species, and they are characterized by a series of unusual traits. (i) They display extreme forms of altruism through the repeated evolution of sterile or near-sterile castes of workers. These workers typically help their mothers to reproduce (eusociality), but sometimes work for their sisters (semisociality) (14) or less related individuals. (ii) The altruism is sex limited: only females are workers, all males are reproductives. (iii) There are striking lapses of altruism, especially worker-queen conflict over the laying of male-producing eggs and worker-male conflict over the amount of investment males receive. (iv) All species are haplodiploid; that is, females develop from fertilized eggs and are diploid, while males develop from unfertilized eggs and are haploid. Hamilton (1) was the first to realize that all four traits might be related and that haplodiploidy could be used to explain the other three. Especially in his 1972 article (6) he demonstrated how 200 years of scientific work on the social insects stood to be reorganized around his kinship theory.

In haplodiploid species every sperm cell produced by a male has all his genes, while each egg produced by a female has (as in a diploid species) only half of her genes. Because any daughter of a male contains a full set of his genes, sisters related through both parents are unusually closely related (r = ¾). The most important r's, under outbreeding, are summarized in Table 1. By pairing relationships that differ in r, a number of predictions have

Table 1. Degrees of relatedness between a female (or
a male) and her (or his) close relatives in a
haplodiploid species, assuming complete outbreeding.
For the effects of inbreeding see Hamilton (6).

Relation	Female	Male
Mother	1/2	1 ⎫ av. = 1/2
Father	1/2	0 ⎭
Full sister	3/4 ⎫ av. = 1/2	1/2
Brother	1/4 ⎭	1/2
Daughter	1/2	1 ⎫ av. = 1/2
Son	1/2	0 ⎭

been advanced, and some of these seem at first to explain the unusual traits
of the social Hymenoptera. (i) A female is more related to her full sisters
than she is to her own children; she "therefore easily evolves an inclination
to work in the maternal nest rather than start her own" (6). (ii) By contrast,
a male is more related to his daughters ($r = 1$) than to his siblings ($r = \frac{1}{2}$).
"Thus, a male is not expected to evolve worker instincts" (6). (iii) A female
is more related to her own sons than to her brothers. "Thus, workers are
expected to be comparatively reluctant to 'work' on the rearing of brothers,
and if circumstances allow, inclined to replace the queen's male eggs with
their own" (6). Since females are more closely related to their sisters than
to their brothers, they are expected to be "more altruistic in their behavior
toward their sisters and less so toward their brothers" (5).

That this system of pairwise comparisons needs refinement is apparent
when both sexes are treated together. For example, a female is related to
her sisters by ¾, but she is related to her brothers by only ¼; if she does
equal work on the two sexes, as expected under outbreeding (6, 9), then
her average effective r to her siblings (½) is the same as that to her offspring.
In short, haplodiploidy in itself introduces no bias toward the evolution of
eusociality. For this reason, Hamilton (6) added the requirement that "the
sex ratio or some ability to discriminate allows the worker to work mainly
in rearing sisters," and he pointed out that inbreeding should be accompanied in haplodiploid species by female-biased sex ratios [or, better put, by
female-biased ratios of investment (9)]. As long as F, the inbreeding coefficient, is larger than 0, a female is more related to a daughter than to a son
by a factor of $(1 + 3F)/(1 + F)$, so that she is selected to produce a similarly
biased ratio of investment (6). Since this unique effect of inbreeding does
not render eusociality more likely—a female's average effective r to her
siblings remains, under inbreeding, the same as that to her offspring (15)—
we suggest that Hamilton's requirement be amended to read: the asymmetrical degrees of relatedness in haplodiploid species predispose daughters

to the evolution of eusocial behavior, provided that they are able to capitalize on the asymmetries, either by producing more females than the queen would prefer, or by gaining partial or complete control of the genetics of male production. The logic for this requirement is given below, along with some of its consequences.

Capitalizing on the Asymmetrical Degrees of Relatedness

In haplodiploid species, a female is symmetrically related to her own offspring (by sex of offspring) by asymmetrically related to her siblings, while a male is symmetrically related to his siblings but asymmetrically related to his own offspring. It is the male parent and the female offspring who can exploit the asymmetrical r's (for personal gain in inclusive fitness); but there is not much scope for such behavior in males (16), while the females can exploit the r's by investing resources disproportionately in sisters compared to brothers or by investing in sisters and sons (or sisters and nephews) instead of sons and daughters.

1. *Skewing the colony's investment toward reproductive females and away from males.*   Imagine a solitary, outbred species in which a newly adult female can choose between working to rear her own offspring and working to rear her mother's (but not both). Assuming that such a female is equally efficient at the two kinds of work, she will enjoy an increase in inclusive fitness by raising siblings in place of offspring as long as she invests more in her sisters than in her brothers—thereby trading, so to speak, r's of ¼ for r's of ¾. For example, by working only on sisters instead of offspring, her initial gain in inclusive fitness would be 50 percent per unit invested. Were this altruism to spread such that all reproductives each generation are reared by their sisters, in a ratio controlled by the sisters, we expect three times as much to be invested in females as in males, for at this ratio of investment (1 : 3) the expected RS of a male is three times that of a female, per unit investment, exactly canceling out the workers' greater relatedness to their sisters. Were the mother to control the ratio of investment, it would equilibrate at 1 : 1, so that in eusocial species in which all reproductives are produced by the queen but reared by their sisters, strong mother-daughter conflict is expected regarding the ratio of investment, and a measurement of the ratio of investment is a measure of the relative power of the two parties (17).

2. *Denying to the queen the production of males.*   Imagine a solitary outbred species in which a newly adult female can choose between working to rear some of her own offspring and some of her mother's. Other things being equal, she would prefer to rear sons and sisters. A second female who had to choose between solitary life and helping this sister would choose the latter, since she would then trade r's of ½ for r's of ¾ and ⅜. The mother

would benefit by this arrangement, since she would gain daughters in place of granddaughters, but she would benefit more if she could induce daughters to work for her without producing any sons of their own, so that strong worker-queen conflict is expected over who lays the male-producing eggs.

Likewise, there should be conflict between the workers over who produces male eggs, but such conflict is expected to be less intense than similar sister-sister conflict in diploid species. Were the arrangement to spread, such that in each generation all female reproductives are daughters of the queen and all males are her grandsons (by laying workers), then if the nonlaying workers control the ratio of investment, we expect a 1 : 1 ratio. Although a worker is twice as related to a sister ($\frac{3}{4}$) as to a nephew ($\frac{3}{8}$), a male is in turn twice as valuable, per unit investment, as a female reproductive. This is because he will father female reproductives ($r = 1$) and males (by a laying worker) ($r = \frac{1}{2}$), while a female will (like her mother) produce female reproductives ($r = \frac{1}{2}$) and males by laying workers ($r = \frac{1}{4}$). Since ($\frac{3}{4}$) ($\frac{1}{2} + \frac{1}{4}$) = ($\frac{3}{8}$) ($1 + \frac{1}{2}$), the workers preferred ratio of investment is 1 : 1 (18). It is trivial to show that a queen also prefers a 1 : 1 ratio of investment, but if laying workers control the ratio of investment, then we expect a 4 : 3 ratio (since a laying worker is related to her sons by $\frac{1}{2}$ and to her sisters by $\frac{3}{4}$). The important general point to bear in mind is that laying workers introduce an extra meiotic event into the production of males, and this extra event automatically raises the value of a male relative to a female reproductive.

3. *The intermediate cases.* When some fraction, p, of the males in each generation is produced by the queen, and the remainder, $1 - p$, by laying workers, then the equilibrial ratios of investment can be calculated as long as one assumes that p remains relatively constant from one generation to the next and that within a colony individuals prefer to allocate resources to the two sexes according to their average r to members of the two sexes. With these two assumptions, it is relatively easy to show (19) that under queen control the equilibrial ratio of investment, x, results when

$$x = \frac{(3 - p)(1 + p)}{(3 + p)}$$

while under worker control it results when

$$x = \frac{(3 - p)^2}{3(3 + p)}$$

and under laying worker control when

$$x = \frac{2(3 - p)(2 - p)}{3(3 + p)}$$

The three competing optimums are presented in Fig. 1. Even if the queen produces as few as one third of the males, there is a substantial difference

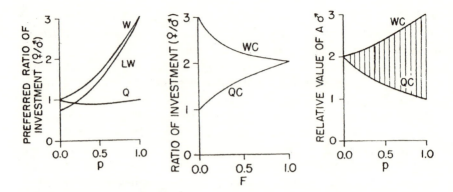

Figure 1. (left). The preferred ratio of investment within eusocial Hymenoptera colonies for the three interested parties, the queen (Q), a laying worker (LW), and the nonlaying workers (W), as a function of the fraction of male-producing eggs laid by the queen (p), where the remainder are laid by a single laying worker. Note that queen-worker disagreement over the ratio of investment increases as p approaches 1. Figure 2. (center). The equilibrial ratio of investment as a function of inbreeding coefficient, F, assuming that the queen lays all of the male-producing eggs. Abbreviations: QC, queen control of the ratio of investment; WC, worker control. Figure 3. (right). The relative value of a male (based on his expected genetic contribution to future generations) compared to the value of a female, per unit investment, as a function of p, depending on whether the queen controls the ratio of investment (QC), the nonlaying workers control the ratio of investment (WC), or the ratio of investment is jointly controlled (shaded area). Only under the unlikely assumption of complete queen control of the ratio of investment (and male production) is the value of a male, per unit investment, equal to that of a female.

between expected ratios depending on who is assumed to control that ratio. Once the queen produces at least two-thirds of the males, workers prefer a ratio of investment of at least 1 : 2.

4. *The effects of inbreeding.*   The above considerations are modified slightly under inbreeding. As long as $F < 1$, the relevant r's remain asymmetrical so that daughters can exploit them as they can under outbreeding. But inbreeding does reduce the asymmetries so that the payoffs associated with the various options become more alike as F approaches 1 (6). This means that the higher the value of F, the less likely is the evolution of eusocial behavior. At $F = 1$ and $p = 1$ (that is, complete inbreeding and complete maternal control of male egg production), both the workers and the queen prefer 1 : 2 ratios of investment, and no conflict is expected over any of the colony's activities. For $p = 1$, the equilibrial ratios of investment are given as a function of differing values of F in Fig. 2. The important point regarding ratios

of investment is that such ratios are never expected to be more female biased than 1 : 2 on the effects of inbreeding alone. All values between 1 : 2 and 1 : 3 must reflect worker performances for sisters over brothers. Values more female biased than 1 : 3 are only expected where extreme patterns of dispersal occur (9).

Although Hamilton has given us an admirable treatment of the possible role of inbreeding in the evolution of the social Hymenoptera (6), we believe its usual role has been negligible, so that the assumption of outbreeding is usually valid. Because the strong selection pressures for producing diverse young act against inbreeding in the same way in which they act against parthenogenesis (20), outbreeding should, like sexual reproduction, have strong positive value in most species. In addition, outbreeding is more easily associated with eusociality than is inbreeding, so that the solitary Hymenoptera should typically show larger values of F than the social species. Most of the evidence we shall later present is consistent with this view of inbreeding.

5. *The early evolution of eusociality.*   Imagine for a moment that daughters are unable to reproduce within their mother's nest, so that they can choose between working there and rearing their own offspring. As pointed out above, they should naturally choose to work for their mother as long as they can preferentially invest in their sisters. Of course the spread of such a preference for sisters should naturally lead to a female-biased ratio of investment, and such a biased ratio raises the value of males, thereby altering the payoffs associated with the daughters' options. The precise genetical analysis is both tedious and complex. Instead, by imagining that the ratio of investment in our incipiently eusocial species is undergoing a steady change, it is easy to give an approximate outline of the relevant selection pressures.

Initially, a female-biased ratio of investment favors mothers who increase the number of their sons, but such behavior should select for workers who respond to sex ratios facultatively, working only when their mothers agree, in effect, to specialize in the production of daughters. As the ratio of investment passes the 1 : 5 mark, a selection pressure appears for workers to work on sons rather than sisters, either on their own or (if we relax our initial requirement) within the maternal nest. If this does not stop the biasing process, the ratio of investment may pass 1 : 2 at which point workers are favored to concentrate on nephews in preference to sisters, intensifying selection for a return to less biased ratios. In short, one never expects a 1 : 3 ratio of investment as an early consequence of a eusocial trend. Instead, one expects that a polymorphism will naturally develop; some large, strifeless nests will specialize in the production of female reproductives and many solitary and small semisocial nests will specialize in the production of males. Such a polymorphism gives to eusocial colonies the evolutionary time to evolve the efficiencies which may eliminate entirely solitary nests in favor of large nests which produce the female reproductives and all the males by a mixture of queen and laying worker contributions.

A second consequence of the imagined early polymorphism is the sharp reduction in the value of males produced in the previous generation. Imagine that fertilized queens overwinter singly and begin new nests in the spring. Some produce daughters who are destined to remain with their mothers to work on rearing female reproductives. Others produce daughters destined to produce sons of their own. Because males are produced from unfertilized eggs, the polymorphism has rendered males superfluous in the first spring generation. As long as the spring queens typically live long enough to produce all the female reproductives, no new sperm is required in the spring and hence all queens can concentrate on the production of daughters to the virtual exclusion of sons. This trivial consequence of haplodiploidy is well documented for so-called primitively eusocial bees (see below).

6. *Summary.*   It appears that there are two, partly overlapping ways by which haplodiploid daughters may be expected to evolve eusocial behavior. Either of these ways tends to bias the ratio of investment toward females, so that this theory can be tested by finding out whether ratios of investment in the eusocial Hymenoptera are typically female biased compared to such ratios in the solitary Hymenoptera. So far as we know, no other theory makes this prediction. In addition, under certain conditions 1 : 3 ratios are expected to be fairly common, so that more precise predictions can be tested along with the main effect. If inbreeding is usually a minor factor, then the most important variable to know is the relative power of the queen and the workers to affect the two parameters over which disagreement is expected: the frequency of laying workers and the ratio of investment.

Fundamental Bias by Sex in Social Behavior of Hymenoptera

With the analysis developed in the previous section, it is possible to present a consistent set of predictions regarding social behavior in haplodiploid species. The most important predictions, along with some of the relevant evidence, are presented here.

1. If and only if workers are assumed to be able to capitalize on the asymmetrical r's in haplodiploid species, does one expect in these species a bias toward eusociality (the evolution of worker castes). If females do not respond appropriately to the asymmetrical r's then their average effective r to their siblings will be the same as that to their own offspring. But if they respond in either of the two ways outlined above, their average r to their siblings will rise above that to their offspring. In short, a bias toward euso-

ciality in the haplodiploid Hymenoptera is contingent upon the discriminatory capacities of the workers. The expectation of this bias does not depend on the assumption that the first workers showed appropriate discriminatory behavior; as long as workers evolved such behavior, their working was more likely to remain adaptive (in the face of fluctuating conditions). Once eusociality appears, it is more likely to endure in haplodiploid than diploid species.

Although species of diploid insects are apparently far more numerous than species of haplodiploid insects, eusociality has evolved only once (the termites) in diploid insects but more than 11 times independently in the Hymenoptera (5, 10). Incipient eusociality, in which an individual helps its parents for one or more years but not usually for life, has evolved repeatedly in mammals but with no bias toward female helpers. In birds, helpers at the nest are usually, but not always, males (21), presumably because the expected RS of a young male is typically less than that of his same-aged sister (22). Helpers in social carnivores may be male or female (23, 24).

2. The same bias toward eusociality can be demonstrated assuming multiple insemination of the queen (compared to multiple insemination in diploid organisms). If the queen is inseminated twice equally, then the average r between a female and her sister will be $\frac{1}{2}$ and between a female and her brother, $\frac{1}{4}$. The average of these two will be $\frac{3}{8}$, which is the same average r between siblings in a diploid species, given the same pattern of insemination. For any multiple insemination, it is trivial to show that a female's average r to her siblings is the same in haplodiploid as in diploid species. But females can still capitalize on the asymmetrical r's. Multiple inseminations remove this possibility only if each daughter is fathered by a different, unrelated male, and if workers are unable to produce any sons. (This same extreme requirement is necessary if the predictions that follow are also to be invalidated through multiple inseminations.) It is, of course, obvious that multiple inseminations render the evolution of worker habits less likely in both haplodiploid and diploid species.

No data exist which would permit one to compare the frequency of multiple inseminations in diploid and haplodiploid insects. What data exist suggest that multiple insemination is infrequent in both groups. In addition, it appears likely that multiple insemination has evolved in the social Hymenoptera as a response to eusociality: a social insect queen may produce tens of millions of workers in her lifetime, overtaxing the spermatogenic capacity of a single male (1, 25). As long as there is a tendency for sperm to clump according to father, as expected (26), there will be a tendency, despite the multiple insemination, for r's between sisters within a colony at any moment to be near $\frac{3}{4}$. The important point is that multiple insemination should not be treated as an independent parameter.

3. Females are more likely to evolve worker habits than are males. Once

females evolve worker habits, a strong bias against the evolution of male workers at once develops. In addition, there develops a bias against males investing in their offspring. A male is unable to exploit the asymmetrical r's to his own advantage. he is equally related to his brothers as to his sisters, so he gains nothing by the over-production of either sex. Likewise, he is unable to produce eggs himself. Since in haplodiploid species a male is no more related to his mate's offspring than to his own siblings, no initial bias in such species (compared to diploid species) is expected either toward or away from male worker habits. (A slight degree of inbreeding introduces a slight bias against male workers). Female workers are expected to exploit the asymmetrical r's and once they do so, in either of the two available ways, the expected RS of a male rises relative to that of a reproductive female, so that the evolution of male workers becomes relatively less likely. If, for example, all males arise from worker-laid eggs, then the expected RS of a male (per unit investment) is twice that of a reproductive female, so that a male would have to be more than twice as effective a worker (gram for gram) as a female in order for selection to favor his helping in the nest (27). In general, when the ratio of investment is controlled by the workers, a male's expected RS is $6/(3 - p)$ times that of a reproductive female (where p is the fraction of males that come from queen-laid eggs). If the ratio of investment is completely controlled by the queen, then the male's expected RS per unit investment is $2/(1 + p)$ times that of a female. For both worker and queen control of the ratio of investment, and for all intermediate cases, the relative RS of a male is given in Fig 3. Only under the unlikely condition of complete queen domination of both male production and the ratio of investment is the expected RS of a male equal to that of a female. Under all other conditions the greater expected RS of a male makes helping behavior and altruism relatively unlikely. In addition, except under complete, or near complete, queen control of the ratio of investment, male parental investment becomes less likely, since a male is expected to inseminate more than one female (per unit investment in him).

In contrast to the termites (all species of which have both male and female workers), there are no species of Hymenoptera that have castes of male workers (5). Indeed, with one or two exceptions (5, 28–31), males have never been seen to contribute anything positive to the colony from which they originate. Again in contrast to the termites, males from social species of Hymenoptera have never been seen to contribute to the colonies that result from their sexual unions, yet rudimentary male parental investment occurs in some solitary species of Hymenoptera (32, 33).

4. No matter who produces the males or who controls the ratio of investment, greater conflict is expected between the workers and the males than between the workers and the reproductive females. Such worker-male conflict is expected to be especially intense where workers control the ratio

of investment. If the queen produces all of the males and also controls the ratio of investment (at 1 : 1), then workers are expected to value their sisters three times as much as their brothers, while each male and each reproductive female values itself twice as much as other reproductives (averaging males and females). Males will then have to work harder to gain appropriate care than will reproductive females. Of course, worker preferences for sisters ought inevitably to lead to a biased ratio of investment. If workers gain their preferred ratio of investment (as a function of p), then they will value reproductives of the two sexes equally; but a male will value himself more relative to his siblings than will a reproductive female relative to her siblings (by approximately the amount shown in Fig. 3) (27), so that selection will more strongly favor male efforts (compared to female efforts) to gain more investment than workers are selected to give, leading to increased worker-male conflict. The argument extends to the intermediate situations as well, but worker-male conflict should be most intense under worker control of the ratio of investment (Fig. 3).

Male-worker conflict appears to be widespread in the social Hymenoptera. For example, male *Mischocyttarus drewseni* mob workers more intensely than do female reproductives, and males are more selfish in their behavior toward larvae (31). Shortly after they enclose, males may be chased from the nest (and killed if they resist), while females are fed in both *Polistes* and *Bombus* (34). In times of food shortage *Camponotus* workers first cannibalize males before turning to female reproductives (28). *Tetramorium* males are apparently starved after they enclose while reproductive females are intensively fed (35).

5. Either laying workers, or a biased ratio of investment in the reproductives, or both, are expected in all eusocial Hymenoptera. Where there are no laying workers, the ratio of investment is expected to approach 1 : 3 (male to female). In other species, the ratio of investment is expected to correlate with p. For reasons outlined earlier, it will be beneficial to the workers if they can produce some or all of the males (but none of the females) or if they can bias the ratio of investment toward their reproductive sisters. Although it is advantageous for the queen to prevent both of these possibilities, there is no reason to suppose that the queen can completely override the maneuvers of her daughters. In the absence of laying workers, one expects a ratio of investment biased toward 1 : 3. As shown earlier, the lower the proportion (p) of males who come from queen-laid eggs, the more nearly the nonlaying workers prefer a 1 : 1 ratio of investment.

A number of species are known to have laying workers (1, 5, 6) but the contribution of these laying workers to the total of males is usually unknown, and most species remain completely unstudied in this regard. It is sometimes supposed that workers must lay male-producing eggs (if they lay any) since they are assumed to be unfertilized, but it is preferable to argue that they remain unfertilized because there is usually no gain in being able to produce daughters. Even wingless, workerlike female ants are fertilized in species lacking winged queens (36, 37), and in some primitively eusocial bees a significant percentage of workers are regularly fertilized; yet fertilized workers have well-developed ovaries no more often than do unfertilized workers (38), suggesting the absence of a selection pressure to produce daughters when the queen is functioning. The ratio of investment in eusocial Hymenoptera is discussed below.

6. The early evolution of eusociality should be characterized by the lengthening of the queen's life so as to produce several generations. Males are expected to be infrequent in the early generations and frequent during the queen's terminal generation. The early evolution of eusociality should be characterized by a polymorphism in which some nests consist of queens and their daughters specializing in the production of female reproductives and other nests consist of daughters, singly or in small groups, producing male reproductives. Such a social grouping actually consists of two generations (in addition to the queen): the generation of adult workers and the generation of adult reproductives whom they rear. If all queens survive to produce the female reproductives, then there will be no value to any males produced along with the generation of workers. Of course, additional generations of workers can be inserted, so that an early eusocial hymenopteran species easily comes to resemble the summer parthenogenetic generations of aphids culminating in the fall production of sexuals.

The correlations proposed are among the most clear-cut in the detailed literature on the early evolution of eusociality in bees (10). For example, the series of eusocial halictine bees, *Lasioglossum zephyrum*, *L. versatum*, *L. imitatum*, and *L. malachurum*, shows "progressively increasing differences in size and in ovarian development between castes, decreasing frequency of worker mating, increasing queen longevity, and decreasing spring and early summer male production" (39).

7. A bias toward the evolution of semisociality (females helping their sister raise her offspring) is expected in haplodiploid species (compared to diploid species). A haplodiploid female is related to her sister's offspring by $r = \frac{3}{8}$ and to her own by $r = \frac{1}{2}$, while a diploid female is related to her sister's offspring by $r = \frac{1}{4}$ and to her own by $r = \frac{1}{2}$, so that, other things being equal, semisociality is more likely in haplodiploid than diploid species. As with eusociality, the bias still persists even if the female is inseminated more than once, as long as each of her daughters is not inseminated by a different male. A male is related to his sibling's offspring by $r = \frac{1}{4}$ and to his mate's by $r = \frac{1}{2}$, so that he is less likely to evolve semisocial habits than are his sisters, but no less likely than males in diploid species. No biased ratio of investment is expected in purely semisocial species.

Semisocial habits (involving females) have evolved independently in the

Hymenoptera even more often than have eusocial habits (5), yet they have not evolved, so far as is known, in the diploid insects. No semisocial behavior is known in haplodiploid males, but their adult behavior is virtually unstudied. Semisocial habits have evolved several times in birds and mammals, more commonly among brothers than among sisters (2, 23, 40).

Ratio of Investment in Monogynous Ants

In the system outlined above, the critical prediction is that workers will bias the ratio of investment toward females whenever some or all of the males come from queen-laid eggs. Since in ants workers feed and care for the reproductives from the time the reproductives are laid as eggs until they leave the nest as adults and since there are usually hundreds of workers (or more) per queen, it is difficult to see how an ant queen could prevent her daughters from producing almost the ratio of investment that maximizes the workers' inclusive fitness. In some ants, such as *Atta* and *Solenopsis* (5), all males appear to be produced by the queen, and in other monogynous ants (single queen per nest) laying workers appear to be a relatively uncommon source of males (compared to eusocial bees and wasps) (41), so that the ratio of investment in ants should often approach 1 : 3. This prediction can be tested by ascertaining the sex ratio of reproductives (alates) commonly produced by a species and correcting these data by an estimate of the relative cost (to a colony) of a female alate compared to a male.

There exist good data on the sex ratio of alates for about 20 ant species, based on complete nests dug up during the time when alates were present in the nest. Ideally, nests should be dug up after all alate forms have pupated (since pupae can be sexed while larvae cannot) but before any of the alates have flown (since one sex may fly earlier than the other). Such data exist, primarily from the pioneering population studies of Talbot (42). Sex ratios so obtained do not differ from sex ratios for the same species based on all nests (42), so data on complete nests dug up anytime were used (43). The number of alates counted, the number of nests from which they came, and the sex ratio for monogynous ants (including two slave-making species) are presented in Table 2. The quality of the data (based on sample sizes) varies widely (44). The sex ratio varies over a 20-fold range (compare *Formica pallidefulva* and *Prenolepis imparis*).

Since in monogynous species workers invest in the reproductives almost exclusively by feeding them, the relative dry weight of a mature male and female alate was taken as a good estimate of their relative cost (45, 46). Dry weights for males and females and the dry weight ratio (female to male) are presented in Table 2. Multiplying the sex ratio by the dry weight ratio gives an estimate of the relative investment in the two sexes (Table 2). This es-

Table 2. The sex ratios of reproductives (males/females) from natural nests of 21 monogynous species of ants and two slave-making species (indicated by s), along with the mean dry weights of male and female reproductives, the dry weight ratio and the inverse of the ratio of investment (inverse of 1 : 3 ratio = 3). Blanks indicate lack of data. Weights are based on dried specimens in the collections of the Museum of Comparative Zoology, Harvard University, except where otherwise stated in the references. The mean weights are based on sample size of five individuals except where noted with the following superscripts: a = 1; b = 2; c = 3; d = 4; e = 6; f = 8; g = 9; h = 10; i = 14; j = 15; k = 20; L = 30; m = 66.

Species	Reproductives counted (No.)	Nests (No.)	Sex ratio	Weight-F (mg)	S.D.	Weight-M (mg)	S.D.	Weight ratio (F/M)	Inverse ratio of investment	Reference
				Subfamily: Formicinae						
Camponotus ferrugineus	1,854	6	1.29	41.18	7.39	6.32	0.50	6.52	5.05	(99)
C. herculeanus	6,300	1*	2.50	56.5[d]	11.1	10.6[d]	2.4	5.33	2.15	(100)
C. pennsylvanicus	1,249	4	0.77	59.5[f]	11.5	8.7[c]	3.3	6.84	8.88	(99)
Formica pallidefulva	2,278	31	0.44	14.4[c]	1.7	7.9[c]	1.2	1.82	4.14	(101)
Prenolepis imparis	1,994	11	8.36	12.7[L]		0.50[g]		25.4	3.04	(102)
				Subfamily: Myrmicinae						
Acromyrmex octospinosus	4,490	10	0.9	19.66[b]	3.5	7.87	2.74	2.50	2.78	(103)
Aphaenogaster rudis	361	14	5.45	6.1[i]		0.48[j]		12.71	2.32	(104)
A. treatae	2,024	12	1.55	9.1[f]		0.9[f]		10.1	6.52	(105)
Atta bisphaerica	35,249	5	3.18					8.00	2.52	(106)
A. laevigata	22,723	6	2.87	263.9[a]		31.5	2.7	8.37	2.91	(106)
A. sexdens	119,936	7	4.90	264.7[h]	100.8	34.5[h]	9.6	7.67	1.57	(106)
Harpagoxenus sublaevis (s)	2,459	58	1.38	0.59[L]		0.34[L]		1.73	(1.25)s	(36.55)
Leptothorax ambiguus	169	12	0.82	0.63[d]		0.10[k]		6.30	7.68	(56)
L. curvispinosus	1,113	82	1.40	0.68[L]		0.15[L]		4.53	3.24	(57)
L. duloticus (s)	1,620	96	2.31	0.20		0.10		2.0	(0.87)s	(54)
L. longispinosus	206	12	0.62	0.54		0.11[e]		4.90	7.90	(56.57)
Myrmecina americana	226	10	1.19	0.55[d]		0.21[d]		2.62	2.20	(56)
M. schencki	795	10	0.31	2.0[d]		1.0[f]		2.00	6.45	(107)
M. sulcinodis	1,114	21	1.15	2.2	0.29	1.2	0.10	1.83	1.59	(108)
Solenopsis invicta	200,491	†	1.00	7.4[g]		2.1		3.52	3.52	(50)
Stenamma brevicorne	235	10	0.90	0.88[d]	0.06	0.36[d]	0.09	2.44	2.71	(56)
S. diecki	391	9	1.30	0.52[c]	0.06	0.15[c]	0.01	3.46	2.66	(56)
Tetramorium caespitum	73,389	126	1.34	6.0		1.5		4.00	2.99	(109)

*Hölldobler (28) also estimated the sex ratio in 15 to 20 additional nests. It ranged between 2 and 3.
†Hundreds of nests (50).

timate should be approximately valid for monogynous and slave-making species but not, as explained below, for polygynous species.

For 21 monogynous ant species, the sex ratio of alates is plotted against their relative dry weight in Fig. 4A. The points tend to scatter around the 1 : 3 line of investment instead of the 1 : 1. The data are fitted by a linear regression in which

$$y = 0.33x - 0.1$$

The slope of this line is not significantly different from a 1 : 3 slope, but it deviates in a highly significant manner from a 1 : 1 slope ($P <<$ .01). In fact, all species are biased toward investment in females, and the least biased species show a 1 : 1.57 ratio of investment. The geometric mean ratio of investment for all species is 1 : 3.45 (range 1.57 to 8.88). The scatter around the 1 : 3 line appears partly to reflect sample size. For example, five of the six species with the best data show a range of only 2.99 to 4.14 (geometric mean = 3.36) (47). The other species (*Acromyrmex octospinosus*) has a ratio of investment of 1.59. It is the only species with a value of p estimated to be lower than 1 ($p = 0.63$), so that its expected ratio, under worker control, is only 1.94. There is a strong inverse relationship ($P <<$ .01; t-test) between the number of males produced and the relative size of a male (compared to a female). This inverse relation is predicted by Fisher's sex ratio theory (9), and, so far as we know, these are the first data—from any group of organisms—demonstrating this relation.

It would be valuable to refine our measure of relative cost. Minor biases are expected from a number of sources. Females contain relatively less water than do males (35, 45), they are richer in calories per gram than are males (35, 48), they are larger than males and therefore consume relatively less oxygen per unit weight (49), and they apparently require less energy (per unit weight) during development than do males (35). Peakin's detailed study permits an overall estimate of the relationship between relative dry weight and relative caloric cost; for *Tetramorium caespitum*, females appear to be three-fourths as extensive as suggested by relative dry weight at the time of swarming (35), so that the ratio of investment based on caloric cost for this species would be 1 : 2.25 (instead of 1 : 2.98 as given in table 2). The need for something like a three-fourths correction also appears likely from the pattern of our investment data: a mean ratio of 3.45 for all species, a mean of 3.36 for the five best studied species, and a 3.54 ratio for the single best studied species, *Solenopsis inpicta* (50), which lacks laying workers and which is certainly typically outbred. In short, real ratios of investment in monogynous ants appear to be near 1 : 3 and certainly larger than 1 : 2.25.

To confirm the contention that the 1 : 3 ratio of investment in monogynous ants results from the asymmetrical preferences of the workers, a series

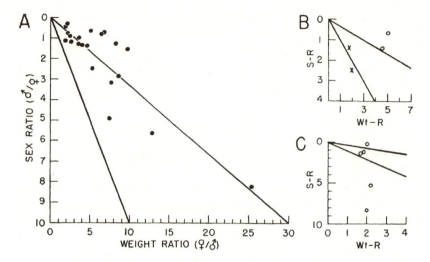

Figure 4. The sex ratio (male/female) of reproductives (alates) is plotted as a function of the adult dry weight ratio (female/male) for various ant species (Table 2). Lines showing 1 : 1 and 1 : 3 ratios of investment are drawn for comparison. (A) All monogynous species. (B) Two slave-making species (x) and three closely related nonslave makers (*Leptothorax*). (C) Five species of *Myrmica* (from top to bottom) *M. schencki, M. sulcinodis, M. ruginodis, M. sabuleti,* and *M. rubra*.

of tests is possible, involving species of ants in which the workers are un-related to the brood they bear (slave-making ants), species of ants in which winged females receive investment in addition to their body weight which males do not receive (polygynous ants), diploid species with workers (ter-mites), haplodiploid species without workers (solitary bees and wasps), and other haplodiploid species with workers (eusocial bees and wasps). Data on the ratio of investment in these species are presented in the following sec-tions.

Ratio of Investment in Slave-Making Ants

In slave-making ants, the queen's brood is reared not by her own daughters but by slaves, workers of other species stolen from their own nests while pupae or larvae. The slave-making workers spend their time slave-raiding, and they typically capture several times their own number in slaves. The slaves feed and care for the slave-making queen and her brood. The slaves are, of course, unrelated to the brood they rear and should have no stake in the ratio of investment they produce. The queen, as always, prefers a 1 : 1 ratio of investment, and in slave-making species she should be able to see her own preferred ratio realized (*51–53*).

The only slave-making ants for which we have found sex ratio data are *Leptothorax duloticus* (54) and *Harpagoxenus sublaevis* (36, 55), two closely related species who prey on other *Leptothorax* species. Fortunately, the data themselves are excellent, being based on large and unbiased samples, and permit a comparison with equally good data from a closely related species that is not slave-making, *L. curvispinosus* (56, 57), and with less detailed data from two other closely related species that are not slave-makers (Table 2). The sex ratio is plotted as a function of relative dry weight for all five species in Fig. 4B. In contrast to these three species, the ratio of investment in both slave-makers is close to 1 : 1 and the geometric mean for the two is 1.00. Each slave-maker has a lower ratio of investment than all other monogynous species shown in Table 2, a highly significant deviation ($P < .001$) toward a 1 : 1 ratio. In *L. duloticus* sexual dimorphism is reduced (through reduction in size of the female) and yet the relative number of males is increased (58).

Leptothorax duloticus enslaves mostly *L. curvispinosus* workers, who in their own nests produce a ratio of investment of about 1 : 3. Since the slaves eclose as adults in a strange nest and go to work caring for the brood as if it were their mother's, why do they not attempt to produce the 1 : 3 ratio of investment typical of their own nests? When *duloticus* first began enslaving *curvispinosus*, the slaves presumably produced a 1 : 3 ratio of investment in the *duloticus* nest, but selection then favored the *duloticus* queen—by whatever means—biasing the ratio of investment back toward 1 : 1, and selection did not favor any countermove by the slaves. In giving up care of the brood in order to raid for slaves, the *duloticus* workers presumably gained sufficient increase in their inclusive fitness to compensate for the loss of their control over the ratio of investment (59).

Ratio of Investment in Polygynous Ants

In polygynous ant species, polygynous nests arise when a queen permits one or more of her fertilized daughters to settle within her nest (60, 61). Large polygynous nests may contain granddaughter queens and even later generation queens. Polygynous nests introduce a bias in the sex ratio because the inclusion of reproductive daughters in the maternal nest increases the relative cost of a female reproductive compared to that of a male (62).

If a reproductive daughter is permitted to settle within or near the maternal nest when unrelated females would not be so permitted, then one must assume that the daughter thereby inflicts a cost on her mother (measured in terms of reproductive success) which her mother permits because of the associated benefit for the daughter. This cost can be treated as a component of investment and raises the relative cost of a reproductive fe-

Table 3. The sex ratio, weight ratio, and inverse ratio of investment for polygynous ant species.

Species	Sex ratio	Weight-F (mg)	S.D.	Weight-M (mg)	S.D.	Weight ratio (F/M)	Inverse ratio of investment	Reference
Crematogaster mimosae	12	4.79[a]	0.62	0.46[a]	0.01	10.4	0.87	(110)
C. nigriceps	6	2.4[b]	0.03	0.57[b]	0.07	4.2	0.70	(110)
Iridomyrmex humilis							0.1	(111)
Myrmica rubra	8.37	2.2	0.21	1.1[c]	0.24	2.02	0.25	(51,111)
M. ruginodis	1.11	1.87	0.29	1.14	0.18	1.61	1.45	(66)
Polygynous	6.71						0.24	
Monogynous	0.92						1.75	
M. sabuleti	5.18	2.2	0.08	1.0	0.23	2.20	0.42	(51,111)
Pheidole pallidula	6.2	3.35		0.6		5.58	0.9	(113)
Tetraponera penzegi	1.8	0.93[a]	0.01	0.48[a]	0.10	1.94	1.1	(110)

a = sample size of 2; b = 3; c = 6.

male. If we assume outbreeding, the male mates with a female who forces the same cost (with its associated benefit) on her mother, so that a male gains the same benefit without inflicting a cost on his mother (63). In short, in polygynous ants we expect the ratio of investment, as measured by relative dry weight, to be biased toward males. This appears to be true for seven polygynous species with the appropriate data (Table 3). There are also several indications of male-biased ratios of investment in polygynous *Formica* (64). Likewise, two polygynous *Pseudomyrmex* have less female-biased ratios of investment than do two monogynous species (65). The more daughters that are permitted to settle in this fashion, the greater will be the relative cost of an individual reproductive female, so that a positive correlation is expected between the degree of polygyny (as measured by the number of queens in a typical nest) and the ratio of investment based on relative dry weight. The most interesting genus in this regard is *Myrmica* (Fig. 4C). Two species are monogynous, *M. schencki* and *M. sulcinodis*, the latter with many laying workers (Table 2). Two are polygynous, *M. rubra* and *M. sabuleti*, with 5 and 15 queens per nest, respectively (Table 3). The third species, *M. ruginodis*, is both monogynous and polygynous (Table 3) (66). The ratios of investment for these species are ordered exactly according to the parameters we have outlined (see Fig. 4C).

Ratio of Investment in Termites

Termites are diploid. In the absence of inbreeding, one expects all colony members, queen and king, female and male workers, to prefer equal in-

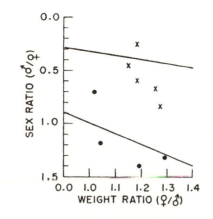

Figure 5. The sex ratio of reproductives is plotted against the dry weight ratio for termites. Data from Roonwal *et al.* (*69*) (closed circles); data from Sands (*67, 70*) (x). The 1 : 1 and 1 : 3 lines are drawn as in Fig. 4.

vestment in reproductives of the two sexes. This is true as long as the colony is monogynous but is not true if the queen is capable of producing some of her daughters by parthenogenesis. Unfortunately, there are almost no data on termite sex ratios and, with one exception (*67, 68*), none based on complete nests. In addition, it is difficult to get specimens to weigh. We have used two kinds of data. (i) Roonwal and his associates gathered sex ratio data, based on naturally occurring swarms, for four species and also ascertained wet and dry weights for male and female alates (*69*). (ii) Sands sampled between two and four nests for five species of *Trinervitermes* and also provided weights (*67, 70*). The data from the two sources are plotted in Fig. 5. The geometric mean ratio of investment for these nine species of termites is 1.62, which is significantly closer to 1 : 1 than are the ratios for monogynous ants ($P < .001$; t-test). There is no significant difference between the termite mean and that of the slave-making ants, a result consistent with the expectation that they be almost equal. However, the termite data are thin enough that they neither strongly support nor contradict our arguments.

Ratio of Investment in Solitary Bees and Wasps

In solitary (nonparasitic) bees and wasps, an adult female commonly builds a cell and provisions the cell with prey or with pollen and nectar. In each cell, she lays either a haploid (male) egg or a diploid (female) egg. In the absence of inbreeding, one expects the typical adult female to invest equally in the two sexes (*9*).

Natural nests. In most solitary bees and wasps, males emerge from and

Table 4. The sex ratio (males/females) from natural nests of solitary species of bees and wasps, along with adult dry weight of males and females of the species. Blanks indicate lack of data. The mean weight for each sex is based on a sample size of five with three exceptions.

Species	Off-spring counted (No.)	Sex ratio	Weight-F (mg)	S.D.	Weight-M (mg)	S.D.	Refer-ence
		Solitary bees					
Agapostemon nasutus	87	2.11	9.7	2.7	6.4	0.6	(114)
Anthophora abrupta	169	1.64	58.0	10.7	36.7	1.6	(115)
A. edwardsii	225	1.48	49.0[a]	2.9	36.6[a]	3.2	(116)
A. flexipes	200	1.50	17.3	2.6	15.1	1.5	(117)
A. occidentalis	241	1.06	89.7	3.7	51.1	6.3	(118)
A. peritomae	70	1.00	23.8[b]	4.1	10.1	1.7	(119)
Chilicola ashmeadi	84	2.82	0.8	0.06	0.5	0.05	(120)
Euplusia surinamensis	297	1.44	148.7	18.7	123.6	31.0	(121)
Hoplitis anthocopoides	351	1.95	11.6	1.6	12.3	2.7	(122)
Nomia melanderi	500	1.01	25.5	5.5	31.9	3.3	(123)
Osmia excavata	2,820	169	24.9	4.8	16.3	1.6	(124)
Pseudagapostemon divaricatus	222	1.61					(125)
		Solitary wasps					
Antodynerus flavescens	200	1.56	22.2	2.5	14.6	2.8	(126)
Chalybion bengalense	183	1.47	19.6	5.9	8.7	2.8	(126)
Ectemnius paucimaculatus	169	1.82	4.0	1.7	3.0	1.2	(127)
Passaloecus eremita	114	0.70	3.4	0.9	1.6	0.3	(128)
Sceliphron spirifex	144	0.95					(83)

a = sample size of 2; b = sample size of 3.

leave their nests earlier than do females (32). In some species the pupal stage itself is known to be shorter in males (71). In addition, female cells are commonly deposited first in twig-nesting species. Because of these sex differences, sex ratio data based on nests collected during the flying season are expected to be biased toward females, as indeed they appear to be (72). By contrast, unbiased data are expected if nests are gathered before any adults have emerged and if the contents are sexed after all larvae have pupated (since larvae can usually not be sexed). We have found such data (with a sample size of 70 or more) for 17 species (Table 4). Since we have no data on cell size or amount of provisions for individuals of either sex, we have again used relative adult dry weight as a measure of the relative cost of a male and a female. Males tend to be smaller than females and more numerous in most species sampled (Table 4). The sex ratio as a function of the dry weight ratio is plotted in Fig. 6. Although there is no tendency for relatively smaller males to be produced in relatively greater numbers, the ratio of investment in solitary bees and wasps is significantly closer to 1 : 1

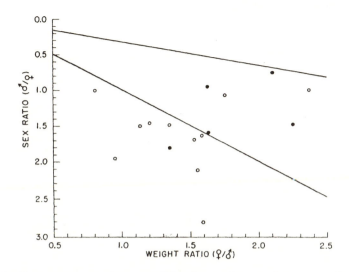

Figure 6. The sex ratio as a function of the dry weight ratio for species of solitary bees (open circles) and solitary wasps (closed circles). Data are from natural nests (Table 4). The 1 : 1 and 1 : 3 lines are drawn as in Figure 4.

than is true in monogynous ants. The geometric mean for all solitary species is 1 : 1.07. The two species which deviate most from 1 : 1 are among the three species with the smallest sample size (115 or less).

These data from natural nests can be supplemented by data from trap-nests (in which artificial nesting sites, usually holes bored in wood, are offered in the field and their contents later reared to maturity). Most such nests consist of a linear series of cells each separated by a partition of mud. The advantage of these data is that they are more numerous than natural data and they can be correlated with direct measures of the relative cost of producing the two sexes. Trap-nests may, however, introduce their own biases, for many bees and wasps prefer to produce the smaller sex (usually males) in smaller diameter holes (73) so that the sex ratio obtained will partly reflect the size distribution of the borings that are presented. Even where the appropriate boring sizes are offered, the data may still be biased if the size distribution of the borings does not exactly match the relative frequency of different size borings in nature. But there is no strong reason in advance to assume that trap-nests will have a systematic bias against any one sex (74) so that strong variance in sex ratio is expected from species to species but no systematic bias.

Trap-nests and the measurement of relative cost.   Krombein (32) has done the bulk of all published trap-nesting work, using the same trapping procedure in a series of localities to capture more than 100 species of solitary bees and wasps. For 27 species (Table 5) with a sample of 70 or more adults

Table 5. The sex ratio (male/female), weight ratio (female/male), cell volume ratio (female/male), and weight per volume ratio (female/male) for the species of solitary wasps and bees studied by Krombein (32). The weight per volume ratio is based only on the cell volumes of those specimens that were weighed. By contrast the cell volume ratio is based on the cell dimensions of all individuals reared. The sex ratio and number of adults reared includes a few individuals whose sex was inferred (75). The species are presented in the order in which Krombein (32) presents them.

Superfamily, family, and species	Adults reared (No.)	Sex ratio	Cell volume ratio (F/M)	Weight-F (mg)	S.D.	Weight-M (mg)	S.D.	Weight ratio (F/M)	Weight per volume ratio
Vespoidea (wasps)									
Vespidae									
Monobia quadridens	227	0.89	1.39	68.2	10.5	35.4	11.8	1.93	1.52
Euodynerus foraminatus foraminatus	96	2.56	1.70	19.0[b]	3.4	12.8	3.4	1.49	1.01
E. f. apopkensis	1,551	2.30	1.69	22.9	2.6	11.0	3.0	2.07	1.33
E. megaera	240	0.67	1.74	28.2	2.9	12.9	2.0	2.22	1.28
Pachodynerus erynnis	240	0.71	0.91	28.0	7.2	12.6[b]	2.3	2.16	3.32
Ancistrocerus antilope antilope	375	1.88	1.48	30.1	4.3	15.8	1.5	1.90	2.0
A. campestris	83	1.77	1.62	18.9	3.6	9.0	2.3	2.10	1.15
A. catskill	189	0.97	1.39	15.5	2.9	6.8	1.2	2.28	1.38
A. tigris	114	0.27	1.53	11.9[b]	2.6	5.0[b]	0.9	2.37	1.85
Symmorphus cristatus	114	1.04	1.20	6.6[a]	0.64	3.4[b]	0.96	1.92	1.12
Stenodynerus krombeini	69	0.86	0.95	9.8	1.5	9.1	1.9	1.07	1.26
S. lineatifrons	92	0.46	0.63	10.5[b]	1.4	5.4[b]	1.4	1.95	1.95
S. saecularis	149	0.69	0.77	17.3[b]	3.8	12.5	2.1	1.39	1.25
S. toltecus	82	0.71	1.34						

186

Pompilidae								
Dipogon sayi	107	0.41	1.05	10.2ᵇ	3.0	0.67	3.46	3.46
Sphecidae								
Trypargilum tridentatum tridentatum	332	0.77	0.90	9.1	10.1	2.3	0.90	0.77
T. clavatum	314	0.89	1.00	11.8	7.9	1.3	1.50	1.01
T. johannis	72	1.18	1.38	17.3	13.8	1.8	1.26	0.82
T. striatum	349	1.60	1.80	24.7	19.0	3.9	1.30	1.03
Trypoxylon frigidum	82	0.71	1.16	2.0ᵇ	1.4	0.37	1.40	0.98

Apoidea (bees)

Megachilidae								
Anthidium maculosum	78	0.3	1.0	34.4	37.0	6.4	0.93	2.38
Prochelostoma philadelphi	85	0.25		3.7ᵇ	3.3	1.0	1.11	1.14
Ashmeadiella meliloti	136	0.64	1.5	6.5	2.8	0.62	2.31	0.97
A. occipitalis	845	0.31	1.29	14.9ᵇ	7.1	2.6	2.10	1.29
Osmia lignaria lignaria	732	2.08	1.49	35.1	14.7	2.3	2.42	
O. pumila	315	0.38	1.24	9.4	5.1	1.6	1.84	1.23
Megachile gentilis	290	5.04	1.0	18.7	12.8	2.5	1.46	
M. mendica	208	2.71	1.0	33.2	16.7	3.3	1.99	

a = sample size of 2; b = of 4.

captured and reared to maturity (75), Krombein provided the sex of the individuals as well as their average cell dimensions. From these dimensions we have calculated the mean relative cell volume (female to male) for each species (Table 5). In addition, Krombein removed from his trap-nesting collection five typical individuals of each sex for each of the 30 species (76) (Table 5). We weighed the specimens and from these weights we have calculated dry weight ratios (female to male) for the same species. Comparison of these data reveals that relative volumes and relative weights are usually greater than one (females occupy more space than males and weigh more).

Since dry weight is partly a measure of the amount of food given and cell volume is a direct measure of the space allotted (77), the relative cell volume and the relative weight of the two sexes (female to male) can be considered partly independent measures of the substitution value of a female (in units of males). In addition, since Krombein's impression is that male and female cells were both stuffed full with prey (78, 79), relative cell volume is probably a good measure of relative amount of food provided. As would be expected, relative dry weight and relative volume are highly correlated, but there is a systematic tendency for females to weigh more than would be expected on the basis of cell volume (Table 5). That this discrepancy is real was confirmed by comparing the weights of our specimens with the volume of the cells they inhabited; weight per unit volume ratios are consistently biased toward females; the mean value for wasps is 1.37, and for bees it is 1.33 (80). Either these wasps and bees allot more space (per unit provisions) for their sons than for their daughters or else development is more expensive in males.

For five species of wasps, Krombein (32) counted the number of caterpillars in a sample of cells that later gave rise to either female or male wasps. On the basis of these data we have calculated the relative number of caterpillars stored in a female cell compared to a male cell. For all five species, this direct measure of provisioning is almost identical to the measure of relative cell volume (81). This is consistent with Krombein's impression that there was no average difference in the size of caterpillars stored in the two kinds of cells (78). To check this impression we weighed the contents of 24 cells of Euodynerus foraminatus apopkensis; these were cells whose wasps failed to develop but for which Krombein could reliably infer the sex of the intended wasp (82). For these 24 cells the mean relative weight of provisions (2.05) is very close to the mean relative cell volume (1.81) (82). These are virtually the only data available permitting a comparison of provisioning ratios with either cell volume ratios or adult weight ratios (83, 84). The only direct measure of developmental cost for any of the Hymenoptera suggests that development may indeed be more expensive in males; during pupation male ants (Tetramorium) lose about 30 percent of their caloric value while females, although similar in size, lose only about 15 percent of theirs (35, 85).

The sex ratio was plotted as a function of relative cell volume for 20 species of wasps (Fig. 7; data from Table 5). Although there is considerable scatter, the species are closer to a 1 : 1 ratio of investment than to a 1 : 3. The data are fitted by the linear regression

$$y = 1.1x - 0.34$$

There is a significant tendency ($P < .01$) for sex ratio and relative cost to be inversely related. The mean ratio of investment based on cell volume is 1.39 for wasps and 1.28 for bees. The ratio based on dry weight is 1.92 for wasps and 2.11 for bees. Similar data have been analyzed from the work of Danks (86), who combined data from natural nests with data from stems of plants made available to wasps and bees. The ratio of investment in these species approximates 1 : 1 (Fig. 8).

Taken together the available data from solitary bees and wasps support the expectation that ratios of investment in the solitary Hymenoptera are typically near 1 : 1 and no greater than 1 : 2. When an individual son is relatively less costly than an individual daughter, relatively more sons tend to be produced. Trap-nests should permit more precise measures of relative cost than presented here (87).

Ratio of Investment in Social Bees and Wasps

Like ants, eusocial bees and wasps are expected to show ratios of investment biased toward females. Since laying workers are known to be an important source of males in some species of bees and wasps (10), ratios of investment in social bees to wasps are not, in general, expected to be biased toward females as in ants (Fig. 0). Unfortunately, it is much more difficult to ascertain the ratio of investment in social bees and wasps than in either ants or solitary bees and wasps. To tell workers from reproductives requires careful, time-consuming behavioral and morphological studies, and these yield too few sex ratio data for our purposes (88). In addition, since female reproductives are hard to distinguish from workers, it is difficult to get an accurate estimate of the relative cost of a reproductive female (compared to a male). We limit ourselves here to detailed data available for bumblebees (*Bombus*) and the closely related parasite (*Psithyrus*).

A temperate bumblebee colony survive for only a season (10), the fertilized females overwintering alone. Reproductive females are produced in late summer at a time when few or no workers are being produced. Young queens remain on the nest for considerable periods where they are readily distinguished from workers. By marking emerging queens and males, Webb

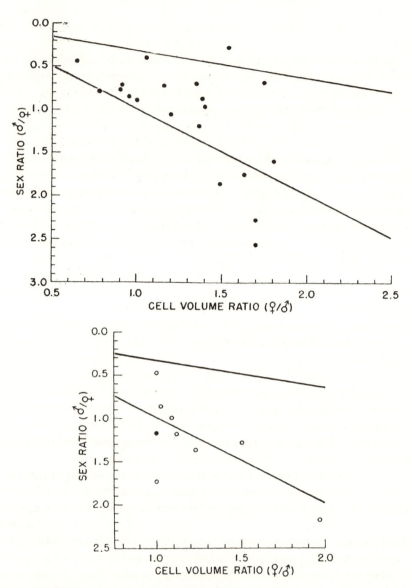

Figure 7 (top). The sex ratio as a function of the cell volume ratio (female/male) for species of solitary wasps trap-nested by Krombein (32) (Table 5, Figure 8). (bottom). The sex ratio as a function of the cell volume ratio for species of solitary wasps (open circles) and bees (closed circle). Natural nests and trap-nests combined by Danks (86).

(*34, 89*) gathered extensive sex ratio data for five species of *Bombus* (and one parasite, *Psithyrus*). These data are presented in Table 6, along with mean weights of male and female reproductives. The sex ratios are all biased in favor of males, and this appears to be general in *Bombus* (*90*). Ratios of investment for the five species lie between 1 : 1.2 and 1 : 3.1.

Psithyrus variabilis is a parasite on *Bombus americanorum* (*34, 89*). A *Psithyrus* queen invades a *Bombus* nest, destroys the host larvae, and rears her own young using the food stores of her host and considerable help from the host workers (*34, 89*). If the *Psithyrus* queen is able to control the ratio of investment then one expects a 1 : 1 ratio and not the 1 : 2 ratio one observes (Table 6). Compared to ratios for the five species of *Bombus*, the ratio in in *Psithyrus* is certainly not biased toward 1 : 1 (as expected), but it is difficult to compare the ratios directly, since mean *Psithyrus* female weight is based entirely on queens caught in the fall while each of the mean *Bombus* female weights is based largely on females caught in the spring after hibernation (and hence weight loss).

Evolution of Worker–Queen Conflict

The information we have reviewed forms an interesting pattern. In monogynous ants the queen appears to produce most or all of the males and the workers apparently control the rate of investment. Where our information is most reliable, this certainly appears to be true (for example, *Solenopsis invicta*, Table 2). The repeated evolution in ants (*5*) of trophic eggs (eggs produced to feed other ants) suggests that in some groups of ants male production by workers was formerly more important than it is now, the queen having regained control of male production and forced a new function on worker-laid eggs. The ant species with the greatest known worker contribution to male production is polygynous (*91*). In some monogynous social bees and wasps, workers contribute heavily to male production (*5, 6*). Why is the queen able to control male production in some species but not in others? Why in monogynous ants is she apparently powerless to affect the ratio of investment (Fig. 4A)? In answering these questions we outline here a theory for the evolution of worker–queen conflict.

1. The asymmetry in aggressive encounters between queen and worker. Aggressive encounters involve violence or the threat of violence. Where two combatants are related, each is expected to adjust its behavior according to the possibility of lowering its inclusive fitness by harming a relative (*92*). In a conflict between a queen and a laying worker, there is an important asymmetry in the way in which each individual is expected to view the possibility of damaging the other. To take the extreme case, early in the life of a large, perennial monogynous ant colony the queen could kill a daughter

Table 6. The ratio of investment in *Bombus* and its parasite, *Psithyrus*. The sex ratio data are from Webb (34). For *Bombus* the weights of females are based entirely on specimens caught in the spring, while *Psithyrus* females were all caught in the fall (before hibernation).

Species	Reproductives counted (No.)	Colonies (No.)	Sex ratio	Weight-F (mg)	S.D.	Weight-M (mg)	S.D.	Weight ratio (F/M)	Ratio of investment
				Species of Bombus					
B americanorum	1780	25	1.52	274.9[c]	32.0	82.4	27.4	4.17	1:2.74
B. auricomus	302	12	1.14	330.3[a]	10.2	136.3	22.8	2.42	1:2.12
B. fraternus	268	4	1.34	315.2[b]	95.28	195.1	32.6	1.62	1:1.21
B. griseocollis	887	20	1.72	207.3[b]	27.1	102.4	8.9	2.02	1:1.17
B. impatiens	351	5	1.42	234.8[d]	106.9	54.9	20.2	4.44	1.313
				Species of Psithyrus					
P. variabilis	290	4	0.91	164.8[a]	29.0	79.9	19.6	2.06	1:2.06

The mean weight for males is based on a sample size of five individuals. 5, 325 (1974) has individuals; b = 3; c = 4; d = 6.

and we would barely be able to measure the resulting decrease in either party's inclusive fitness. By contrast, a worker who kills her mother harms her own inclusive fitness in three different ways. She destroys the one highly specialized egg layer in the colony. She destroys the one individual capable of producing reproductive females to whom the worker is related by 3/4. And she destroys the one individual capable of producing new workers (and hence keeping the colony alive). In short, the worker inflicts a catastrophic loss on her own inclusive fitness.

In such situations there is a large bias in favor of the queen winning any aggressive encounter with her workers. The bias in favor of the queen is largest where the colony is expected to reproduce again in the future (perennial colonies), where there is no alternate, closely related reproductive to whom the workers can attach themselves (for example, monogynous ants), where the queen is strongly specialized as an egg layer, and where the ratio of investment is controlled by the queen (since this decreases the expected RS of the males produced when the queen is destroyed, assuming only males can still be produced). Aggression, as we shall see, is expected to have an important influence on male production but little or none on the ratio of investment.

2. The relevance of aggression to the production of males.   Within a colony a small number of acts result in the laying of the male-producing eggs for a season. If the queen can be present, at or soon after these events, then her advantage in aggressive encounters should permit her to destroy worker-laid eggs, provided that she can recognize such eggs. Alternatively, if she can detect other potential egg layers, she may be able to attack them directly. To discriminate worker-laid eggs from her own, the queen must see them being laid, find them in places or circumstances where her own eggs are not, or learn to discriminate the two kinds of eggs. There is evidence for all three kinds of discrimination (5, 10, 34, 93). In particular, Gervet (93) has shown that *Polistes* females antennate the first several eggs they lay and may eat one or two. When deprived of this experience, a female does not develop the capacity to discriminate strange eggs from her own. Regardless of experience, the female does not destroy eggs that are more than 3 hours old. In short, the mechanisms are known by which a queen can easily discriminate against many worker-laid eggs. In addition, it is difficult to see how a worker could become an effective laying worker yet conceal this fact from the queen. West-Eberhard (34) has suggested that in *Polistes* a female will fail to destroy an egg if she does not have one herself to lay, so that both the capacity to produce male eggs and to destroy those of the queen must depend on how often a laying worker is fed (and how rarely she feeds others). This, in turn, ought to depend on how attractive such an individual is to other workers or how aggressively she begs from them (without herself being altruistic). Queens should be selected to be aggressive toward workers attractive to others and to be aggressive toward begging workers who are themselves not altruistic. In summary, deception cannot save either the laying worker or most of her eggs.

3. *Annual versus perennial colonies.*   In an annual bumblebee colony, killing the queen at the time when male eggs are being laid should not in itself lower the eventual output of the colony by much because, once male production begins, no more workers are produced anyway, and the colony does not survive beyond the production of reproductives. Of course if workers are unfertilized, the entire production of the altered colony must consist of males; but if workers control the ratio of investment then the males' expected RS must equal that of a similar mass of females, so that initially there is only a slight selection pressure against killing the queen. However, it cannot be assumed that the total production of the altered colony will equal that of the colony with the queen intact, for more conflict is expected in the altered colony (94), especially if it is large and no one individual can dominate all others. In summary, worker production of males is much more likely in annual colonies than during most of the life of perennial colonies, but there still remains some bias in favor of the queen.

It is noteworthy that worker–queen conflict and reciprocal egg-eating have been known in the annual colonies of bumblebees since the 19th century, but male production is only known to occur largely by laying workers in one species, *Bombus atratus*, which is perennial. According to our theory, male production by the queen in ants is associated with the perennial colonies typical of this group.

4. *The relationship between polygyny and laying workers.*   If workers easily lose in fights with the queen because of her unique reproductive role, then polygynous societies where the several queens are close relatives should be characterized by workers who are much more willing to risk injuring their mother than workers in monogynous colonies are (95). Although it would be preferable to gain one's way without harming one's mother, injuring her is associated with less drastic effects for the workers' inclusive fitness since they can at least transfer their work to close relatives. We thus predict that polygyny should be associated with laying workers. Arguing from a hypothesized association between inbreeding and polygyny, Hamilton (6) came to the opposite conclusion (96); but he admitted that the only available evidence shows an association between being polygynous and having laying workers produce many of the males (91).

5. *Conflict over the ratio of investment.*   Laying the male-producing eggs can be achieved by a small number of acts, but the ratio of investment (which includes all that goes into rearing the reproductives) results from thousands upon thousands of acts. In addition, the queen—via egg destruction—can often aggressively dominate male production, but it is much more difficult aggressively to impose a ratio of investments. By laying more male eggs than

the workers would prefer, the queen may begin with a sex ratio that would, without intervention lead to equal investment. But the workers care for the eggs and with care goes the power to destroy. The queen may guard her male eggs, but once they hatch they will need care from workers (97). As the larvae grow, execution of excess males becomes increasingly inefficient and underfeeding more likely. Consistent with this argument is the discovery that adult male ants lose weight while their reproductive sisters are being fattened up (35).

If workers can evolve the ability to estimate the ratio of investment within the colony, then they will be able to counteract the queen's maneuvers more efficiently and more precisely. In other words, the capacity to measure and produce a given ratio of investment (which may involve coordinating the activities of millions of workers) must lie within the workers. Perhaps the special cognitive strains of being a haplodiploid worker account in part for the enlargement of the brain in the social Hymenoptera in contrast to its diminution the termites (6).

6. *The concept of offspring power.*   The data that we have gathered are inconsistent with the notion of complete parental domination (13). The female daughters of monogynous ant queens appear to completely dominate their mothers where the ratio of investment is concerned, while she enjoys in the same species nearly complete domination of the genetics of male production. The queen's royal status (highly protected, completely cared for, and the recipient of much altruism and deference) flows from her unique genetic role, but this role does not give her royal powers—at least not where care for her offspring is concerned (the ratio of investment). Instead, the relevant principle is more like: to those who do the work shall be delegated the authority over how such work is allocated. But our slave-making data show that this cannot be a general principle. Likewise, we have no data that would show whether the reproductives in the system get more investment than either the queen or her working daughters prefer. However, Brian has made the remarkable discovery that workers in *Myrmica rubra* have to actively bite larvae in order to decrease the number that develop into reproductives (98), and this is reminiscent of the inefficiencies of weaning conflict in mammals. Instead of supporting a general principle predicting who shall dominate situations of conflict, our work supports the notion that there is no inherent tendency for evolution to favor any particular party in situations of conflict.

Summary

Hamilton (1) was apparently the first to appreciate that the synthesis of Mendelian genetics with Darwin's theory of natural selection had profound

implications for social theory. In particular, insofar as almost all social behavior is either selfish or altruistic (or has such effects), genetical reasoning
suggests that an individual's social behavior should be adjusted to his or her
degree of relatedness, r, to all individuals affected by the behavior. We call
this theory kinship theory.

The social insects provide a critical test of Hamilton's kinship theory.
When such theory is combined with the sex ratio theory of Fisher (9), a
body of consistent predictions emerges regarding the haplodipoid Hymenoptera. The evolution of female workers helping their mother reproduce is
more likely in the Hymenoptera than in diploid groups, provided that such
workers lay some of the male-producing eggs or bias the ratio of investment
toward reproductive females. Once eusocial colonies appear, certain biases
by sex in these colonies are expected to evolve. In general, but especially in
eusocial ants, the ratio of investment should be biased in favor of females,
and in ants it is expected to equilibrate at 1 : 3 (male to female). We present
evidence from 20 species that the ratio of investment in monogynous ants
is, indeed, about 1 : 3, and we subject this discovery to a series of tests. As
expected, the slave-making ants produce a ratio of investment of 1 : 1, polygynous ants produce many more males than expected on the basis of relative dry weight alone, solitary bees and wasps produce a ratio of investment
near 1 : 1 (and no greater than 1 : 2), and the social bumblebees produce
ratios of investment between 1 : 1 and 1 : 3. In addition, sex ratios in monogynous ants and in trap-nested wasps are, as predicted by Fisher, inversely
related to the relative cost in these species of producing a male instead of a
female. Taken together, these data provide quantitative evidence in support
of kinship theory, sex ratio theory, the assumption that the offspring is capable of acting counter to its parents' best interests, and the supposition that
haplodiploidy has played a unique role in the evolution of the social insects.

Finally, we outline a theory for the evolution of worker–queen conflict,
a theory which explains the queen's advantage in competition over male-
producing workers and the workers' advantage regarding the ratio of investment. The theory uses the asymmetries of haplodiploidy to explain how
the evolved outcome of parent–offspring conflict in the social Hymenoptera
is expected to be a function of certain social and life history parameters.

REFERENCES AND NOTES

Dr. Trivers is an associate professor of biology at the Museum of Comparative Zoology, Harvard University, Cambridge, Massachusetts 02138. Mrs. Hare is his research assistant at the museum. This article is dedicated to Ernst Mayr on the occasion of his retirement from the Harvard faculty.
    1. W. D. Hamilton, *J. Theor. Biol.* 7 1 (1964); J. B. S. Haldane, *Nature (London)*

New Biol. **18**, 34 (1955); G. C. Williams and D. C. Williams, *Evolution* **11**, 32 (1957).

2. J. Maynard Smith and M. G. Ridpath, *Am. Nat.* **106, 447** (1972).

3. G. E. Woolfenden, *Auk* **92**, 1 (1975); I. C. R. Rowley, *Emu* **64**, 251 (1965).

4. E. O. Wilson, *Sociobiology* (Harvard Univ. Press, Cambridge, Mass., 1975).

5. ———. *The Insect Societies* (Harvard Univ. Press, Cambridge, Mass., 1971).

6. W. D. Hamilton, *Annu. Rev. Ecol. Syst.* **3**, 193 (1972).

7. R. L. Trivers, in *Sexual Selection and the Descent of Man 1871–1971*, B. Campbell, Ed. (Aldine, Chicago, 1972), pp. 136–179.

8. ———. *Am. Zool.* **14**, 249 (1974).

9. R. A. Fisher, *The Genetical Theory of Natural Selection* (Clarendon, Oxford, 1930). In diploid organisms (in the absence of inbreeding) natural selection favors equal investment in the two sexes. Where investment in a typical male produced equals investment in a typical female, natural selection favors the production of a 50 : 50 sex ratio—regardless of differential mortality by sex after the period of parental investment. For a definition of parental investment, see Trivers (*7*). For preferred sex ratios under inbreeding, see W. D. Hamilton, *Science* **156**, 477 (1967). In the absence of inbreeding, Fisher's argument applies to haplodiploid species. For the effects of inbreeding on the parents' preferred sex ratio in haplodiploid species, see Hamilton (*6*). Hamilton's argument on this point must be modified for species with laying workers, as pointed out to us by J. Pickering, Harvard Biology Department. For the offspring's preferred sex ratio (and the male parent's) in typical diploid species, see Trivers (*8*).

10. C. D. Michener, *The Social Behavior of the Bees* (Harvard Univ. Press, Cambridge, Mass., 1974).

11. Slightly more precise formulations are found in Hamilton (*6*). Under inbreeding, *r* must be redefined to take into account the probability that an individual will have two copies of an allele that is identical by descent.

12. Several of the degrees of relatedness in Hamilton (*1*) and Wilson (*5*) are in error. For corrections, see W. D. Hamilton, in *Group Selection*, G. C. Williams, Ed. (Aldine, Chicago, 1971); R. H. Crozier, *Am. Nat.* **104**, 216 (1970); and (*6*).

13. By contrast, R. D. Alexander [*Annu. Rev. Ecol. Syst.* **5**, 325 (1974)] has argued that offspring should naturally act in their parents' best interests, but in our opinion, neither the arguments presented nor the evidence available support this viewpoint. For example, the spread of alleles conferring selfish behavior on offspring may reduce the eventual reproductive success of both parent and offspring but this is not an argument against the spread of such alleles (although it is an argument for choosing a mate who was selfless when young). Similarly, the ability of the parent to respond to offspring selfishness by harsh retaliation is limited by the parent's growing investment in an offspring, as well as by the offspring's growing independence of the parent (*8*). Alexander interprets the apparent conflict in parent-offspring relations as an efficient system by which the two parties communicate regarding the optimal parental strategy, but this view fails to answer three questions (*6, 8*): (i) Why is such a system of communication not vulnerable to the kinds of deceit already described for vertebrates and social insects? (ii) In what way are kicking, biting, and screaming, or reciprocal egg-eating, efficient systems of communication? (iii) And finally, if conflict occurs because of different estimates of the same parameters, why are parent and offspring estimates predictably biased in favor of

the estimator? Total parental domination has also been argued by M. T. Ghiselin, *The Economy of Nature and the Evolution of Sex* (Univ. of California Press, Berkeley, 1974). See also R. L. Trivers, *Science* **186**, 525 (1974); W. D. Hamilton. Q. *Rev. Biol.* **50**, 175 (1975). For an introduction to kinship theory, see M. J. West Eberhard, *ibid.*, p. 1.

14. C. D. Michener, *Annu. Rev. Entomol.* **14**, 299 (1969). For a review of the literature on the social insects, see Wilson (*5*). For a discussion of the important "primitively eusocial" bees, see Michener (*10*).

15. This point was first brought to our attention by D. M. Windsor of the Smithsonian Tropical Research Institute. Windsor independently derived many of our conclusions.

16. Because a male typically invests nothing in the offspring, his only avenue to a biased sex ratio of investment is through his sperm. These are selected to wriggle through to fatherhood more often than the potential mother would prefer, thereby selecting for a spermatotheca under subtle sphinctral control. Since any success of such sperm automatically raises the value to mothers of producing sons, it is difficult to imagine how spermal ingenuity could compete with mother-power for long.

17. This assumes that such species are typically outbred and that queens are actually or effectively singly inseminated (as when sperm of successive fathers are highly clumped). All full sisters agree among themselves on the preferred ratio of investment, so that the mother's power is pitted against that of a united three-fourths sisterhood.

18. J. Pickering, personal communication.

19. It is desirable to solve the equation $r_M \times$ the expected $RS_M = r_F \times$ the expected RS_F, where r_M is the r of an interested party (workers) to reproductive males and the expected RS_M is the expected RS of a male (per unit effort), measured by both the number of females he is expected to inseminate and by his average r to his mates' offspring. From the nonlaying workers' standpoint (where x is the ratio of investment in males compared to females) the equation reads

$$(\frac{3-p}{8})\ (1 + \frac{1-p}{2}) = \frac{3x}{4}\ (\frac{1}{2} + \frac{1+p}{4})$$

20. See G. C. Williams, *Sex and Evolution* (Princeton Univ. Press, Princeton, N. J., 1975). Inbreeding also decreases heterozygosity. The large degree of genetic variability that is found within sexual species supports Williams' arguments as well as the claim that inbreeding is usually trivial. For further evidence, see R. C. Lewontin, *The Genetic Basis of Evolutionary Change* (Columbia Univ. Press, New York, 1974).

21. A. F. Skutch, *Condor* **63**, 198 (1961); J. L. Brown, *Am. Zool.* **14**, 63 (1974).

22. C. J. Ralph and C. A. Pearson, *Condor* **73**, 77 (1971).

23. G. B. Schaller, *The Serengeti Lion* (Univ. of Chicago Press, Chicago, 1972).

24. L. D. Mech, *The Wolf* (Natural History Press, New York, 1970); J. Van Lawick-Goodall and H. Van Lawick, *Innocent Killers* (Houghton Mifflin, Boston, 1971).

25. Selection would not necessarily favor the production of a male capable of fertilizing all of a queen's daughters, because the cost of such a male to those that produce him may outweigh his benefits to them.

26. G. A. Parker, *Biol. Rev. Cambridge Philos. Soc.* **45**, 525 (1970); S. Taber, *J. Econ. Entomol.* **48**, 552 (1955).

27. A male's r to reproductives that he might help rear (his siblings) will also

change as a function of p, as will a female's; but the difference between the two is slight, especially if there is more than one laying worker per nest. When only one laying worker produces all of the males in each nest, a male is more related (than is a female) to other reproductives in the nest by a factor of 10:9. (With additional laying workers, this factor approaches 1.0.) In general, with a single laying worker per nest the average r of a male to reproductives divided by the average r of a female to reproductives is

$$2(2p^2 - 3p + 5) / (9 - p)$$

This value varies between 1.11 and 0.93.

28. B. Hölldobler, personal communication.

29. ———. Z. Angew. Entomol. **49**. 337 (1962).

30. ———. Z. Vgl. Physiol. **52**, 430 (1966). Males apparently feed each other more often than they feed reproductive females or workers (28), and this bias is exactly consistent with the greater expected RS of males assuming workers are able to bias the ratio of investment or lay some of the male-producing eggs (Fig. 3). Hamilton (6) is apparently mistaken in supposing that Camponotus tend to inbreed (29).

31. R. L., Jeanne, Bull. Mus. Comp. Zool. Harv. Univ. **144**, 63 (1972).

32. K. Krombein, Trap-nesting Wasps and Bees: Life Histories, Nests and Associates. (Smithsonian Press, Washington, D.C., 1967).

33. D. J. Peckham, F. E. Kurczewski, D. B. Peckham, Ann. Entomol. Soc. Am. **66**, 647 (1973); R. M. Bohart and P. M. Marsh, Pan-Pac. Entomol. 36, 115 (1960); C. G. Hartman, Entomol. News 55, 7 (1944); M. N. Paetzel, Pan-Pac. Entomol. **49**, 26 (1973). For Trypargilum, see also Krombein (32).

34. M. J. West Eberhard, Misc. Publ. Mus. Zool. Univ. Mich. 140 (1969); M. C. Webb III, thesis, University of Nebraska (1961). Alternatively, as both authors suggest, the workers may be guarding against inbreeding.

35. G. J. Peakin, Ekol. Pol. **20**, 55 (1972).

36. A. Buschinger, Insectes Soc. **15**, 89 (1968).

37. ———, Zool. Anz. **186**, 242 (1971); C. P. Haskins and R. M. Whelden, Psyche **72**, 87 (1965). Thus, there is no inherent block to sexual reproduction by workers.

38. Reviewed in S. F. Sakagami and K. Hayashida, J. Fac. Sci. Hokkaldo Univ. Ser. VI Zool. **16**, 413 (1968). Comparison of 12 primitively eusocial bees reveals no correlation between the percentage of workers with ovarial development and the percentage of workers that are inseminated. Within Lasioglossum duplex at least, there is a weak tendency for inseminated females to show some ovarial development. Insemination of workers may be useful if the queen weakens or dies.

39. Each of the subgenera of primitively eusocial bees reviewed by Michener (10) shows some or all of these trends, but quantitative data are still too weak to permit detailed correlations between numbers of males produced, degree of worker insemination, and frequency of worker supersedure.

40. C. R. Watts and A. W. Stokes, Sci. Am. **224**, 112 (June 1971).

41. For example, in Aphaenogaster rudis workers are capable of producing males when the queen is dead, but when she is alive she apparently produces most or all of the males [R. H. Crozier, Isozyme Bull. 7, 18 (1974)].

42. This is true for Leptothorax curvispinosus. L. duloticus, and Prenolepis imparis (data separated by M. Talbot).

43. For one monogynous species (*Camponotus herculeanus*) we used data on alates counted while leaving the nest. For another species (*Tetramorium caespitum*), reproductives were removed from the nests each year, leaving the nests intact so that they could be sampled in succeeding years (each year's sampling of a given nest is counted as a separate nest).

44. There is strong variance between nests in the sex ratio produced. Small nests within some species apparently tend to produce males (35). Producing one sex at a time may be a ploy to reduce male selfishness, assuming that males are sensitive to the sex ratio of reproductives within a nest (since they will typically value other males more than female reproductives).

45. Relative dry weight was chosen (instead of relative wet weight), on the assumption that water is relatively inexpensive to an ant colony. For several species, the dry weights were found by collecting live alates, killing, drying, and weighing them. (Alates were dried for 1 hour at 275°F; additional drying had no effect on dry weight.) For most of the rest of the species, we weighed dried specimens in the collection of the Museum of Comparative Zoology (Harvard), and we normally weighed five individuals of each sex. For five species (Table 2), specimens preserved in alcohol were dried and weighed. Although two authorities on ants guessed that females would lose relatively more weight while being dried, the reverse was invariably true: males commonly lose about two-thirds of their weight, while females lose somewhat less than half of theirs. Thus the wet weight ratio is larger than the dry weight ratio by a mean factor of 1: 1.7 (geometric mean for seven species: range, 1.4 to 1.9). Since it is unlikely that water has no cost to the colony, the true relative cost may lie between the dry weight ratio and the wet weight ratio but closer to the dry weight ratio.

46. A number of methodological safeguards were used. All sex ratio data were gathered in ignorance of the relative cost of the two sexes. For each species weighed, H.H. weighed male and female specimens (usually singly) on a Mettler balance scale, type B6 (precision to ±0.01 mg). Whenever we chose specimens to weigh from a larger sample, they were chosen without knowledge of the relevant sex ratio. More than half of the specimens weighed were sent to us by other scientists in response to our request for five specimens of each sex that were typical by size. None of the scientists knew of the predictions being tested.

47. In our opinion the six monogynous species with the best data are *Formica pallidefulva*, *Leptothorax curvispinosus*, *Myrmica sulcinodis*. *Prenolepis imparis*, *Solenopsis invicta*, and *Tetramorium caespitum*.

48. Bomb calorimetry was performed for us by the Warf Institute (Madison, Wis.) with an estimated accuracy of ± 10 percent. Each female sample was slightly higher in calories per gram than the comparable male sample. Two species of *Camponotus*: male, 4.8 kcal/g, female (incomplete burns), 5.15 kcal/g. Composite males of *Myrmica emeryana*, *Aphaenogaster rudis*, and *Pogonomyrmex barbatus*, 5.16 kcal/g. Female *Myrmica emeryana*, 5.19 kcal/g. *Aphaenogaster rudis* (incomplete burns), 6.62 kcal/g, and *Pogonomyrmex barbatus*, 6.29 kcal/g.

49. R. G. Wiegert and D. C. Coleman, *BioScience* **20**, 663 (1970).

50. W. L. Morrill, *Environ. Entomol.* **3**, 265 (1974); personal communication; G. P. Markin and J. H. Dillier, *Ann. Entomol. Soc. Am.* **64**, 562 (1971).

51. M. V. Brian, *Ekol. Pol.* **20**, 43 (1972).

52. L. Passera, personal communication.

53. This prediction should also hold for parasitic ants (5) except that such species are likely to practice adelphogamy, resulting in a strong bias toward females (6, 9). Ratios of investment in parasitic ants do appear to be biased toward females (6). *Sifolinia laurae* females are larger than males and more numerous (51). The same is true of *Plagiolepis xene* (52). As is consistent with the hypothesis of adelphogamy, nearly every nest of *P. xene* that was examined by Passera (8 of 11) produced at least one or two males.

54. As in *Leptothorax curvispinosus*, nests consisted of individual acorns and other nesting places in which *Leptothorax duloticus* were found. For additional information on *L. duloticus*, see M. Talbot, *Ecology* 38, 449 (1957); L. G. Wesson, *Bull. Brooklyn Entomol. Soc.* 35, 73 (1940); E. O. Wilson, *Evolution* 29, 108 (1975).

55. A. Buschinger, personal communication. Buschinger sent us specimens stored in alcohol which we dried and weighed. For additional information on the biology of the species, see A. Buschinger, *Insectes Soc.* 13, 5 (1966); *ibid.*, p. 311; *Zool. Anz.* 187, 184 (1971).

56. M. Talbot, unpublished data gathered at the E. S. George Reserve, Pinckney, Mich. All specimens were sent in alcohol by Talbot and dried and weighed by us.

57. A. E. Headley, *Ann. Entomol. Soc. Am.* 36, 743 (1943) (10 nests); M. Talbot (56) (72 nests). For additional observations on the species, see E. O. Wilson, *Ann. Entomol. Soc. Am.* 67, 777 (1974); *ibid.*, p. 781.

58. The dry weight of the slave-making queen may be reduced because she founds a colony by expropriating a nest of the slave species; hence she may need little in fat reserves.

59. In nests of both *Leptothorax duloticus* and *Harpagoxenus sublaevis* the slaves greatly outnumber the slave-makers (36, 55, 56). The slaves gather all of the food and do all of the nest and brood care.

60. D. H. Janzen, *J. Anim. Ecol.* 42, 727 (1973).

61. Polygynous nests, usually temporary, may also arise when young queens (perhaps sisters) work together to found a new colony [C. Baroni-Urbani, *Zool. Anz.* 181, 269 (1968)]. Queens and workers are not expected to agree on relative merits of monogyny and polygyny. Workers are more likely to favor a reproductive sister joining the nest than is the queen, and in at least one ant species, *Camponotus herculeanus*, polygynous queens are tolerated by workers but act aggressively among themselves (29). By contrast, two related females are more likely to agree to continue the polygyny, and polygyny between founding females often ends when the first workers eclose [N. Waloff, *Insectes Soc.* 4, 391 (1957)]. For detailed studies of any polygyny, see A. Buschinger, thesis, Würzburg University (1967).

62. Inclusion of reproductive daughters in the nest also changes the average *r* between workers and the male and female reproductives that they finally rear. Insofar as workers are rearing the offspring of someone other than their mother, they will tend to prefer a 1 : 1 ratio of investment.

63. This is a general argument. In any species in which the offspring of one sex, after the end of the period of parental investment, inflict a cost on their parents not inflicted by offspring of the opposite sex, then the inflicting sex will be produced in smaller numbers than expected on the basis of parental investment. The argument holds in reverse for altruistic behavior performed by offspring of one sex. This suggests a simple way to measure whether helpers at the nest (for example in birds) are really helping or are inflicting a net cost on their parents [A. Zahavi, *Ibis* 116, 84

(1974); J. Brown, *ibid.* **117**, 243 (1975)], or are helping only enough to make up for costs they are inflicting. The sex ratio that is produced should show appropriate biases. This argument may explain some of the sex ratio variation in bird species with helpers.

64. The following polygynous *Formica* appear to have male-biased ratios of investment: *F. obscurventris* [M. Talbot, *Anim: Behav.* **12**, 154 (1964)]; *F. ulkei* and *F. obscuripes* [————. *Am. Mid. Nat.* **61**, 124 (1959)]; *F. opaciventris* [G. Scherb *J. N.Y. Entomol. Soc.* **69**, 71 (1961)]. See also M. Ito and S. Imamura, *J. Fac. Sci. Hokkaldo Univ. Ser. VI Zool.* **19**, 681 (1974). Relative dry weight (F/M) in *Formica* is typically lower than 2: 1.

65. D. H. Janzen, *Science* **188**, 936 (1975). The ratio of investment was based on a single nest for each species.

66. G. W. Elmes, personal communication: 21 nest were monogynous (2130 reproductives), four were polygynous (293 reproductives). Polygynous queens were microgynes (average, eight to a nest).

67. W. A. Sands, personal communication.

68. *Neotermes connexus* (sex ratio 2791: 912; 55 nests were examined): H. A. Bess, in *Biology of Termites*, K. Krishna and F. M. Weesner, Eds. (Academic Press, New York, 1970), vol. 2. We have been unable to get specimens to weigh, but it is unlikely that females are more than 1.5 times as heavy as males (67).

69. M. L. Roonwal and S. C. Verma, *Ann Arid Zool.* **12**, 107 (1973); M. L. Roonwal and N. S. Rathor, *ibid.* **11**, 92 (1972). For one species, *Microcertermes raja*, the dry weight ratio was calculated from Roonwal and Verma's linear measurement of the alates. A termite swarm consists of individuals who are just departing their nest.

70. W. A. Sands, *Insectes Soc.* **12**, 117 (1965). We thank W. A. Sands and M. T. Pearce of the Centre for Overseas Pest Research, London, England, for the use of their unpublished data on *Trinervitermes*. Specimens stored in ethanol were weighed after being dried for 1 minute on filter paper. For the five species of *Trinervitermes*, the sex ratios and the dry weight ratios are as follow:

	Sex ratio	Weight ratio
T. trinervius	61:136	1.15
T. germinatus	230:347	1.25
T. togoensis	60:101	1.18
T. oeconomus	95:114	1.27
T. occidentalis	52:214	1.18

Although the sex ratio data are sparse they a consistent with other indications of a female biased sex ratio. For nomenclatural changes see W. A. Sands, *Bull. Br. Mus. (Nat. Hist.) Entomol. Suppl.* **4** (1965).

71. S. D. Jayakar and H. Spurway, *J. Bombay Nat. Hist. Soc.* **61**, 662 (1964).

72. Data from the following studies are likely to be biased in this fashion: K. A. Stockhammer, *J. Kans. Entomol. Soc.* **39**, 157 (1966); R. P. Kapil and S. Kumar, *J. Res. Punjab Agric Univ.* **6**, 35 (1969): C. D. Michener, W. B. Kerfoot, W. Ramirez, *J. Kans. Entomol. Soc.* **39**, 245 (1966).

73. The depth of the holes available may also affect the sex ratio produced, as

in *Megachile rotundata*. [H. S. Gerber and E. C. Klostermeyer, *Science* **167**, 82 (1970)].

74. There is usually considerably greater mortality trap-nests than in nature. If the mortality is differential by sex, then the sex ratio of adults when enclosing misrepresents the sex ratio at the time parental investment (when the eggs were laid). In general, species with little or no male parental investment (such as bees and wasps) are expected to show differential male mortality (7).

75. Plus *Stenodynerus krombeini* ($N = 69$). For some species, Krombein (32) was able to infer sex for some of the individuals who failed to develop and for these species the sex ratio we have used is based on all individuals reared plus those whose sex was inferred. Inclusion of individuals whose sex was inferred has only a slight (and nonsystematic) effect on the sex ratio measured.

76. All individuals were reared by Krombein (32) in trap-nests and were killed and pinned shortly after eclosing.

77. Weighted equally by the two dimensions, cell length and cross-sectional area.

78. K. V. Krombein, personal communication.

79. Also relevant to relative cost are features such as cell position and hence time of parental investment, cell partition width, and associated intercallary cells. Neither partition width nor associated intercallary cells appear to differ strikingly by sex (32), but female cells are often deeper in the nests; that is, they are the first to be occupied and the last to be vacated. The effect of these factors on relative cost is not clear.

80. This is also true for a larger sample (15 individuals of each sex) of *Euodynerus foraminatus apopkensis* (Table 5). Cell dimensions were not available for many of the specimens on which mean weights were based, thus Krombein picked out additional specimens to permit the measure of weight per unit volume.

81. For the five species, the provisioning ratio (sample size given in parentheses) and the cell volume ratio are: *Monobia quadridens* 1.38 (9), 1.39; *Euodynerus foraminatus apopkensis* 1.75 (77), 1.69; *E. megaera* 1.76 (9), 1.74; *E. schwarzi* 2.29 (11), 2.37; and *Ancistrocerus antilope* 1.21 (17), 1.48.

82. The caterpillars were removed from alcohol and dried for 18 hours before being weighed. Because the contents of some cells were lumped, it is not possible to calculate variances in the measures. Sample size: 8 females, 14 males, and 2 uncertain individuals that are likely to be males. For the same species (sample of 2 males and 2 females) the weight ratio of provisions (1.41) is very near the volume ratio of provisions (1.44) [J. T. Medler, *Ann. Entomol. Soc. Am.* **57**, 56 (1964)].

83. E. White, *J. Anim. Ecol.* **31**, 317 (1962). The weight ratio is based on White's data.

84. In *Osmia rufa* the provisioning ratio (1.3), based on the number of provisioning trips required to fill up two male cells and two female cells, was the same as the volume ratio (1.3), but not the same as the adult weight ratio (for all individuals, 1.7). The sex ratio was 1.4 [A. Raw, *Trans. R. Entomol. Soc. London* **124**, 213 (1972)]. In *O. lignaria* the peak larval weight ratio (1.4) is the same as the weight ratio of provisions consumed [M. D. Levin, *J. Kans. Entomol. Soc.* **39**, 524 (1966)]. In *Sphecius speciosus* the adult weight ratio (2.4) is almost identical to the provisioning weight ratio (2.3) but the latter was only estimated [R. Dow, *Ann. Entomol. Soc. Am.* **35**, 310 (1942)]. In *Sceliphron spirifex* (83) the provisioning ratio is about 1.6 while male and female cell volumes are nearly identical, but this species is a mud dauber and does not build cells end to end in a limited space.

85. In *Megachile rotundata* the dry weight ratio for adults is only slightly greater

(3 to 8 percent) than the wet weight ratio of larvae for three independent samples (73).

86. H. V. Danks, *Trans. R. Entomol. Soc. London* **122**, 323 (1971). Relative cell volumes are based on relative cell length alone since this is the only cell dimension that Danks supplies. For one of his species (*Cemonus lethifer*) Danks also provides weights of prepupae. The prepupal weight ratio (*N* = 252) is 1.29, slightly greater than the cell volume ratio (1.12).

87. In particular, prey that is stored must be weighed as well as counted. In addition, by watching females provision cells, one can measure the time spent in provisioning cells of different sizes.

88. For example, Jeanne (*31*) and West Eberhard (*34*) give sex ratio limited by small samples. C. D. Michener [*Bull. Am. Mus. Nat. His.* **145**, 221 (1971)] gives detailed sex ratio data for allodapine bees but it is very difficult to separate workers from reproductives in these bees. For other data, see Michener (*10*).

89. Since males leave the nest before young queens do, Webb's data probably slightly underestimate the number of males produced.

90. J. B. Free and C. G. Butler [*Bumblebees* (Collins, London, 1959)] give an estimate of 2: 1 for the sex ratio in *Bombus* generally.

91. *Myrmica rubra* (Table 3) [M. V. Brian, *Insectes Soc.* **16**, 249 (1969)].

92. W. D. Hamilton, in *Man and Beast: Comparative Social Behavior* (Smithsonian Press, Washington, D.C., 1971). For the expected effects on aggressive behavior of asymmetries in the payoffs, see G. A. Parker, *J. Theor. Biol.* **47**, 223 (1974).

93. J. Gervet, *Insectes Soc.* **9**, 343 (1964).

94. Conflict is expected because each worker would prefer to produce sons rather than nephews. This conflict is also expected when the queen is alive, but workers then agree on the production of sisters which is the major part of their work.

95. The acacia trees that are host to ant colonies of *Pseudomyrmex* show a similar bias. Queens of monogynous species are protected within heavily fortified thorns, while queens of polygynous species occupy less protected ones (*60*).

96. For species such as *Myrmica rubra* polygyny apparently results when fertilized daughters are permitted to return to the colony of their origin. It is not known whether such females mate with close relatives. That inbreeding should be associated with laying workers is not a strong argument. Queens will be less antagonistic toward eggs laid by workers, but there will also be less gain for the workers.

97. The queen may be able to influence the ratio of investment in mass provisioning social bees because investment occurs at the time of egg laying.

98. M. V. Brian, *Colloq. Int. C.N.R.S.* **173**, 1 (1967).

99. J. L. Pricer, *Biol. Bull. (Woods Hole)* **14**, 177 (1908). *Camponatus-ferrugineus* specimens were captured in Delaware and killed and weighed by us (after drying) on 10 April 1975. *Camponotus pennsylvanicus* females were captured and killed in Stoughton, Mass., on 10 June 1974, dried, and weighed. (Males came from the Museum of Comparative Zoology collection.)

100. B. Hölldobler and U. Maschwitz, *Z. Vgl. Physiol.* **50**, 551 (1965).

101. M. Talbot, *Ecology* **29**, 316 (1948).

102. ———, *ibid.* **24**, 31 (1943). *Prenolepis* reproductives overwinter as alates in the nest. Specimens were collected in mid-May while swarming in Lexington, Mass.

103. T. Lewis, *Trans. R. Entomol. Soc. London* **127**, 51 (1975).

104. M. Talbot, *Ann. Entomol. Soc. Am.* **44**, 302 (1951); A. E. Headley, *ibid.* **42**,

265 (1949). Specimens were collected from the nest 21 July 1974 at Blue Hills, Mass.

105. M. Talbot, *Contrib. Lab. Vertebr. Biol. Univ. Mich.* **69**, 1 (1954). Specimens collected 18 July 1974 at Wellfleet, Mass.

106. M. Autuori, *Arq. Inst. Biol. São Paulo* **19**, 325 (1950). No specimens were available for *A. bisphaerica*, so an approximate weight ratio of 8.0 was inferred from weights of the other two *Atta*.

107. M. Talbot, *Ann. Entomol. Soc. Am.* **38**, 365 (1945).

108. G. W. Elmes, *Oecologia (Berlin)* **15**, 337 (1974). At least 223 of the 596 males were apparently produced by laying workers, so that the ratio of investment is expected to lie near 1: 2 (rather than 1: 3).

109. M. V. Brian, personal communication; ———, G. Elmes, A. F. Kelly, *J. Anim. Ecol.* **36**, 337 (1967). Weights (at swarming) are from Peakin (35).

110. B. Hocking, *Trans. R. Entomol. Soc. London* **122**, 211 (1970).

111. G. P. Markin, *Ann. Entomol. Soc. Am.* **63**, 1238 (1970). Each month of the year four nests were sampled at random and the contents were weighed.

112. G. W. Elmes, *J. Anim. Ecol.* **42**, 761 (1973).

113. Twenty-three nests, 1852 reproductives (52).

114. G. C. Eickwort and K. R. Eickwort, *J. Kans. Entomol. Soc.* **42**, 421 (1969).

115. T. H. Frison, *Trans. Am. Entomol. Soc. (Phila.)* **48**, 137 (1922).

116. R. W. Thorp, *Am. Midl. Nat.* **82**, 321 (1969).

117. P. F. Torchio and N. N. Youssef, *J. Kans. Entomol. Soc.* **41**, 289 (1968).

118. J. C. Porter, *Iowa State J. Sci.* **26**, 23 (1951).

119. P. F. Torchio, *Los Ang. Cty. Mus. Contrib. Sci.* **206**, 1 (1971).

120. G. C. Eickwort, *J. Kans. Entomol. Soc.* **40**, 42 (1967).

121. D. H. Janzen, personal communication.

122. G. C. Eickwort sexed pupae and prepupae in nests found prior to adult emergence in May 1972. He was unable to sex 27 prepupae. For additional data, see G. C. Eickwort, *Search* **3**, 1 (1973).

123. P. F. Torchio, personal communication.

124. Y. Hirashima, *Sci. Bull. Fac. Agric. Kyushu Univ.* **16**, 481 (1958).

125. C. D. Michener and R. B. Lange, *Ann. Entomol. Soc: Am.* **51**, 155 (1958).

126. S. D. Jayakar and H. Spurway, *Nature (London)* **212**, 306 (1966).

127. K. V. Krombein, *Proc. Biol. Soc. Wash.* **77**, 73 (1964).

128. O. Lomholdt, *Vidensk. Medd. Dan. Naturhist. Foren. Kbh.* **136**, 29 (1973).

129. Supported by the Harry Frank Guggenheim Foundation. For the use of specimens we thank M. V. Brian, A. Buschinger, G. Eickwort, Y. Hirashima, D. H. Janzen, K. V. Krombein, O. Lomholdt, C. D. Michener, M. Talbot, P. Torchio, and E. O. Wilson. For the use of unpublished data we thank M. V. Brian, A. Buschinger, G. Eickwort, G. W. Elmes, D. H. Janzen, K. V. Krombein, O. Lomholdt, L. Passera, W. A. Sands, M. Talbot, and P. Torchio. We thank K. V. Krombein, M. Talbot, and E. O. Wilson for generous and unstinting aid. For additional help we thank W. L. Brown, B. Hölldobler, S. Hyman, P. Hurd, J. Pickering, M. L. Roonwal, J. Scott, L. S. Trivers, and K. Strickler Vinson. For detailed comments on this article we thank R. D. Alexander, G. Eickwort and K. Eickwort, C. D. Michener, and P. Torchio.

Postscript

When Ed Wilson was working on *Sociobiology* in the early 1970s, he used to joke that having mastered the more difficult literature on the social insects for his *Insect Societies* (1971), he thought it would be easy to master the less impressive literature on vertebrate social behavior and therefore treat the whole subject together. Of course, there was the usual asymmetry: students of the social insects usually knew a lot about vertebrate behavior, but students of vertebrates rarely knew anything about social insects beyond the famous waggle dance of honey bees. It was kinship theory that brought the social insects to the attention of many other biologists, because when Hamilton outlined his theory in 1964, the best evidence came from the haplodiploid Hymenoptera. What was true then is still true now. For detailed tests of kinship theory, for precise measurements of degrees of relatedness, for a host of advances in understanding the interaction with sex ratio, the social insects provide the most detailed and extensive comparative data available.

There is a very extensive literature now on the subject, with major comparative studies available for various sugroups, e.g., vespine wasps (Foster and Ratnieks 2001). To illustrate the sophistication of this literature here are a few recent titles, dealing only with wasps, and mostly the work of only one group of scientists:

> "Insurance-based advantage to helpers in a tropical hover wasp" (Field et al. 2000)
> "Conflicts of interest in social insects: Male production in two species of *Polistes*" (Arevallo et al. 1998)
> "Kin selection, relatedness, and worker control of reproduction in a large-colony epiponine wasp" (Hastings et al. 1998)
> "Control of reproduction in social insect colonies: Individual and collective relatedness preferences in the paper wasp, *Polistes annularis*" (Queller et al. 1997)
> "Lack of kin discrimination during wasp colony fission" (Solis et al. 1998)
> "Colony life history and demography of a swarm-founding social wasp" (Strassmann et al. 1997)

For an excellent review of the general subject of kinship and sex allocation in social insects, see Crozier and Pamilo (1996). For ants, see Bourke and Franks (1995). For recent work on slave-making ants, see Herbers and Stuart (1998); for social parasitism in ants, Aron, Passera, and Keller (1999). For the importance of parasites in selecting for colony genetic diversity, see Liersch and Schmid-Hempel (1998), Baer and Schmid-Hempel (1999), and Schmid-Hempel 1994). The best way to gain access to the latest work is to do a "cited reference search" for Trivers and Hare (1976) on a computer search engine. On Web of Science, this will spit forth almost 600 references, in reverse chronological order.

6

SIZE AND REPRODUCTIVE SUCCESS IN A LIZARD

I first visited Jamaica in August 1968, in the company of my advisor, Professor Ernest Williams, Harvard's Curator of Herpetology. Dr. Williams was an authority on *Anolis* lizards, about 200 species spread from South America to the southern United States and including many species in the West Indies. I took one look at the women of Jamaica and a second look at the island itself and decided that if I had to become a "lizard man" to pay for frequent visits to the island, then by God I would humble myself and become a lizard man.

My arrival in Jamaica with Professor Williams was captured in a dedication I wrote in Hicks and Trivers (1983):

Ernest W. Williams took me to Jamaica in 1968. He was on a collecting expedition; I was the driver. We arrived in Kingston in the evening and drove in a blinding rainstorm to the Maryfield Guest House, a decaying English great house set on three acres. The beautiful old trees and well-tended garden attracted a big population of lizards. We began our fieldwork over breakfast on the veranda watching *Anolis lineatopus*. The males warmed themselves in the sun and then engaged in display and aggressive encounters as they reoccupied their territories. These bright, active little lizards reminded one almost of puppies or kids, enjoying a little social play in the early morning hours.

Ernest soon drew my attention to a more sinister species, which seemed to hide in the background. This was *Anolis valencienni*, which moved in a very distinctive fashion. . . . Individuals of this species seemed unusually abundant at the Guest House and since an *Anolis* of this type had not been studied, I soon concentrated on figuring out its social system. Ernest impressed me very warmly on that trip to Jamaica. He traveled in a very calm, quiet, unpretentious style. When we filled out our immigration cards and were asked to state our occupation, my natural impulse was to jack up the description as high as I could. I expected Ernest to do likewise. Nothing less than "Alexander Agassiz

Professor of Zoology at Harvard University" seemed appropriate to the occasion. Instead he wrote "Teacher," simple, unpretentious, and when need be, nonspecific or ambiguous. He also seemed fearless in his travels through Jamaica, and I much respected him for this. He neither swaggered nor scraped, but held himself calmly at all times. (pp. 571–57)

I had initially thought to become a monkey man much in the mold of Irven DeVore, and when granted $800 on admittance to graduate school in the summer of 1968 to study whatever I wished, I proposed to fly to Panama and spend a few weeks watching howler monkeys on Barro Colorado Island. Ernest Williams was a middle-aged Harvard bachelor who had never learned how to drive and preferred a bright and diligent graduate student for company, so he prevailed upon me to study lizards instead. I remember he used every possible argument in favor of the lizards. He even said that if I was obsessed with primate behavior, why, there was a *fossil* monkey on Jamaica! He also said that any fool could study a monkey (which, I am sure, is true) but that it took a *real* biologist to study a lizard (of this I am less certain). Professor Williams was on a collecting expedition for only a week but I chose to stay for another twelve days in order to begin work on *Anolis valencienni*, which could be studied on the hotel grounds where I was staying.

I returned in 1969 for a summer of research on the species and found an apartment near the hotel grounds, which were now abandoned. I slowly studied the species, capturing two, three, or perhaps even five individuals in a day with a slip knot tied to the end of a stick. But in retrospect, working by myself really made little sense since Anolis lizards are arboreal and I have a fear of heights that makes the second rung on a ladder seem like an unsafe perch. So I did not climb the trees and had to wait until the lizards came to the ground to feed or to move on to a neighboring tree before I could catch them. Then I was invited for a weekend in the countryside, and there a fourteen-year-old boy, upon hearing that I was eager to hold one of the large green lizards, *Anolis garmani*, climbed up the tree and quickly noosed and caught a large male. I saw at once that I could greatly increase my sample size by employing teenage youths to catch my lizards for me, and this became the basis for my project on *garmani*. In one way I learned to regret the change. When I worked on my own, I caught very few lizards but spent many hours watching the lizards through binoculars. I thus learned much more about their behavior and spent a more interesting day asking functional questions that I did with my scientific data-gathering operation, in which much of my day was taken up looking at the undersides of lizards while measuring, marking, describing, and releasing them.

The great advantage of the *garmani* project was that I could run a mark-recapture study, *requiring* my presence on the island three times a year. At the same time Professor Williams was fortunate to get one of the very first NSF grants in ecology with, by the standards of the times, substantial funds.

I set out to measure natural selection acting on the species and ended up concentrating on size, growth rate, survival, and reproductive success in *both* sexes. This was the first or one of the very first studies to measure how a single parameter (size) affected reproductive success differently in the two sexes in nature. A description of our lizard team in action is provided by a piece I wrote for the *Jamaican Sunday Gleaner* on the greatest lizard catcher I ever employed, Mr. Glenroy Ramsey, later of Alligator Pond, St. Elizabeth.

Glenroy Ramsey Master Lizard Catcher

Glenroy Ramsey was—and is—undoubtedly the greatest lizard catcher I have ever met. This is saying more than it sounds like because I have handled, I believe, more Jamaican lizards than any human alive (more than 3,000, including more than 1,000 of the dreaded green lizard alone). I have caught lizards also in the United States, in Haiti, in Panama, in Europe, and in East Africa, and I have never seen the likes of Glenroy Ramsey. If there were an All-Jamaican Lizard-Catching Team which competed internationally (as do cricketers), Glenroy would certainly be on the team, might be team captain, and would very likely be one of those rare individuals whose exploits are heralded for decades.

Glenroy was five to ten times better than the best I employed. That is, on his best days, if I went out out with Glenroy and three other of my top lieutenants, he would catch three times as many lizards as the other three combined.

Let me give you two examples of Mr. Ramsey's special powers. Glenroy once caught the largest male green lizard I ever handled, and he also caught the largest female. This was not terribly surprising since he probably caught about one-third of all the green lizards I ever handled. What was truly exceptional was that he caught the two at the same moment, while copulating with each other, slipping a nylon noose of fishing line around both their heads before jerking them off the tree in one sweet motion.

To have the largest two lizards sexing with each other is such a statistically rare event as to constitute a valid scientific fact. In other words, as a scientist I could have published a short, but meaningful, note in a journal such as *Herpetologica*, based almost entirely on this single capture!

A second rare achievements suggests that the Lord Herself may be intervening on Mr. Ramsey's behalf. Two scientists from Florida visited me at my work Southfield, St. Elizabeth. They were studying crabs in bromeliads but wished to see the lizards I was studying, especially the green lizard and the coffee lizard (also called, in some parts of Jamaica, "the white croaking lizard"). Too involved in conversations to go out myself, I asked Glenroy to capture a large male of either species. In less than 15 minutes, he returned with a large male green lizard hanging by its neck from his noose. Protruding from the male's mouth were the hind feet and tail of another lizard. When we pulled this lizard out, it turned out to be a large male coffee lizard, dead for some time and enveloped in slime! Once again, Mr. Ramsey had performed way beyond ordinary expectation.

Difficult

Incidentally, it was not beyond Glenroy to have concocted the entire exhibit himself, capturing each male separately and forcing one down the throat of the other. However, it is very difficult to force one large lizard down the throat of another: both lizards tend to resist your efforts at every turn. Glenroy could have grabbed the coffee lizard, killed it, and stuffed it down the throat of a live green lizard, which he had then caught, but Glenroy rarely chose to handle the lizards himself and, in any case, the direct physical evidence indicated otherwise: the male coffee lizard showed too many signs of having been inside the green lizard too long. It was, in fact, the green lizard's considerable difficulty swallowing his prey that permitted a relatively easy capture.

Glenroy and I often worked together as a team. When green lizards were frightened by our approach, they sometimes took to the very top leaves of their tree. Although Glenroy scaled the tree, he could not always sight the lizard himself (flattened, for example, on top of a large mango leaf). When this happened, I and others directed the attack from the ground, I through binoculars, giving instructions to maneuver his lizard stick close to the prey. When finally the noose slipped over the lizard's head, someone would yell, "Jig!" and Glenroy would sweep the lizard into the air. Sights such as these convinced some country people that we were involved in some very serious *obeah*, indeed—a suspicion confirmed for them when we painted numbers on the lizards' backs and released them back to nature. This permitted us to study individuals in nature and we also clipped two toes to give each lizard a permanent number since they shed their skin every three or four weeks, along with any paint.

I would like to say that Glenroy was smarter than any lizard he ever met, but this is not, in fact, quite true: he was only smarter than all the males. A few females eluded him for good. Being smaller, females are more difficult to catch, but they are also "wilder." That is, they run from us quicker and stay away longer. They seem to apprehend more quickly than males that we are up to no good, that we are in fact, serious lizard catchers and intend to hold our prey. Males, by contrast, have their brains addled by their feistiness. They are continually turning back to face us, as if reasserting their authority in their own tree and this is usually their undoing.

In our later years we sometimes joked that we were the only two people on the island who were smarter than at least half the lizards! (That the lizards have well more than 90% of the adult human population firmly under their control is, of course, obvious from even a short visit to the island.)

What He Looks For

We once asked Glenroy what he looked for when he headed out to capture lizards. Most people look for the head jutting out from a branch or trunk. Sometimes we look for the tail or the general outline of the lizard. But Glenroy said he only looked "for the eyes." We all had a good laugh on that one. After all, most of the lizards are quite small as it is without restricting our vision only to the eyes!

As the years rolled by I came to see this as a very significant fact. The sighting of eyes plays an extremely prominent role in predator-prey relations in nature. Even when prey play dead—play possum, for example—they invariably keep their eyes wide open. Eyes glisten with the consciousness and attentiveness of their possessors. Predators also key on the eyes of their prey because a strike at the head will tend to kill or disable. This bias had led to the evolution of numerous eye-mimics: fake eyes, located, for example, on the wings of butterflies so as to deflect the predator's attack onto a relatively harmless part of the anatomy.

Who knows exactly how the interaction works? For all we know, some lizards seeing Glenroy coming their way say to themselves (so to speak) "Rasta George! I can't look this devil in the eye, he'd recognize me! I'd better get the hell out of here," thus making a movement and attracting the attention of the very person it seeks to avoid.

Do not imagine for a moment, though, that Glenroy's skill was just a matter of some eye trick or trick with lizard eyes. To appreciate the brain-power behind his success you had to watch him mature as a tactician to take on the most difficult challenges. In our later years together a capture might go as follows: upon sighting a lizard he would see at a glance that this was going to be a difficult capture: the lizard was a small adult female, already nervous at our presence, in a tree with too much canopy and too many escape routes. He would then devise a strategy, closing off escape routes, positioning others to watch from key angles, maneuvering the lizard through his own climbing approach and stick movements.

Some epic struggles between man and lizard lasted a half an hour or longer and when Glenroy finally emerged from the tree, with lizard in hand, the lizard knew that it had been had, that it had run up against a superior force, been worn down by a greater mind. Of course, Glenroy used others in his campaign so that the impertinence of some of these lizards thinking they could run and hide from the Greater Southfield Co-operative Lizard-Catching Team, permanently postponing their encounter with science, gave us considerable amusement.

I knew that they had less to fear from us than they supposed. We rarely killed a lizard through rough capture, Glenroy least of all. The lizard would lose two toes, to be sure, but so would everyone else in the study site, so that no particular injustice was implied. (We never cut their large toe on the hind leg out of respect for its obvious utility.) But the lizards never grew to love us, this is certain, and later recaptures only made matters worse. It eventually became apparent that the lizards could tell us apart from other people in the area. For some time I imagined that this might be due to my light skin color, and easy cue, but one day I learned that this cue was certainly not necessary. I was watching a large, male green lizard from a great distance, through binoculars, hidden out of clear sight behind a bush. Jamaican after Jamaican walked by in both directions, some passing quite close without any clear response from the lizard. Suddenly he took one quick look up the street, turned and bolted for the foliage above.

I looked up the street and saw two of my lizard catchers with bun and box juice in hand, returning from the shop. Neither carried a lizard stick, but both had heads tilted slightly back and, through long habit, frequently raked the trees with their eyes. Perhaps this was the lizard's cue or perhaps by now he knew the odious sight of each and every one of us, regardless of the way in which we approached!

Blood Sample

In a recent visit to the island (December 1985) Glenroy and I went up to Hardware Gap, north of Newcastle, to find the elusive "water lizard" *Anolis reconditus*. This time we were taking a blood sample from each lizard, in a study of their blood parasites (especially malaria). We spent two hours searching for the lizard when we finally spotted a large male resting on top of a dead wild pine. When the lizard saw the group of people assembling beneath him, he darted down into the hollow trunk of the wild pine, the better to avoid what lay in store for him.

Mr. Ramsey leapt to the base of the tree and conked it a couple of times with his machete so as to freeze the lizard inside. He assigned each of us to posts so that if the lizard chose to dart from his hiding place, his every move would be spotted. Glenroy then skillfully stripped the tree of its vegetation and began to hack at its base.

When the tree was ready to be felled he gave it one mighty swipe and three of us heaved the tree onto the road. The plan was to isolate the lizard out in the open where we would make a relatively easy capture. The lizard obliged by remaining within the hollow trunk. Now surrounded on all sides, with plenty of open road to traverse before safety, the lizard watched from inside while its home was systematically destroyed.

An odd assortment of creatures emerged from the tree, including a large cricket, a baby lizard, and a large, plump toad. Each was grabbed up in turn. Finally the large lizard bolted. I knew from long experience that this was *my* moment. You see, no matter how long people have caught lizards for me and no matter how high their motivation (for example, in expectation of immediate financial reward), a certain distaste for actual physical contact prevents the hand from reaching where the lizard will be. I had watched myself for too many years putting on truly impressive displays of attempted lizard catching, while unconsciously maneuvering my hand so as to avoid actual contact, not to know this fact as it applied to all of us.

Motivated

In this situation, however, I was the most motivated to catch the lizard—permitting me to be the first person to bleed this species of lizard in the name of science!—and as the lizard bolted, I flung myself on the road, grabbing him in one swift move. A full round of congratulations and praises to Jah were enjoyed by all of us. And I indulged myself to shout at the poor male in my hand, "Wha do you, man! You are going up against Marse Glenroy and myself, in the name of 'science' and all that is holy!" The poor lizard could only moan inwardly at his fate to have run into two serious lizard catchers that day.

On the way home that evening I couldn't help wondering how that poor lizard was spending his first night after getting to know us. He could hardly be expected to have any notion of the fine use to which we would put this drop of blood (later revealed to contain neither malaria nor any other blood disease). What he knew was that he had been spotted by humans, had darted to safety as he had innumerable times in the past, but then had his home destroyed, been grabbed and yelled at, measured, had two of his toes cut off, and a drop of blood extracted and then was tossed back near his old tree. One thing was certain. He would not be happy to see us again!

Romanticizing Glenroy

One comment above rings false: "We rarely killed a lizard through rough capture, Glenroy least of all." This was not true. Glenroy did not show any special reverence for life. Except for the fact that I did not pay for the dead lizards, he would have been happy to deliver all of them dead to me. He also had larcenous tendencies, perhaps with a heritable component. His grandfather was known for his ability to milk the cows of others, tying grass to his body to act both as a camouflage and as food to keep the cows happy. It still sickens me slightly to remember the sight when I swung my binoculars outside of my seven-acre study site and spotted Glenroy catching a lizard. This could only mean that he was violating the most sacred trust on which our work rested: no catching lizards outside the area and pretending they were within! This happened only late in the study, when burdened with remeasuring more lizards during ever shorter visits, I often sat in one place for ten or fifteen minutes, instead of always, for the sake of pure data, accompanying my workers during their rounds.

Little did I know. It was only years after this paper was published that my lizard workers relaxed enough to tell me the full range of Glenroy's perfidy. He in fact often caught lizards on his way to work in the morning, secreting them in empty juice cartons until such time as I was encumbered measuring and marking several lizards at once. Then he would go to this secret stash and bring in a lizard at a time (of course, with a story as to which tree in the study area had produced the lizard). I learned that he even attempted to induce a copulation by releasing a fat individual of each sex onto the same tree, even shooing the female toward the male. Indeed, the statistically significant copulation described above may have been the result of just this sort of prearrangement.

These stories helped explain some very interesting observations that kept me up without sleep for several nights after I first made them. I would catch a lizard in the morning at the eastern corner of my study area and recapture the same male that same afternoon two hundred meters away on the western edge. After the second such observation of large-scale migration in a very short time, I imagined that I had uncovered a new category of male, a new strategy, if you will: individuals who moved several hundred meters every day, cruising through large numbers of territories, presumably copulating as they went. Discovery of this new category helped explain why we often caught large, unmarked males late in the study. It seemed unlikely that they had remained undetected within the study area for so long, but if they were supermales passing through only briefly, then I would only detect them after the local, stationary ones had been caught. Of course, "nutting don't go so," in the local patois. The migrations were always east to west because Glenroy lived to the west of the study area. The lizards were only trying to get back home. The late captures of large, unmarked males resulted, of course, from the same cause. Out of a modest sense of scientific duty, I reanalyzed my data but none of the major findings, e.g., on copulation frequency as a function of size, changed during the course of the study, so I decided, with a sigh of relief years after the fact, that it was not necessary to publish a small "Erratum" in *Evolution*.

"Rasta, How Much Would You Pay to See Two Humans Sex?"

One day three of my lizard workers (including Glenroy Ramsey) stood eagerly in front of me with a question: how much would I pay to watch two human beings have sex? I had to laugh. They must have reasoned to themselves, If this freak will pay $2 to watch two green *lizards* have sex, he'll pay a fortune to watch two humans. What is more, the copulation was in my study area! Kenneth "Sexy" Wray—known and admired for his large piece of man-wood—was having sex with a young woman in the abandoned tailor's shop by the roadside. For one wild moment, I thought, Yes, by God, a random sample of humans copulating! No more of this social "science" in which you arrived and asked people who was doing the copulating. Absence of language in other species holds us to a higher standard than we reserve for ourselves.

Writing the Paper: The Two-Hour Minimum

I had finished the first two parts of my doctoral thesis (chapters 1 and 2) by spring 1971 and needed only to write up my lizard work to earn my Ph.D. Irv DeVore helped me get a one-year NIH postdoc with him, awarded in the spring of 1971, but the fellowship required that the thesis be submitted within a year, that is, spring of 1972. It was not a difficult task since I had been analyzing the data for the thesis while the work went on. It required only that I sit down and do the writing, at most a month of work. Finally, the day of reckoning arrived. It was early January 1972 and if I failed to finish by March I would lose my fellowship. I sat myself down and convinced myself that I would work a minimum of two hours a day until the paper was finished. Two was so little that you could hardly run from it, but the key was to work two every day, no matter how many hours you had worked the day before. After about 18 days of this, I was sick of lizard data and had a flood of new ideas on parent–offspring conflict. A few days later I ran to Irv DeVore and suggested to him that I quickly push along my manuscript on parent–offspring conflict and present my advisor, Ernest Williams, with three theoretical papers instead of two plus a lizard paper. Professor DeVore urged caution. He felt certain that Professor Williams, whose very life revolved around *Anolis* lizards, who had funded all of my work, and who fully expected to see it in my thesis, would not be happy with parent–offspring conflict, no matter how brilliantly rendered. How much work had I done on the lizards, he asked. Quite a bit, I explained. What was the problem? I had only analyzed about three-fifths of my data. Did Ernest know how much data I had? Not exactly. Professor DeVore asked to see what I had written.

Ernest will be delighted with this, he assured me. A few loose ends tied up, another table or two, and I could call it quits—no need to tell Professor Williams about the large mass of unanalyzed data on lizard tail fractures.

And so it was that I learned a valuable trick for getting myself to work when depression lay heavy on me or a necessary but onerous task blocked my path: the two-hour, daily minimum, no exceptions.

Sexual Selection and Resource-Accruing Abilities in *Anolis garmani*

ROBERT L. TRIVERS

In 1969, I began a field study of the social behavior of two Jamaican lizards, *Anolis garmani* and *A. valencienni*. This study has developed into an attempt to measure how sexual selection has molded adult dimorphism in size and associated dimorphism in growth rates, mortality rates, and in sexual and aggressive behavior. Preliminary data for *Anolis garmani* were presented as part of a general theory for the evolution of sex differences (Trivers, 1972). Meanwhile, several important theoretical advances (Maynard Smith, 1971; Williams, 1975; Zahavi, 1975) have convinced me that female choice is a central factor molding male sexual dimorphism and systems of male–male competition. The data presented here for *Anolis garmani* tend to support the view that those males who do most of the mating are those who have tended to demonstrate the superiority of their non sex-linked genes.

Methods

Social behavior was studied for eight months (between July, 1969 and January, 1972) during seven separate visits to a study area which eventually comprised 9 acres in Southfield, St. Elizabeth, Jamaica. (The study area was expanded during the 2nd visit and again during the 3rd; the last visit concentrated on only a portion of the study area.) The study area consisted of scattered fruit trees typical of Jamaica: *Mangifera indica, Blighia sapida, Pimenta officinalis, Cocos nucifera, Guazuma ulmifolia, Bursera simaruba* and three species of *Annona*, among others. During each visit to the study area an effort was made to catch all resident adults.

The lizards were captured, sexed, measured, marked and released where captured, usually within five minutes of capture. Toe-clipping gave a unique permanent mark to each individual, and, in addition, a number was painted on each individual's back for short-term identification without recapture. To reduce error in measurement, each individual was stretched by hand while being measured. A number of individuals were also weighed. Unless otherwise indicated, all measurements are snout-vent lengths given in mm.

During each visit the sighting of any marked lizard was recorded, as well as its behavior. Few lizards were observed for any length of time unless they were engaged in display or social behavior. A particular effort was made to spot copulating pairs. If one or both partners were unmarked the copulation was interrupted to catch the unmarked individuals. This was almost invariably successful in the case of males but often unsuccessful in the case of females (see below).

Sexual Dimorphism

Dissections of males and females in the collection of the Museum of Comparative Zoology, Harvard University, indicate that males reach sexual maturity at about 85 mm in size (based on the presence of seminiferous tubules) and females at about 70 mm in size (based on the presence of an egg). This accords well with field observations in which the smallest female seen copulating was 70 mm long, and the smallest male, 87 mm. The mean size of all adult males captured in the study area was 109.7 mm, and that of adult females, 83.5 mm. An analysis of weight vs. snout–vent lengths showed, as expected, that weights varied as the cube of snout-vent lengths. Adult males weigh typically, therefore, 2.25 times as much as adult females. This degree of sexual dimorphism is unusual among anoles as large as or larger than *A. garmani*, but it is typical of several other Jamaican anoles.

Males also have proportionally larger jaws than females. Measuring the entire length of the lower jaw of animals in the field showed that males had jaws 1.06 times the size of jaws of females whose snout-vent length was identical. In addition, males have several external structures absent or poorly developed in females: a bright yellow dewlap used in display, a series of spikes along the back resembling the teeth of a saw (Fig. 1), and a stouter, more muscular neck and body.

I assume that males and females hatch at about the same size, and that the adult dimorphism is achieved by the male's faster growth rate, as documented below. In a small sample of the closely related and equally sized dimorphic *Anolis lineatopus*, hatching males do not differ in size from hatchling females (A. S. Rand, unpublished data); and the same is true of a large sample of hatchling *A. valencienni* in the Museum of Comparative Zoology (Harvard) collection (Trivers and Hicks, in preparation). *Valencienni* is a

Figure 1. Male and female A. *garmani* engaged in a typical copulation face down four feet up the trunk of a coconut tree (photo: Joseph K. Long).

closely related Jamaican anole in which adult males are 1.65 times as heavy as adult females. Both *lineatopus* and *valencienni* show the same sexual dimorphism in growth rates documented below for *garmani* (Rand, unpublished data; Trivers and Hicks, in preparation).

The Measurement of Reproductive Success

Anolis lizards are ideally suited to the measurement of reproductive success because only a single egg matures at a time and females typically copulate only once in order to fertilize that egg (Crews, 1973). Thus, reproductive success can be measured by measuring frequency of copulation.

In all *Anolis* studied, females lay a single egg at a time (Smith, Sinelnik, Fawcett and Jones, 1973). I never saw a *garmani* female lay an egg, but dissection of ten females accidentally killed in the study area always showed one mature egg or none, although large females often showed one or two smaller eggs in earlier stages of development. For Puerto Rican anoles, each female caught copulating by George Gorman (personal communication) had one mature oviducal egg.

I assume that females typically copulate once in order to fertilize each egg. If a pair is permitted to copulate undisturbed ($N \cong 50$), the female (with one exception) is not seen to copulate again for at least one month. But if the copulation is interrupted (in order to catch one or both lizards), the female may copulate within a day with the same male. By interrupting successive copulations, I observed the same couple copulating as often as four times in one week. Once permitted an undisturbed copulation, however, at least a month elapsed before such a female was seen copulating again. (Only the first copulation of these interrupted series is counted in the data below.) These facts suggest that females copulate infrequently, presumably to fertilize an egg, and that fertilization inhibits further copulation within an ovulatory cycle (see Crews, 1973). The exceptional case referred to above concerns a large female (91 mm) who copulated eight days apart on the same tree, first with the territory holder, then with an interloper (the fourth copulation in Table 2) at a height of nearly 50 feet. (Both copulations are included in the data below.) The apparent tendency of *garmani* females to copulate once per egg contrasts sharply with the non-territorial female *valencienni* who often copulate more than once on the same day, usually with different males (Trivers and Hicks, in preparation).

Sperm storage has been demonstrated in *Anolis* (Fox, 1963) but if female *garmani* store sperm I assume they do so without regard to their own size (or to the size of the copulating male). If females copulate more than once to fertilize a single egg, I assume they do not do so as a function of their size (or that of the male). Finally, I assume that length of copulation is independent of the sizes of the participants.

Biases in sighting copulations are thought to be minimal for the following reasons. (1) *Garmani* appears to be unusual among lizards (and other animals) in that individuals choose highly conspicuous places in which to copulate. Although over half of over 3000 observations of the lizards showed them to be in foliage where they are difficult to see, only one of 90 copulations took place in the foliage, and it was a homosexual copulation. (It and a second homosexual copulation are excluded from the data that follow.) All heterosexual copulations took place on highly visible perches, the pair almost invariably facing down on the lower exposed trunk of a tree (Fig. 1). (2) In the congeneric Jamaican A. *valencienni*, by contrast, whose normal perch heights parallel those of *garmani* (Rand, 1967; Schoener and Schoener, 1971), copulations take place from the ground to the outer leaves of the foliage with no preferred orientation, facing up or down, and do not differ in height or visibility from typical non-copulating perches (Trivers and Hicks, in preparation). (3) *Garmani* copulations appear to last a considerable time. Only one was observed from beginning to end and it lasted

Table 1. Mean size (x) of individuals seen copulating (and standard error of the mean, s_x) compared to mean size of individuals never seen copulating, analysed by sex, for six separate samples. The first sample consists of data from the first two visits combined; all other samples represent separate visits.

	Summer '69 and Winter '69–70			Spring '70			Summer '70		
	x	s_x	N	x	s_x	N	x	s_x	N
Copulating males	117.5	0.8	6	114.8	2.5	23	119.5	2.7	14
Non-copulating males	100.0	5.4	42	106.9	1.2	77	108.4	1.3	79
Copulating females	83.4	3.5	5	79.6	7.0	16	82.1	2.8	9
Non-copulating females	78.8	0.8	55	78.7	2.6	102	81.7	0.8	104

	Winter '70–71			Spring '71			Winter '71–72		
	x	s_x	N	x	s_x	N	x	s_x	N
Copulating males	122.2	2.5	18	124.0	3.5	11	126.0	3.5	7
Non-copulating males	112.3	1.3	134	112.1	1.6	89	105.6	3.0	23
Copulating females	88.8	1.5	17	91.3	1.0	6	86.7	2.2	7
Non-copulating females	86.1	0.5	177	87.3	0.6	120	85.0	1.1	33

25 minutes. The median length of time of copulations first observed when already under way was about 10 minutes. Couples were extremely reluctant to disengage, even when approached, so that couples were sometimes noosed while still copulating. Individuals who had frequently been recaptured, however, would sometimes disengage when sighted or approached. Frequently recaptured individuals tend to be the largest, so the bias, if any, would tend to make copulations of large animals less frequently observed. (4) Once sighted, the male of a copulating pair was almost invariably marked, or, if not, was captured during or shortly after the copulation. Females were sometimes unmarked and remain uncaptured. The magnitude of this potential bias can be seen by comparing the number of marked females seen copulating (Table 1). The first, fourth and seventh samples are nearly unbiased in this respect and they differ in no obvious way from the other samples, so that the bias introduced, if any, appears to be trivial. (5) Smaller adult males (or females) copulate on perches which do not differ in either height or visibility from perches on which larger individuals copulate. (6) The bright green copulating pairs are so easy to spot against the brown trunks of trees that it is unlikely that many pairs were overlooked while a tree was being searched. All trees were searched regularly and thoroughly each day, but most work was done between 7:30 AM. and 3: 00 PM. In these data, there is no relationship between the time of day and the sizes of those who are copulating.

Size And Reproductive Success

Copulations in which one or both individuals were marked were recorded and the sizes of those seen copulating at least once were compared with the sizes of those adults never seen copulating for five separate samples (Table 1; data from the first two visits were combined to form the first sample). In each sample the mean size of copulating males is significantly larger (P < .05) than the mean size of adult males never seen copulating. For each sample, the mean size of copulating females is also larger than the mean size of non-copulating females, but in only the last sample is this difference significant (P < .05). Ranking females by size, however, shows a significant tendency in all but the third sample for larger females to copulate more often than would be expected by chance (Kolmogorov-Smirnoff one-sample test; P < .05).

Taking all the data together and comparing the frequency of copulation for each non-overlapping 5 mm size category (Fig. 2), the tendency for females to copulate more often with increasing adult size is significant (*t* test; P < .05), and the tendency for males to copulate more often with increasing adult size is *significantly stronger* than that tendency for females (*t* test; P < .05). A male 110 mm or larger, for example, is 4.8 times more likely to copulate than a smaller adult male, while a female in a comparable category (85 mm or larger) is only 1.53 times more likely to mate than a smaller adult female. Large size is more important for male reproductive success than for female, and other things being equal, one would predict larger male adult size and higher growth rates for males when young in order quickly to reach the large sizes at which copulation becomes frequent.

Although there is an extremely regular tendency for frequency of copu-

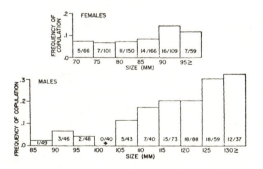

Figure 2. Reproductive success in males and females as a function of size. Reproductive success is measured by the number of copulations observed per number of individuals (male or female) in each non-overlapping 5 mm size category. Data from all visits combined.

lation to increase with each increase of 5 mm in male size, this tendency is violated for males 95 mm–104 mm in size. In fact, the frequency of copulation for males 100 mm–104 mm in size is significantly lower than that predicted by a linear fit to the frequencies for the other categories (t test; $P < .005$). As shown below, this is a time of transition in the lives of most males, a time in which they leave the territories of larger males and establish new ones of their own. A trough in male reproductive success at this time is not completely unexpected.

Growth Rates: Age and Size

In Fig. 2, size and age are confounded, the larger lizards tending to be the older ones. Indeed, in theory the entire correlation shown in Fig. 2 could result from the effects of age (and hence experience) on reproductive success. If this were true, however, then no sexual dimorphism in growth rates would be expected (unless, for some reason, size affected survival differently in the two sexes, which does not appear to be true: see Figs. 14 and 15). In short, higher male growth rate is uniquely predicted by the assumption that large size is more important for male reproductive success than for female reproductive success.

To gather data on growth rates, animals were recaptured on successive visits to the study area, roughly at four month intervals. At all seasons of the year males grow faster than similar-sized females, as documented in Figs 3, 4, 5, and 6, in which are plotted growth rates as a function of size at initial capture for three different seasons. Note that during the two periods in which growth rates are greatest (summer, fall) there is no overlap between male and female growth rates. During the period of slowest growth rates, however (see Fig. 5) there is substantial overlap between the growth rates of males and females between 85 and 95 mm in size. Indeed, there is no evidence that males between 85 and 95 mm grow on average any faster than males 95 to 110 mm do.

These data appear to show that during the dry season (early spring) when growth rates, in general, are depressed, males who overlap the size of adult females suffer especially reduced growth rates (compared to larger males). Less numerous data from the spring of 1970 (Fig. 6) show some overlap in male and female growth rates, but less than in the spring of 1971, in which growth rates, in general, were much lower. In short, variance in male growth rates is greater in the poor growing season (the spring) and greatest when that season is especially harsh.

Ages of lizards were never exactly known because no lizards were marked as hatchlings so it is not possible to compare directly the effects of age and size on reproductive success. One can, however, compare the reproductive

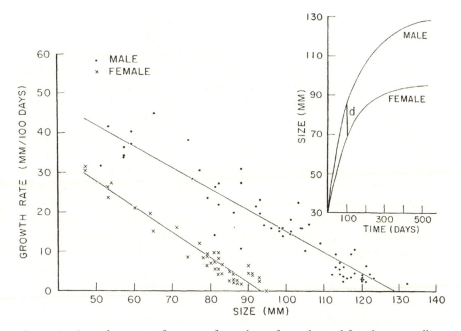

Figure 3. Growth rate as a function of initial size for males and females originally captured in April–May 1970 and recaptured in August 1970. Inset shows size as a function of age for both sexes if individuals grew throughout the year at the summertime growth rates. The regression line fitting the male data is given by y= 0.53x + 68.4 and for the female data y=0.63x + 59.2. The mean time between recapture for males was 107 days and for females, 103.7 days.

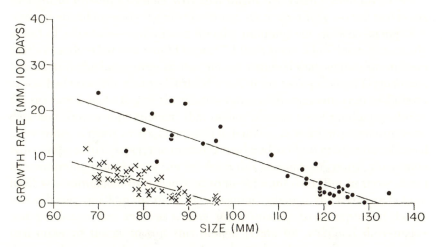

Figure 4. Same as in Fig. 3 except that initial capture took place in August 1970 and recapture in December–January 1970–71. Male data: y = −0.35x + 46.2; 144.6 days between captures. Female data: y = −0.29x + 27.8; 143.6 days between captures.

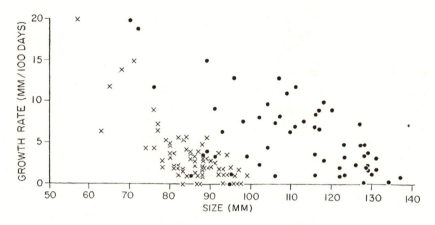

Figure 5. Same as in Fig. 3 except that initial capture took place in December–January 1970–71 and recapture in April 1971. Average period between recapture: males, 95.0 days; females 96.8 days. Note that growth rates are strongly reduced compared to other seasons, that variance in male growth rates are increased and that there is overlap between male and female growth rates.

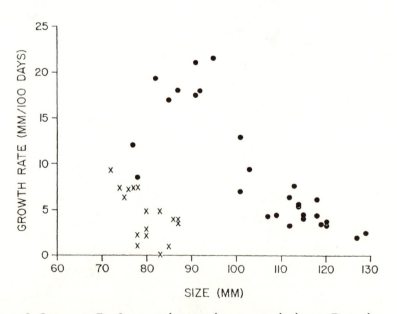

Figure 6. Same as in Fig. 3 except that initial capture took place in December–January 1969–70 and recapture in April 1970. Mean time between recaptures for males, 107.5 days; for females, 100.3 days. Note that growth rates for the two males smaller than 80 mm are markedly lower than expected based on growth rates of larger individuals.

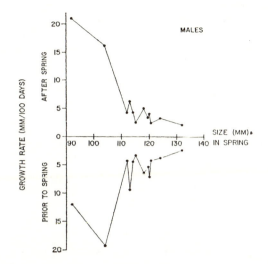

Figure 7. Growth rates from winter 1969–1970 to spring, 1970, and from spring, 1970, to summer, 1970, are plotted for the same males as a function of their size in spring. Note that the lines have a tendency to be mirror images of each other.

success of individuals who were the same size four months earlier but who are now of different sizes. One's interpretation will depend on whether one assumes that individuals who were once the same size are the same age or that the individual who grew more slowly afterward also grew more slowly earlier and hence was older when the two were the same size. In short, one must know how growth rates during one four month period are associated with growth rates in other periods.

When growth rates of the same individual in two successive periods are plotted together against size of the individual at the intermediate recapture, then it is clear that there is a strong, positive correlation between an individual's growth rate in the two successive periods: individuals who grew fast in the first period are more likely to grow fast in the second (Figs. 7, 8 and 9). This means that we can safely assume that of individuals initially the same size, the ones which subsequently grew more slowly are older. This, in turn, permits us to measure the effect, if any, of age on male reproductive success (frequency of copulation). If age makes an independent contribution, then there should be a tendency during any one visit for those males to copulate who later grew slowly (as corrected for body size). No such tendency is evident: in Figure 10 the males who copulate during a visit (open triangles) do not tend to have smaller growth rates thereafter. By contrast, in one of the three samples (after the lowest growth rates), those who grew fastest tended to copulate more often in the visit after these growth rates (open circles, Fig. 10).

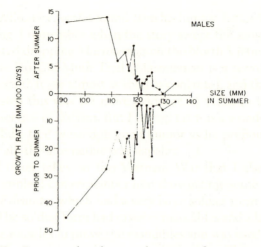

Figure 8. As in Fig. 7, except that the growth rates are from spring to summer, 1970, and from that summer to winter, 1970–71, as a function of size at summer, 1970. Note that the lines have a tendency to be mirror images of each other.

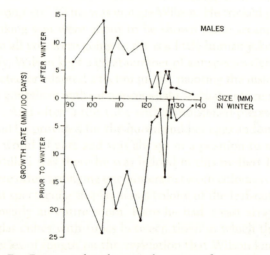

Figure 9. As in Fig. 7, except that the growth rates are from summer, 1970, to winter, 1970–71, and from that winter to spring, 1971. Note the same mirror image as in Figures 7 and 8.

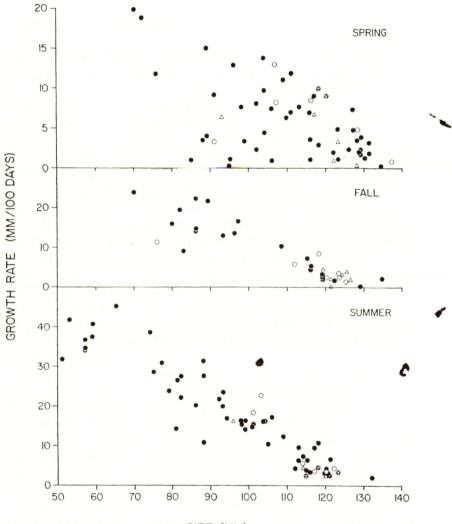

Figure 10. Copulation as a function of growth rate in males for three separate periods. Open circles = males who copulated during the visit after the growth rate shown. Open triangles = males who copulated during the visit before the growth rate shown. Dark circles = males not seen to copulate during the visits before and after the growth rate shown. See text for interpretation.

Male Aggressiveness

With the exception of the two homosexual copulations discussed below, two adult males were never seen near each other unless they were displaying aggressively (as described for *A. lineatopus* by Rand, 1967) or fighting. Only three fights were observed in the study area between adult males. In each case, both males attempted to bite the other on the jaw, neck or body. In one case, one male secured with his jaw a grip on the other's neck.

To confirm the part aggression may play in reducing territorial overlap, 37 experiments were conducted in which a male, captured elsewhere, was tied down on a tree for at least 30 minutes or until a response was elicited. The tethered male was tied around his waist with nylon fishing line which was secured (with about a yard's free reign) to the tree, usually about 10 feet up the trunk. Five tethered males were used, ranging in size from 97 mm to 118 mm. None of the experiments was continually monitored; instead, each experiment was visited several times before being terminated. These experiments can be classified into those eliciting a strong, aggressive response, those eliciting a cautious response and those apparently eliciting no response at all. Twenty-four experiments elicited no reaction, but 13 of these were conducted on trees on which no male was ever seen during that visit (of four weeks) to the study area.

In six of the experiments, strong aggressive responses were elicited. Typically, within 30 minutes, a second male appeared, bright green, nuchal crests raised, throat area inflated, body fully extended, and approached the tethered male. If his approach was not interrupted (by an attempt to capture him) a fight soon broke out, the second male lunging repeatedly at the tethered male and attempting to bite him on the jaws, neck and side of the body.

Seven experiments elicited cautious responses. In these, a second male appeared but remained at some distance from the tethered male and approached only very slowly, if at all. For example, a 102 mm male was tethered on a thatch tree (*Cocos nucifera*) within sight of the resident (190 mm) who was perched down, fully green, on a brown frond, 17 feet from the ground. For the first several minutes the resident male responded only by eyeing the tethered male closely. He then moved onto the trunk and hid himself under a frond. Twelve minutes after the tethered male was introduced, the resident male had turned completely black and was nearly invisible under the frond. Meanwhile, the tethered male had moved part way around the trunk away from the resident male. Although still green, some darkening had begun to appear. Several minutes later he escaped from his tether and abandoned the tree. Fifteen minutes after his escape, the resident male was bright green again and headed back up the tree.

One might suppose that the relative size of the tethered male would control the type of response (aggressive or cautious) shown by the resident male: if the resident is larger he should attack. The few data do not bear this out. The absolute size of the tethered male may be important in determining a cautious response in the resident, especially if the chance of serious injury is greater when larger males fight each other than when smaller males fight. I assume that in *garmani*, as has been shown in *A. lineatopus*, the larger of two males usually wins in a naturally occurring fight (Rand, 1967).

To observe a more natural aggressive sequence two experiments were conducted in which males were released onto trees known to be occupied by other males. In the first, a 100 mm male was released onto a tree occupied by a 119 mm male. The smaller male slowly made his way to 10 feet up on the tree when he spotted the large male, who was perched two feet further up on the same branch. The larger male started toward the smaller one, who quickly turned and ran down to a height of six feet from which he leapt to the ground. He was never seen again. In the second experiment, a 109 mm male was released onto a tree occupied by a 114 mm male. The latter spotted him almost at once but reacted very cautiously, displaying in sight of the introduced male but not moving toward him. For an hour both moved cautiously within sight of each other, neither attacking, but the intruder moved out into the foliage while the resident remained low on the tree. Two hours later both were seen fighting on the trunk of the tree. Both fell to the ground, and the intruder then escaped back up the tree into the foliage. Later that day, the resident was seen chasing the intruder through the foliage. After this, the intruder was never seen again while the resident was seen on successive days occupying the same tree.

Male Territoriality

During the first, second and seventh visit to the study area, I attempted to map male territories by concentrating on a small portion of the study area. Males are sighted too infrequently to measure territory size in the usual way, that is, to construct a volume fitting such sightings. But fortunately males 105 mm and larger show a strong tendency to occupy trees which they may share with a male smaller than 105 mm but which they do not share with any other large male (105 mm and larger) (Fig. 11). Typically, during a given visit, a large male will be sighted between five and ten times in a large tree (or, in several smaller contiguous trees) without any other large male being sighted in that tree (or trees). When followed, such males wander throughout the tree (or trees). In the data presented here, a tree was assigned to a male if he was seen three or more times in it without any other adult male being seen therein. If, as happened several times, a large

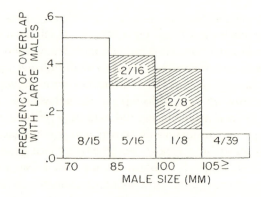

Figure 11. Percentage of males in each size category known to be overlapping home ranges with large adult males (≥ 105 mm). Data taken from visits one, two and seven. Shaded areas refer to males who dispersed *during* the visit from the territory of a large male to a tree unoccupied by any other adult male. Two of the four cases of overlap in the ≥ 105 mm category were cases of temporary overlap which resulted in copulations with resident females.

tree was also known to be occupied by a small adult male (85 mm-104 mm), both males were excluded from the data, since too few data were available to partition the tree between them. If a large male occupied several contiguous trees, of which one was also occupied by a small adult male, ½ of that tree was arbitrarily assigned to each male.

Territory size was measured by foliage volume of trees exclusively occupied. The foliage volume of a tree was calculated by estimating the height of the tree and multiplying that by the area of a rectangle approximating the average horizontal dimensions of the foliage. This very crude estimate is obviously incapable of discriminating between very similar territories, but it is adequate to differentiate unambiguously between categories of trees (e.g., large, medium and small). A better way of estimating foliage volume was not devised partly because it soon became apparent that the lizards were not responding to foliage volume as such. Pimento trees (*Pimenta officianalis*), for example, never reach a large size but they are very attractive to the lizards, and a large adult male will often occupy a pimento tree containing several females. To prevent any subjective bias, the foliage volume of each tree in the study area was estimated at the beginning of the study *before* the sizes of any of the occupants were known.

There is a strong tendency for male territory size to increase with the size of the adult male (Fig. 12). Furthermore, there is a strong tendency for the number of females resident within a male's territory to increase with the male's size (Fig. 12). Since about 90% of all copulations take place within the exclusive territory of the male observed copulating (see below), it is

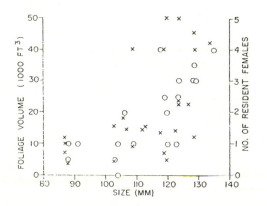

Figure 12. Territory size (X) and the number of females resident within the male's territory (O) as a function of male size. Territory size was measured by estimating foliage volume (see text). One-half a female was arbitrarily assigned to each of two males if she overlapped both of their territories.

plausible to assume that the primary correlation between male reproductive success and male size results from the correlation between territory size and male size.

Males show a decreasing tendency to overlap with large males as a function of increasing size (Fig. 11). To avoid overlap growing males must usually disperse to an unoccupied tree, and five males were seen to do so *during the visit* in which they were first seen overlapping with a large male (shaded areas, Fig. 11). Likewise, dispersal data gathered from recapture work show a peak in dispersal for males 100–109 mm in size (Fig. 13).

Male Territories and Copulations

For 49 of the observed copulations it is possible to say whether the male was copulating within his territory (in the sense of copulating in the tree in which he alone, among large males, was observed). Of these 49, 44 (or about 90%) were performed by the territory holder. Of the additional five, three were performed by temporary interlopers, and two by small males resident within a large male's territory.

In each of the three cases involving an interloper, the interloper was smaller than the resident male (see Table 2, copulations 1, 2 and 4). In the first case, the interloper occupied one corner of the larger territory for at least two days; he was not seen before or after. When first seen he was copulating with the one female known to be resident in that portion of the territory (a mango tree). Two hours later the larger male entered this portion of his territory, but no encounter was observed between the two males. It

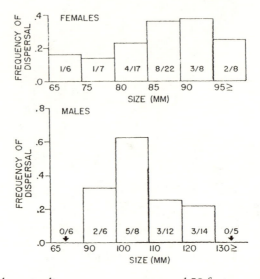

Figure 13. Tendency to disperse across open ground 50 feet or more as a function of initial size in males and females. Animals were measured and marked in December 1970–January 1971 and recaptured in April 1971. The percentage of those recaptured who had dispersed is plotted against size of the animals in December–January. Mean distance of dispersal for the females, 123 feet; for the males, 119 feet.

is possible that the large resident male was unaware of the copulation while it was going on. In the second case, the interloper was first seen while copulating and was never seen thereafter. The copulation took place on the main tree of the resident's territory (a pimento tree), and it seemed unlikely at the time that the resident could be unaware of the copulation. The resident was seen many times before and after in the same territory. In the third case, the interloper entered from his own territory 40 feet away and returned to it sometime after the copulation. It is very likely that the resident was not in this tree at the time but in one of eight small contiguous trees which he frequented (and in which he copulated).

Two of the 49 copulations were performed by small males regularly resident within a larger male's territory (Table 2, copulations 3 and 5). In the first case, the smaller male occupied a portion of the larger male's territory. In the second case, the larger resident male was observed approaching the copulating couple in a very aggressive posture: bright green, nuchal crests raised, body extended. Both the small male and his copulating partner had turned black, which in aggressive encounters seems to indicate fear. Unfortunately, the incident was interrupted in order to capture the copulating couple (who were unmarked).

The choice of the place to copulate appears to be the female's. I have

Table 2. Data on five copulations within the territory of a large male in which the territory owner (the large male) did not perform the copulation.

Date	Size of Territory Owner	Size of Copulating Male	Behavior of Copulating Male
9.7.69	124	116	Temporarily occupied one tree of 5-tree territory.
1.7.70	119	118	Seen only once on main tree of territory.
4.20.70	115	96	Resident within larger territory.
5.9.71	124	117	Entered from and quickly returned to his own neighboring territory.
8.6.71	121	91	Resident within territory; approached by owner.

only observed the beginning of two copulations. In both cases, the female was already perched face down at the spot where the copulation took place while the male approached her from several feet away. In addition, I watched a female move down to the exposed trunk of a tree and remain perched nearly motionless for 40 minutes while a male repeatedly tried to approach her from about 20 feet further up the tree, but he was frightened back by passing humans (the male had been captured many times before). A local naturalist (Mrs. Dorel Staple) has often seen females perch at places where they soon copulate, and she believes that while they wait they adopt a special posture, with the rear slightly elevated. Crews (1973) has shown that female *Anolis carolinensis* have a copulation invitation pose.

If females choose conspicuous perches in which to copulate, why do they do so? It seem difficult to believe that the perches are safer than inconspicuous ones. It would be most interesting if, in this highly territorial species, females were advertising their availability in order to increase the number of males from which to choose. The only clear case of female rejection I observed involved a female who had been caught by the tail by a small adult male. For over a minute he attempted the impossible, that is, to copulate while maintaining a hold on her tail. He finally released the tail, moved quickly to cover her, but she moved even more quickly to escape him.

Female Aggressiveness

No two adult females were ever seen near each other unless they were fighting or unless one was clearly hiding from the other. Only nine experiments were conducted with tethered females but two of these induced attacks from resident females. One induced an attack from a small resident male (91 mm) but none induced any sexual displays or any other response from any other male.

If adult females defend territories these certainly overlap more than male territories. It was common to see as many as four different adult females at different times on the same part of the trunk of a large tree. Unfortunately, since females do not tend to occupy trees exclusively and since I have less than ⅓ as many resightings per adult female as per adult male, I am unable to present any data on the sizes of female home ranges, their precise degrees of overlap, and whether they are sufficiently distinct to constitute territories, i.e., defended areas. Rand (1967) has shown for *A. lineatopus* that females do defend territories and that these show greater overlap than do male territories. T. Jennsen (unpublished data) has similar very detailed data for another Jamaican anole, *A. opalinus*.

Homosexual Copulations

Of the 62 copulations in which the individual underneath (see Fig. 1) was identified, two were homosexual copulations. The animal on top is invariably a male. The two males underneath were within the size range of adult females (83 and 76 mm in length), while the two males on top were large fully adult males (121 mm and 125 mm). One of the two copulations was unusual in that it took place on a small branch (less than two inches thick) high in the tree surrounded by foliage. The homosexual copulations differed in no other observable way from heterosexual copulations except that the small male underneath had turned partly black, which, in aggressive encounters, indicates fear. In only one heterosexual copulation had the female turned black, and that was one in which the male had also turned black, both animals being in the presence of a much larger male, who was approaching aggressively (see Table 2). Intromission appeared to have been achieved in both homosexual copulations.

The two males underneath were never seen again on subsequent visits to the study area, but one of the males on top was followed on two successive visits to the study area, during which he was observed in two heterosexual copulations. These two and his homosexual copulation occurred on the same tree, a territory the larger male apparently occupied for at least eight months.

In the very similar *A. lineatopus*, Rand (1967) has shown that small adult males compete with adult females for territories within large male territories. There is no evidence of the large male ejecting a small adult male in order to open up space for an additional female. Arguing from natural selection, one would predict that large males ought to eject small males in order to open up space for additional adult females. Natural selection might favor small males remaining where they were, if their territories increased survival (or growth rate) above that expected elsewhere. An occasional buggery might be a small price to pay for the advantages of remaining within the large male's territory. Such a hypothesis can be tested by comparing the survival and growth rates of small males within large males' territories against the rates of small males found living on their own. Unfortunately, there were too few data from this study to make the appropriate test.

Adult Sex Ratio

Biases in capture strongly favor males: they are larger, are more often found on conspicuous perches, and are easier to catch once spotted. During the first two visits a marked male was resighted more often than a marked female by factors of 3 and 3.8 respectively. Despite this bias, more adult females were captured than adult males for each of the visits (Table 3). As each is an independent sample, the difference is highly significant. The difference cannot result from males maturing at a greater age than females, since the sexes reach sexual maturity at about the same age (see Fig. 1).

If, as argued above, eggs destined to produce males are about the same size as eggs destined to produce females, then the cost to an adult female of producing a male will about equal the cost of producing a female. According to Fisher (1958), the sex ratio of conception should therefore be about 50/50. Natural selection would only favor a biased production of females if the population were strongly inbred (Hamilton, 1967), but dispersal data strongly argue against this assumption since in a typical three-month period approximately 28% of the adult of each sex disperse across ground 50 feet or more to new trees (Fig. 13). The alternate explanation is differential male mortality, something that data from other lizards, as well as theoretical considerations, would predict (Trivers, 1972).

Table 3. Adult sex ratio.

	Males	Females	Sex ratio (M/F)
1	10	15	0.67
2	38	45	0.84
3	100	118	0.85
4	93	113	0.82
5	152	194	0.78
6	100	126	0.79
7	30	40	0.75
Total	523	651	0.80

Size and Survival

Recapture data were used to test the relationship between size and survival or, chance of being recaptured within the study area for each of three four-month periods. If a lizard was not recaptured four months later but was known to be alive then (because of recapture still later), it was counted as being alive. Data on chance of survival four months later for lizards captured during three successive visits (winter 1969–70, spring 1970, summer 1970) showed no clear differences from visit to visit, and these data were aggregated to give a general view of chance of recapture as a function of initial size for both males and females (Figs. 14 and 15).

In order that such recapture data reflect actual chance of survival, one must know how many lizards are dispersing out of the study area as a function of their size. Fig. 10 gives dispersal (within the study area) as a function of size. Since larger females appear to disperse slightly more often than smaller females, I conclude that at least up to 90 mm in size females appear to survive better with increasing size. For males, it appears safe to assume that they survive better with increasing size up to 110 mm but they show no increase in survival after this (and they may show a decrease). One must remember, of course, that size is confounded

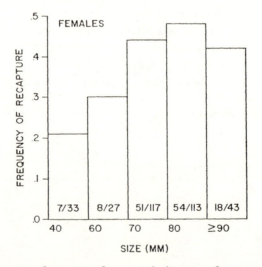

Figure 14. Frequency of recapture four months later as a function of initial size for females captured in the winter 1969–70, the spring 1970 and the summer 1970. Females were counted as recaptured even if they were not recaptured four months later as long as they were known then to be alive because of later recapture.

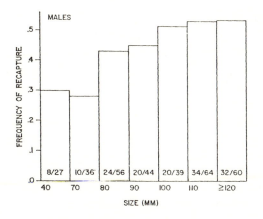

Figure 15. Same as Figure 14, except for males.

with age, and part of the relationships observed may be attributed to differences in age.

The relationship between size and survival appears to be similar for the two sexes, especially over the common size range of 40 to 90 mm in size.

There was no evidence that survival was correlated with growth rate. That is, there was no relationship between growth rate in one four-month period and survival in the next. Likewise, the growth rate of individuals who survived eight months (or twelve months) did not differ from the growth rate of comparable individuals who were only known to survive four months.

There was no relationship between weight at capture (as corrected by snout–vent length) and survival four months later.

Sexual Selection and Sexual Reproduction in *Anolis garmani*

In species with no male parental investment (such as *A. garmani*) half the species investment is thrown away each generation on males (Fisher, 1958) so that there appears to be a constant two-fold advantage every generation in favor of asexual mutants (Maynard Smith, 1971). Williams (1975) has drawn the inference that the advantages of genetically diverse offspring must be far greater than hitherto imagined. Despite considerable success with specific life history models, Williams (1975) believes that sex must, in general, be at a disadvantage in low-fecundity organisms such as many vertebrates and insects. An attractive alternative is to imagine that female choice is sufficiently discriminating in such species so that those males actually mating each generation have genes which are about twice as good at producing successful daughters as are the genes of the mothers themselves.

Because only daughters invest, a block to the spread of asexual mutants exists only when the reproductive success of *daughters* is doubled by mating with preferred males (compared to asexual reproduction). In short, the advantage of genetically diverse offspring combined with a system of female choice could provide a block to the spread of asexual mutants in low-fecundity organisms.

Since environments are constantly changing, the ideal male genotype must also be changing. How can females come up with general criteria that work regardless of environment (see Zahavi, 1975)? A daughter is selected to maximize survival and resource accrual (where these resources are translated into investment) so that males who appear to have maximized this criterion are males whose genotype is preferable regardless of the environment within which this choice is being made. What is striking about our data on *A. garmani* is that those males who have maximized growth rates times survival are those who inseminate the adult females. Although I have no data on female choice, our other data suggest that females naturally mate with males whose daughters will have the best chance of maximizing resource accrual times survival. That there is a strong system of male-male competition for females does not imply that females are merely mating with males good at besting other males. Instead, systems of male-male competition are expected to evolve under the influence of female choice, so that males compete among themselves in such a way as to reveal to females that their genes are good at maximizing survival times resource accrual.

Summary

1. The social behavior of *A. garmani* was studied for a total of eight months during seven separate visits to a study area of nine acres in Jamaica.

2. Adult males are 2.25 times larger by weight than adult females. At all times of the year and at all sizes for which there are data males grow faster than females. Growth rates for the same individuals in different four month periods are positively correlated.

3. By measuring reproductive success as frequency of copulation, it was discovered that in adults of both sexes increasing size increases reproductive success, but that this tendency is *significantly stronger* in males than in females. There is no evidence that age makes an independent contribution to a male's frequency of copulation.

4. The steep slope of the function relating reproductive success to adult male size results apparently from sexual selection: males compete aggressively to occupy exclusive territories containing females. There is a strong tendency for large males to occupy correspondingly large territories which

in turn contain large numbers of females. About 90% of copulations were performed by a territorial male within his territory.

5. As young males mature they tend to be found less frequently within the territories of large males. Males 105 mm and larger are almost never seen within the territories of other males in this category. Males 100-104 mm must disperse if they are still within a large male's territory and males within this size range show a significantly lower frequency of copulation than would be expected from the function relating size to reproductive success for other adult males.

6. Two homosexual copulations were observed and are described.

7. The adult sex ratio is biased in favor of females.

8. Recapture data appear to show that females survive better with increasing size up to 90 mm. Males appear to survive better up to 110 mm but probably not at larger sizes.

ACKNOWLEDGMENTS

I thank E. E. Williams for advice, encouragement and support throughout. I thank Mike Sutherland (Harvard Statistics Department) for extensive help with the statistical tests. I thank Allen Greer for help with determining sexual maturity in lizards dissected. I thank Glenroy Ramsey, Verne Staple, Raymond Simpson and Harper Parchment for extensive, expert help in the field. I thank especially Mrs. Dorel Staple for seven months of hospitality in Southfield. This work was supported by a grant from Sigma Xi, by NSF Grant B 019801X to E. E. Williams, and by an NSF prédoctoral fellowship.

LITERATURE CITED

Crews, D. 1973. Coition-induced inhibition of sexual receptivity in female lizards (*Anolis carolinensis*). Physiology and Behavior 11: 463–468.

Fisher, R. A. 1958. The Genetical Theory of Natural Selection. Dover Publications, 291 pp.

Fox, W. 1963. Special tubules for sperm storage in female lizards. Nature 198: 500–501.

Hamilton, W. D. 1967. Extraordinary sex ratios. Science 156: 477–488.

Maynard Smith, J. 1971. What use is sex? J. Theoretical Biology 30: 319–335.

Rand, A. S. 1967. Ecology and social organization in the Iguanid lizard *Anolis lineatopus*. Proc. U.S. Nat. Mus. 122: 1–79.

Schoener, T. W., and A. Schoener. 1971. Structural habitats of West Indian *Anolis* lizards. 1. Lowland Jamaica. Brevoria 386: 1–24.

Smith, H. M., G. Sinelnik, J. D. Fawcett, and R. F. Jones. 1973. A survey of the chronology of ovulation in anoline lizard genera. Transactions Kansas Academy Science 75: 107–120.

Trivers, R. L. 1972. Parental investment and sexual selection. *In* B. Campbell, ed.,

Sexual Selection and the Descent of Man, 1871–1971. Chicago: Aldine, pp. 136–179.

Trivers, R. L., and R. Hicks. *In Prep*. The social behavior of *Anolis valencienni* in relation to its feeding ecology.

Williams, G. C. 1975. Sex and Evolution. Princeton University Press.

Zahavi, A. 1975. Mate selection—a selection for a handicap. Journal of Theoretical Biology 53: 205–214.

Postscript

There is now a very substantial literature on sexual selection in lizards but I have not kept up with it beyond the summary of general trends found in the ninth chapter of my 1985 book and a recent update of these trends by Stamps, Losos and Andrews (1997). Failure of *Anolis* lizards to conform to predictions of Hamilton and Zuk (1982) have recently been shown by Schall and Staats (1997).

I wish I had not overstated the role of sexual selection in maintaining sexual reproduction, namely that it could cover the entire cost of sex. The conventional view then was that sexual selection made no contribution, or only a trivial one, or perhaps even a negative one. Since then, a much greater appreciation of the positive power of mate choice and sexual selection has developed and two recent papers have shown that sexual selection can, in theory, make a strong contribution to maintaining sex by eliminating negative mutations (Agrawal 2001, Siller 2001).

7

SELECTING GOOD GENES FOR DAUGHTERS

When I had published my social insect and lizard work in 1976 I saw two topics ahead of me worth developing: one was a theory of mate choice, and the other, a theory of self-deception. For various reasons I was slow to develop either topic. In chapter 8 I describe my thinking on self-deception. Here I introduce the only paper I managed to publish on a theory of mate choice, a widely neglected paper with Jon Seger, which I believe demonstrates something fundamental, namely, that systems of female choice will naturally evolve with a bias toward the interests of daughters.

Upon finishing my work on parental investment and sexual selection, it was obvious that there was a need for a theory of mate choice based on genetic quality. Trivers (1972) had only managed to clear out the undergrowth on this topic, preparing the land to plant, so to speak, but with very little planted. A case in point is the notion, briefly pushed forward, that females may in some cases prefer older males because their age indicates ability to survive. On second thought, this seems unlikely on its face. For one thing, daily experience in our own species argues against it. Very few women say about me, "Well, Bob is not good-looking, he has an unpleasant personality, and he is difficult to live with, but God, is he *old*—I'm turned on!" Logic also suggests that this argument is wanting. In one sense, we are all equally old, about four billion years, that is, as old as life itself. The difference between an individual who is sixty years old and one who is twenty is how far back into the past their unique genotypes extend. The *genes* inside a twenty-year-old have survived just as long as those in a sixty-year-old. It is just that in the twenty-year-old the genes were united just 21 years ago, from two separate individuals, each conjoined with other genes. In short, they have undergone recombination more recently than have the genes of the sixty-year-old. Since evolution is ongoing, the unique genetic combinations of older individuals are apt to become increasingly irrelevant as time passes. In short, there is no clear and obvious logic by which age should be associated with mate choice, and any mature theory of mate choice and age requires some underlying understanding of recombination.

I set about working on a theory of mate choice by 1977. My intuition was that if it could be shown that female choice would evolve with a bias toward daughters (in species lacking male parental investment), then males would evolve to show off how good a female they would have been had they had the good fortune of being made female. The reason for this bias grew in part out of an understanding of selection acting on sexual reproduction itself. As I have already mentioned, the most important advance in our thinking about the evolution of sex differences was the appreciation of the problem of evolving sexual reproduction itself. The key to this is the advantage of producing genetically variable offspring, but part of the two-fold cost of sex can also be paid if sexual selection is working adaptively from the female's standpoint, that is, if the genes of the males breeding are on average superior to those of the females. This will be most likely where female choice for genes is operating but only if *daughters* are genetically improved. That is, if sexual selection is operating adaptively from the *female's* standpoint, those males actually breeding would have superior genes for her offspring than the average female genes themselves. If these superior genes improved daughter survival and/or reproduction, then an asexual mutant would have less than a twofold advantage each generation. You could indeed imagine a series of species in competition with each other differing in the degree to which sexual selection benefited females and their daughters. Those that did so more extensively would be those that should outcompete the other species.

But I was eager to see if I could find, at the *individual* level, an argument for sexual selection benefiting daughters. There was no immediate hope on the male–male competition side of the equation since what was initially favored were genes that benefited males. Was it possible that female choice was biased in favor of daughters? The conventional approach at the time was either to see female choice improving the male lineage, as in the "sexy son hypothesis" (females prefer males who are attractive to other females in order to have sons that will be attractive to the daughters of other females) or, to follow Fisher's sex ratio theory and give equal weight to sons and daughters, in calculations of the effects of female choice genes. I wanted a third way, to have effects on daughters count more heavily in decisions of female choice.

I tried and I tried but I could not see the solution. I was caught in an unusual space. Looking at the matter at the level of the species, sexual selection that failed to benefit daughters made no sense, but lacking an argument at the individual level left me with nothing but a species selection process enriching the survival of species that happen to have sexual selection oriented in the right way. Any kind of force *within* the species in the opposing direction would carry far more weight.

I do not remember exactly when it was, but sometime in the early 1980s I saw that female choice benefiting daughters would be more quickly self-reinforcing than female choice benefiting sons, because female choice genes expressed themselves only in females; thus, when benefiting a son, you would have to wait two generations before seeing an increase in the number of females choosing with this bias. By contrast, female choice genes that benefited daughters would increase immediately in the choosing sex in the next generation.

In those days I often presented the argument by reference to sex-antagonistic genes. These are genes that have a beneficial effect when they appear in one sex and a negative effect when they appear in the opposite one. I said imagine there were two females, A and B, choosing two kinds of males, a and b. Imagine that the sons of male a survived 30% better than average, while the daughters survived 30% less well than average. Imagine the offspring of male b had the opposite tendency, 30% improvement in female survival at the cost of a 30% loss in male survival. Now imagine these two were in competition with a third female choice gene, C, which expressed no preference. At first glance, all three appear to have equal fitness. What is lost on the female side for A is gained on the male side, and vice versa for B, while of course C has zero effect. But, I argued, this was an illusion, because the female choice gene B increased immediately in frequency among females, which in turn intensified the bias in favor of male b offspring, which over time will contain a disproportionate share of female choice B genes since they are likely to be the offspring of females with that bias. This female choice gene would begin to spread, while the female choice gene A would disappear, even in competition with the no-choice genes (C), since females with the appropriate bias *decrease* in frequency in one generation. I expressed this idea in my 1985 book thus:

Selecting Good Genes for Daughters

Imagine a gene in a female leading her to choose a male whose genes improve her daughters' chance of survival by 10%, while decreasing her sons' survival by a similar amount. At first glance, genes for this kind of female preference appear to be neutral in their evolutionary effects: what is gained in daughters is lost in sons. But since the genes for this kind of female choice will have increased in one generation among females, more females will choose in this way, giving a benefit to the males they prefer. These males will tend to contain a disproportionate number of the very same female choice genes, since these males will—more often than randomly—be the sons of males preferred by such females one generation previously. Thus, the net effect of such genes is positive: they enjoy an increase in survival in females, a comparable decrease in survival in males, but an offsetting increase—however small—in reproduction among males due to the increased number of females who prefer males likely to have these genes.

Notice also that genes in males that increase survival in sons make life slightly more difficult for these sons (fewer females per male), while genes in males that increase survival in daughters make life slightly easier for males carrying these genes (more females per male). Thus, female choice genes for sons inhibit their own spread, while choice genes for daughters do not.

If these arguments are valid, it suggests that genes in males that reveal quality of the remaining genes *for daughters* will more rapidly be selected for by female choice than similar genes that merely reveal quality in sons. Of course, beneficial genes need not be as sex-antagonistic as we have described. The genes most favorably selected for through choice will be those that improve survival in both sons and daughters, although it must inevitably be true that males will be selected for in ways that are not advantageous to females. Our theory of female-biased female choice suggests that genes in females for protecting themselves from this will easily gain a toehold. They are self-reinforcing within one generation. Male–male competition, in turn, may begin to adjust itself appropriately. In the hamadryas baboon, lower-ranking males are more likely to attack a dominant in consort with a female if she lacks a special relationship with him, presumably because their payoff—via choice—is higher.

There is another reason for expecting that this neglected form of female choice may be important in nature. We have reasons to believe that the type of female choice in a species may affect its chances of survival. Consider two species: one chooses to benefit daughters in every generation; the other benefits sons. Imagine that these two species are in close competition. Since only females invest in young, their numbers control the rate of increase. Increased survival of sons will temporarily increase population size but will result in no increase in the number of offspring produced. Thus, the species that improves daughters should have a competitive advantage. It has the capacity to increase its numbers in each generation compared to the second species. Thus, species with female choice for daughters will tend to increase at the expense of other species (pp. 353–354).

I Talk Jon Seger into Working on the Problem

I thought this was an important insight. It suggested renewed attention to the factors controlling variation in *female* reproductive success as a key to understanding variation in male morphology. The conventional view was that female distributed themselves in nature primarily with reference to food—but also kin-directed networks and predators—and this distribution in space and time determined male–male competition, with female choice being an added force with no particular orientation. In my view, it was precisely the factors that affected *female* reproductive success that were expected to be important in female choice; thus, males might easily evolve to show off to females how good a female they would have been had they had the good fortune of being made one! Were the antlers of a male deer really meant to represent the offspring he would have carried—through roughly the same annual cycle—were he a female instead? All sorts of interesting possibilities beckoned, but the argument itself required a quantitative underpinning. The verbal argument does not permit you to answer the question, Is this an important force or a weak force? What kind of (heritable) correlation between the attributes of the father and the daughter is required to permit selection to mold an appropriate response? For all of these questions I needed a quantitative treatment of the idea, but as I have indicated many times, I had given up any mathematical work myself.

I may have already tried to interest one or two others in solving the problem mathematically when I had the good fortune of again running into Jon Seger, a biologist who had been my graduate student while I taught at Harvard. I first met Jon when he was a graduate student in education at Harvard, I believe preparing to be a high school teacher. He had taken Professor DeVore's primate behavior course when I was a teaching assistant. He came to my office with a paper he had written earlier, which, as was usual in those days, was implicitly group selectionist. It was a measure of Jon's quickness that it took only one lesson to beat this style of thinking out of him and as I remember it, he returned a week later with the whole paper rewritten in an individual selection vein. Like myself, Jon decided to enter biology relatively late in life, but unlike me he also forced himself to learn some more mathematics.

Jon Seger has published several brilliant papers, but the one I always love most is "Partial Bivoltinism May Cause Alternating Sex-Ratio Biases That Favour Eusociality" (1983), as obscure a topic as it was well treated by Jon. We met again at a conference at Irven DeVore's home in Cambridge, Massachusetts, perhaps connected with a birthday celebration for Irv. In any case, I remember catching his attention with the problem while we were seated together on a couch, with me criss-crossing my arms to convey the essential logic of sex-antagonistic effects benefiting females being given more weight than similar effects benefiting males. I pointed out to Jon that Mark Kirkpatrick had already produced equations that explored a trivial part of the nonadaptive landscape. That is, his equations showed that if enough females liked a male whose genes decreased male survival, such a male's genes and the female choice gene linked to him would spread rapidly to fixation. I pointed out that all Jon needed to do was to reverse the signs, so to speak, and imagine a male whose genes decreased daughter survival and then compare the fate of the choice genes and their matching male phenotype genes. This is exactly what Jon did, and when he did so he showed that a female choice gene harming daughters was inhibited in its spread compared to a female choice harming sons and that the quantitative effect was appreciable.

Jon then wrote up the paper pretty much as you see it here, expect that I tried to include verbal argumentation wherever I could, and Jon discovered a weak kind of group selection that would operate if you imagined different populations frozen along different equilibria for the relevant female choice and male genotype genes.

Asymmetry in the Evolution of Female Mating Preferences

JON SEGER AND ROBERT TRIVERS

Trivers[1] has suggested that where genetic or developmental constraints on the expression of a trait prevent male and female fitnesses from being maximized simultaneously, female mating preferences should evolve to favour males who exhibit variants of the trait that confer relatively low fitness on males but relatively high fitness on females. This asymmetry is expected because alleles that affect mating preferences are expressed only in females, but are genetically correlated with alleles that differentially affect the fitnesses of the two sexes. Here we describe a two-locus population-genetic model that embodies this idea. The model's qualitative behavior is exactly like that of previous models[2-10] for the joint evolution of male traits and female mating preferences: evolution is equally likely to proceed in either direction along (or away from) a line of neutral equilibria that relates given frequencies of the preference alleles to corresponding frequencies of the trait alleles. But there is a quantitative asymmetry, of the expected kind, in the shape of the line of equilibria. When we extend the model to include migration between partially isolated demes (breeding groups), 'selective diffusion'[11,12] moves the demes along the line of equilibria in the direction that increases average female fitness while lowering average male fitness.

The model species is a sexual haploid with two unlinked loci and discrete, non-overlapping generations. Alleles at the trait locus (T) determine male and female viabilities according to the following scheme:

	T_1	T_2
Males	α_1	α_2
Females	β_1	β_2

In males, T_1 and T_2 also determine a phenotypic difference that is visible to females. At the beginning of each generation the frequencies of T_1 and T_2 and t_1 and $t_2 = 1 - t_1$. After viability selection, the frequencies of T_1 in adult males and females are

$$t_1^m = \alpha_1 t_1 / (\alpha_1 t_1 + \alpha_2 t_2) \tag{1}$$

and

$$t_1^f = \beta_1 t_1 / (\beta_1 t_1 + \beta_2 t_2) \tag{2}$$

with $t_2^m = 1 - t_1^m$ and $t_2^f = 1 - t_1^f$.

Female mating preferences are determined by alleles at the preference locus (P). Females carrying P_1 prefer to mate with T_1 males, and females carrying P_2 prefer to mate with T_2 males. There are many different behavioural models of female choice[3-10,13,14]; we used two simple ones that give qualitatively different results in some contexts[8]. Under the fixed-relative-preference model[6,8], the probability that a P_1 female mates with a T_1 male is

$$U_{11} = a_1 t_1^m / (a_1 t_1^m + t_2^m) \tag{3}$$

where (a_1 ($\geqslant 1$) is a parameter that sets the strength of the female preference, which vanishes at $a_1 = 1$. The probability that a P_1 female mates with a T_2 male is $U_{12} = 1 - U_{11}$. The behavioral model is one in which P_1 females encounter T_1 and T_2 males sequentially, at random, accepting them as mates with fixed probabilities whose ratio is $a_1 : 1$. Similarly,

$$U_{22} = a_2 t_2^m / (a_2 t_2^m + t_1^m) \tag{4}$$

is the probability that a P_2 female mates with a T_2 male, and $U_{21} = 1 - U_{22}$ is the probability that she mates with a T_1 male.

Under the better-of-two model of female choice,

$$U_{11} = t_1^m + c_1 t_1^m t_2^m \tag{5}$$

and

$$U_{22} = t_2^m + c_2 t_1^m t_2^m \tag{6}$$

where c_1 and c_2 ($0 \leqslant c_1 \leqslant 1$) are parameters that set the strength of the mating preference. Here the behavioural model is that of a lek. Each lek is composed of two males, sampled at random from the population after viability selection. If the lek contains one T_1 and one T_2 male, the female chooses her preferred type with probability $\frac{1}{2}(1 + c_1)$. Thus the preference vanishes at $c_i = 0$.

The frequencies of P_1 and P_2 are p_1 and $p_2 = 1 - p_1$, at the beginning of each generation. Although P_1 and P_1 have no effects on male or female viabilities, their frequencies change slightly in each sex during viability selection at the T locus, because of the genetic covariance (phase disequilibrium) that exists between the two loci. This covariance is maintained, even at gene-frequency equilibrium, by the female mating preferences[2-10].

Given the assumptions and definitions described above, analysis of the model proceeds in a straightforward way, along the lines described more fully elsewhere[6,8]. Additional details are given in the figure legends.

The asymmetry mentioned above can easily be demonstrated by comparing models in which the fitnesses $(\alpha_1, \alpha_2, \beta_1 \beta_2)$ have been symmetrically rearranged. Consider the case $(1, \alpha_2, 1, 1)$, with $\alpha_2 < 1$. Here females are unaffected by their T-locus genotypes, but T_2 males suffer reduced viability relative to T_1 males. Kirkpatrick's[6] first model is this fitness scheme with fixed-relative-preference female choice, where $a_1 = 1$ and $a_2 > 1$ (P_1 females mate randomly, while P_2 females prefer the less viable T_2 males). Figure 1A shows the line of stable equilibria illustrated in Kirkpatricks's Fig. 1. If we reverse the pattern of sex limitation, so that T_2 females suffer reduced viability, then the fitness array is $(1, 1, 1, \beta_2)$ and we obtain the line of stable equilibria shown in Fig. 1B. Although T_2 invades at lower frequencies of P_2 when its deleterious effect is limited to females than it does when limited to males, much higher frequencies of P_2 are required to take it to fixation. Of course, T_2 can have any equilibrium frequency between 0 and 1, under either of these fitness schemes, given an appropriate frequency of P_2. But under almost any probability distribution of $\hat{p}_2$ centred on $\hat{p}_2 = \frac{1}{2}$, the average equilibrium frequency of T_2 is higher when its deleterious effect is limited to males than when the same deleterious effect is limited to females.

In a more fully symmetrical model, both T_1 and T_2 would affect the fitnesses of their carriers, and both P_1 and P_2 females would exhibit mating preferences. Consider the fitnesses $(1, \alpha_2, \beta_1, 1)$. Here T_1 males are fitter than T_2 males, as in Kirkpatrick's model. But now T_2 females are fitter than T_1 females. If we set $\alpha_2 = \beta_1$, then the fitness effects are completely symmetrical, and in the absence of any female mating preferences, T_1 and T_2 have a strongly attracting equilibrium[15-19] at $t_1 = t_2 = \frac{1}{2}$. If we also set $a_1 = a_2 > 1$, then P_1 and P_2 females have equally strong preferences for T_1 and T_2 males, respectively, and the model is fully symmetrical. But the line of stable equilibria is not symmetrical. T_2 is taken to high frequencies more easily than is T_1 (Fig. 2).

The interior equilibria are always stable under the fixed-relative-

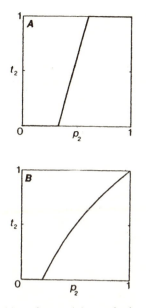

Figure 1. Lines of stable equilibria for models in which only one sex is affected by T_1 and T_2. A, Only males are affected. This is Kirkpatrick's[6] model in which $\alpha_1 = \beta_1 = \beta_2 = 1$, $\alpha_2 = 0.6$ $a_1 = 1$ and $a_2 = 3$. B, Only females are affected. Here $\alpha_1 = \alpha_2 = \beta_1 = 1$, $\beta_2 = 0.6$, a_1 and $a_2 = 3$. T_1 goes to fixation when the frequency of P_1 is greater than 0.8 ($p_2 < 0.2$), but T_2 goes to fixation only when P_2 is also at fixation ($p_2 = 1$). If a_2 (the preference parameter) were larger than 3, or if β_2 (the fitness of T_2 females) were larger than 0.6, then the upper end of the line of equilibria would intersect the upper boundary ($t_2 = 1$) at $p_2 < 1$. Conversely, if a_2 were smaller than 3 or if β_2 were smaller than 0.6, then the line would intersect the right-hand boundary ($p_2 = 1$) at $t_2 < 1$, and T_2 would have an interior equilibrium even at fixation for p_2. In all of the models discussed here, the interior equilibria satisfy the equation

$$p_2^f = \frac{t_2(2 - 1/F) - U_{12}}{U_{22} + U_{11} - 1} \qquad (8)$$

where p_2^f is the frequency of P_2 in adult females and $F = \beta_1 t_1 + \beta_2 t_2$ is the average female fitness. Equation (8) was used to draw the lines of equilibria shown above and in Figure 2. If female fitnesses are not affected by T_1 and T_2, then $F = 1$, and $p_2^f = p_2$. But if the T locus is expressed in females ($\beta_1 < 1$ or $\beta_2 < 1$), then $F < 1$, and p_2^f is slightly different from p_2. the change in the frequency of P_2 in females arises during viability selection at the T locus, owing to the phase disequilibrium (covariance) between the two loci. Thus, the line of equilibria shown in A (Kirkpatrick's model, in which $\beta_1 = \beta_2 = 1$) is exact, but the lines shown in B and Figure 2 are approximate. (If the horizontal axes were relabelled as p_2^f, then all of the lines would be exact.) The error involved in this approximation (or misrepresentation) was estimated by numerical iteration of the recurrence equations in the genotype frequencies, which model the full dynamics of the two-locus system. Even with strong selection and strong female mating preferences, as in B, p_2^f differs from p_2 by only a few per cent for intermediate gene frequencies, becoming identical to p_2 at the boundaries. (The exact end points of the lines of equilibria can easily be derived analytically.) With weaker selection and weaker mating preferences, as in Figure 2, the approximation is excellent everywhere.

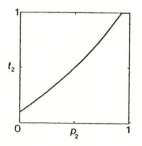

Figure 2. Line of stable equilibria for a fully symmetrical version of Kirkpatrick's model. Here $\alpha_1 = \beta_2 = 1$, $\alpha_2 = \beta_1 = 0.8$ and $a_1 = a_2 = 1.05$. Note that the strengths of natural selection and female choice are weaker here than in the examples shown in Figures 1 and 3. T_1 and T_2 have an interior equilibrium at fixation for P_1 $(p_2 = 0)$, but T_2 is taken to fixation when P_2 is above a frequency of 0.925. Qualitatively similar lines of equilibria exist even if $\alpha_2 \neq \beta_1$ and $a_1 \neq a_2$.

preference model of female choice[6,8], but they can be unstable under the better-of-two model[8]. Figure 3 shows a line of equilibria for the symmetrical fitnesses described above, under symmetrical better-of-two female choice $(c_1 = c_2 = 1)$. Again the asymmetry is obvious, and is illustrated dramatically by the dynamical behaviour of a population that begins at the centre of the gene-frequency plane $(t_2 = p_2 = \frac{1}{2})$. The population evolves rapidly to a boundary equilibrium at $t_2 = 1$, $p_2 > \frac{1}{2}$ (Fig. 3).

The causes of the asymmetry are easy to see. P_2 females who choose T_2 males instead of T_1 males thereby increase the survival of their daughters. Of course, they also decrease the survival of their sons, so there would seem to be no net advantage associated with the preference for T_2. But in the next generation it will be females, not males, who do the choosing, and allele P_2 increases its frequency in females by associating itself nonrandomly with T_2.

The stable equilibria are points at which a population would sit forever, if affected only by the deterministic forces of natural and sexual selection. But in a world of finite, partially isolated demes, differential migration may give rise to a weak directional force that moves an average deme along the line of equilibria, towards higher frequencies of P_2 and T_2. Consider a set of n demes, each of which has the same carrying capacity, K. The expected number of adult females in the ith deme is $KF_i/(F_i + M_i)$, where F_i and M_i are the average viabilities of females and males born in that deme. If males do not help females to rear offspring, then the productivity of each deme is proportional to its number of adult females, which is proportional to its frequency of T_2, under the fitness scheme discussed above [$(1, \alpha_2, \beta_1,$

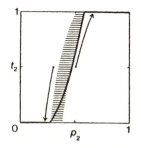

Figure 3. Line of equilibria for the symmetrical model with better-of-two female choice. Here $\alpha_1 = \beta_2 - 1$, $\alpha_2 = \beta_1 = 0.6$ and $c_1 = c_2 = 1$. From points to the left of the line, populations move down and to the left. From points to the right of the line, they move up and to the right. From points within the hatched region, populations are attracted to stable interior equilibria on the lower region of the line (approximately, the region below $t_2 = 0.55$). From points outside the hatched region, populations move to the stable boundary equilibria at $t_2 = 0$ or $t_2 = 1$. The illustrated trajectories begin at $p_2 = 0.3$, $t_2 = 0.5$ and at $p_2 = t_2 = 0.5$. Each of these trajectories is less than 100 generations in length. The upper region of the line (between approximately $t_2 = 0.55$ and $t_2 = 1$) is unstable. From points just to the right of the line, populations move to the boundary at $t_2 = 1$, and from points just to the left, they move to stable interior equilibria on the lower region of the line. The line of equilibria, the regions of attraction and the illustrated trajectories were all found numerically, by iteration of the recurrence equations in the genotype frequencies. The approximate line of equilibria (equation 8); see Figure 1 legend) has the same end points as the line illustrated here, but it passes through the central point ($p_2 = t_2 = \frac{1}{2}$).

1) with $\alpha_2 = \beta_1$]. Let each deme contribute to the migrant pool in proportion to F_i, and let it then receive mK migrants from the pool. It follows that

$$\Delta \bar{t}_2 = \frac{m}{1 + m}\left[\frac{(1 - \beta_1)\ \text{var}\ (t_2)}{\beta_1 + (1 - \beta_1)\bar{t}_2}\right] \tag{7}$$

For example, if $m = 0.01$, $\beta_1 = 0.8$, $\bar{t}_2 = 0.5$ and var$(t_2) = 0.01$, then $\Delta \bar{t}_2 = 2.2 \times 10^{-5}$. The overall frequency of T_2 increases because demes with higher than average frequencies of T_2 make larger than average contributions to the migrant pool, which therefore has a higher frequency of T_2 than does the population at large.

Suppose that at the beginning of a generation the demes are at various points along an approximately straight segment of a line of stable equilibria. Then after migration, selection and reproduction, the demes, will still be on the line, because the migrants that entered each deme will have changed its frequency of P_2 by an amount proportional to the amount by which they changed its frequency of T_2, the constant of proportionality being the reciprocal of the slope of the line. The changes of gene frequency caused each generation by differential migration are not opposed by selection within demes, because within each deme the new frequency of T_2 is stabilized by a new and appropriate frequency of P_2. Even where the line of equilibria is moderately curved, migration moves the demes along paths that are almost tangent to the line. Then, depending on the direction of the curvature, there are either slight decreases or slight additional increases in the average frequencies of T_2 and P_2, as selection pulls the demes in toward the line.

This process of 'selective diffusion'[11,12] or 'group selection'[20–23] depends on the variance of gene frequencies among demes, but the variance is reduced by migration. In group selection models for the evolution of reproductive altruism or female-biased sex ratios of parental investment, the overall direction of gene-frequency change is reversed when the variance falls below a critical minimum, because within each deme there is selection against the trait that is favoured by differential migration between demes. This problem does not arise in the present model, because within each deme female choice stabilizes the current frequency of T_2. Thus, $\bar{t}_2$ continues to increase as long as the balance between migration and sampling drift maintains even a small interdemic variance of gene frequencies. Female mating preferences change the frequencies of T_2 only in demes that are away from the line of equilibria. In demes that are on the line they do not directly change frequencies, but they can act as a catalyst for changes induced by other causes.

The main implication of these results is that females in polygynous species are expected to prefer mating with males who, other things being equal, show evidence that they would have been reproductively successful had they been females rather than males. We are not the first to argue that female choice should evolve to favour, in part, traits that promote ecological fitness[1,2,24–34], but the present model seems to be the first to show explicitly how female choice could evolve to place special emphasis on traits that promote female fitness[1,26].

Many well-known patterns of sexual dimorphism are consistent with an interpretation of the kind implied by this model. For example, rates of senescence are higher in the males of most species than in the females[35,36]. Unfortunately, this dimorphism has plausible alternative explanations based on male-male competition rather than female choice[1,35,37]. The same ambiguity seems likely to appear in all comparative tests of the model, as the following example will illustrate.

Male *Anolis* lizards tend to be much larger than conspecific females on islands, but not on the mainland[38]. Island populations appear to be more strongly food-limited than mainland populations. Individuals belonging to island populations grow relatively slowly, mature at relatively late ages, feed relatively often on small items, and show sharp increases in growth rates when additional food is supplied experimentally[38]. If the fitnesses of island females depend relatively strongly on the efficient acquisition and assimilation of food, we would expect island females to prefer mating with males who had demonstrated special competence in these abilities. The greater size dimorphism of island populations is consistent with this expectation. But it is also consistent with the argument that female-defence polygyny is a more successful male strategy on islands than it is on the mainland, because island populations live at higher densities (with smaller territory and home-range sizes) than do mainland populations. Direct male-male competition would then be the selective force that favours relatively large male sizes on islands[39]. The two models are not mutually exclusive, in this example or in general, so considerable ingenuity will be needed to devise tests that clearly distinguish between them.

We thank J. Bull, P. Harvey, R. Lande and M. Kirkpatrick for helpful advice and discussion.

REFERENCES

1. Trivers, R. *Social Evolution* (Benjamin/Cummings, Menlo Park, 1985).
2. Fisher, R. A. *The Genetical Theory of Natural Selection* (Clarendon, Oxford, 1930).
3. O'Donald, P. *Genetic Models of Sexual Selection* (Cambridge University Press, 1980).
4. Lande, R. *Proc. Natl. Acad. Sci. U.S.A.* **78**, 3721–3725 (1981).
5. Lande, R. *Evolution* **36**, 213–223 (1982).
6. Kirkpatrick, M. *Evolution* **36**, 1–12 (1982).
7. Kirkpatrick, M. *Am. Nat.* **125**, 788–810 (1985).
8. Seger, J. *Evolution* **39**, 1185–1193 (1985).
9. Engen, S. & Saether, B.-E. *J. theor. Biol.* **117**, 277–289 (1985).
10. Lande, R. & Arnold, S. J. *J. theor. Biol.* **117**, 651–664 (1985).
11. Wright, S. *Ecology* **26**, 415–419 (1945).
12. Wright, S. *Evolution and the Genetics of Populations* Vol. 3 (University of Chicago Press, 1977).
13. Janetos, A. C. *Behav. Ecol. Sociobiol* **7**, 107–112 (1980).
14. Parker, G. A. in *Male Choice* (ed. Bateson, P.) 141–166 (Cambridge University Press, 1983).
15. Owen, A. R. G. *Heredity* **7**, 97–107 (1953).
16. Haldane, J.B.S. *Nature* **193**, 1108 (1962).
17. Li, C. C. *Evolution* **17**, 493–496 (1963).
18. Bodmer, W. F. *Genetics* **51**, 411–424 (1965).
19. Kidwell, J. F. *et al. Genetics* **85**, 171–183 (1977).
20. Wade, M. J. *Q. Rev. Biol.* **53**, 101–114 (1978).
21. Uyenoyama, M. & Feldman, M. W. *Theor. Popul. Biol.* **17**, 380–414 (1980).
22. Wilson, D. S. *A. Rev. Ecol. Syst.* **14**, 159–187 (1983).

23. Wilson, D. S. & Colwell, R. K. *Evolution* **35**, 882–897 (1981).

24. Mayr, E. in *Sexual Selection and the Descent of Man 1871–1971* (ed. Campbell, B.) 87–104 (Aldine, Chicago, 1972).

25. Zahavi, A. *J. theor. Biol.* **53**, 205–214 (1975).

26. Trivers, R. L. *Evolution* **30**, 253–269 (1976).

27. Bell, G. *Evolution* **32**, 872–885 (1978).

28. Eshel, I. *J. theor. Biol.* **70**, 245–250 (1978).

29. Borgia, G. in *Sexual Selection and Reproductive Competition in Insects* (eds Blum, M. S. & Blum, N. A.) 19–80 (Academic, New York, 1979).

30. Andersson, M. *Biol. J. Linn. Soc.* **17**, 375–393 (1982).

31. Hamilton, W. D. & Zuk, M. *Science* **218**, 384–387 (1982).

32. Dominey, W. J. *J. theor. Biol.* **101**, 495–502 (1983).

33. Thornhill, R. & Alcock, J. *The Evolution of Insect Mating Systems* (Harvard University Press, Cambridge, 1983).

34. Heisler, I. L. *Evolution* **38**, 1283–1295 (1984).

35. Williams, G. C. *Evolution* **11**, 398–411 (1957).

36. Comfort, A. *The Biology of Senescence* 3rd edn (Elsevier, New York, 1979).

37. Rose, M. R. *Am. Zool* **23**, 15–23 (1983).

38. Andrews, R. M. *Copeia* 1976, 477–482 (1976).

39. Stamps, J. A. in *Lizard Ecology* (eds Huey, R. B. Pianka, E. R. & Schoener, T. W.) 169–204 (Harvard University Press, Cambridge, 1983).

Postscript

This paper has fallen like a stone to the bottom of the discipline, leaving not even a ripple at the surface. By summer of 2001, the paper had only been cited 20 times in the scientific literature and most of these were only glancing references. Moller (1997) has made the valuable suggestion that a bias in the primary sex ratio produced by parents might serve as a measure of the bias of female choice, since if genes primarily good for sons were passed on in a family then there might be a selective advantage in producing more sons in that family, and vice-versa. Freed (2000) has drawn attention to the theory, especially as it applies to explaining female traits correlated with male.

I know of no follow-up paper on the logic itself, nor in applying the logic to patterns of sexual dimorphism. Indeed, the only examples of the latter can be found in chapter 9 of my 1985 book. At the time I thought it likely that a literature would grow up that would seek to see whether factors that controlled variation in female reproductive success predicted male characters, but this has not happened. To me it is still true that it would be very valuable to classify breeding systems according to the degree to which they elevated to breeding status genes good for daughters.

8

SELF-DECEPTION IN SERVICE OF DECEIT

I was interested in self-deception well before I became interested in evolutionary biology, and I was conscious of deception as a problem from a very early age. At the end of my freshman year at Harvard in 1962, when I left mathematics in despair and disrepute (stripped of my scholarships, placed on academic probation, etc.), I decided to devote my life—not to science—but to fighting the good fights in society, toward justice instead of truth. Toward that end, I saw myself becoming a lawyer, concentrating on civil rights, poverty, and criminal law. I asked someone what you majored in if you were going to become a lawyer, and they said U.S. history—the Federalist papers, the Bill of Rights, Supreme Court decisions, that kind of thing. So, I studied U.S. history. It was at once apparent that U.S. history was not so much an academic discipline as an exercise in self-deception. They were title of the books gave away the game, for example, *The Genius of American Democracy*, a popular book in the early 1960s by Daniel Boorstin. The chief problem in U.S. historiography was, why are we the greatest nation that ever existed and the greatest people who ever strode the face of the earth? The competing schools of thought in U.S. history were competing answers to those questions. For example, the existence of a frontier was said to inspire such virtues, the benefits of upper-class Englishmen designing a new government, or the wave of immigrants that populated the United States, and so on. So, I saw the field as an exercise in self-deception.

During the Vietnam War, one could see, perhaps, some of the negative consequences that accompanied the self-glorification that was the study of U.S. history. I also remember being astonished, then, at how often in defense of a particular administration official or the president himself, someone would say, "Yes, but he sincerely believes that we are acting in the best interests of the Vietnamese." Such a person was believed to be superior to one who cynically and consciously maintained the same viewpoint. Why, I wondered, should we so prize the degree of perfection of our leaders' self-deception? At the same time, I remember reading my first and only piece by Karl Marx, *The Communist Manifesto*. The first half was a searing indictment of the self-deceptions and lies of bourgeois society, while the second was a brand new set of self-deceptions regarding the future under communist leadership. This was progress?

Regarding deception, one of my most vivid childhood memories is my first realization of how pervasive and stupid patterns of human deception could be. I was about six or seven years old, as I remember it, when I had spotted a choice item in a store window display that I wanted to buy. I believe it may have been a knife, if not some extra special toy, and it cost $6.00, so the sign in large numbers declared. I saved my coins until some weeks later I had $6.00 plus the few pennies then required to cover taxes. I went into the store and asked for the knife. The man brought me the knife, I paid, and he told me that I was $1 short. I begged to differ. I informed him that the proper price was $6.00. He remained firm. I started to become desperate. If the price was really $7.00, I wanted to know, why did the sign in the window say that the price was only $6.00? He assured me that the price in the window said $7.00. We went to the display and he drew my attention to the small .98 written next to the big 6, so that the true price for the knife was $6.98, in effect $7.00. I was flabbergasted. What sense did it make, I asked the man, to misrepresent the true price of an item by subtracting two pennies and then writing the remainder in small numbers? He assured me that the practice was widespread. I left the store without the knife but seething with indignation and outrage. I soon confirmed that the practice was widespread. Virtually all prices being advertised for goods, whether they were cars, a suit of clothing, gasoline, or a child's knife, were misrepresented as being one unit below their real value by the trick I have just described. I went around in a daze for weeks scarcely believing what was in front of my eyes. How on earth, I thought, could an entire society arrange prices in such a way that you were forced, whenever you saw a price, to do a little addition in order to figure out the true price? The amount of mathematical computation thereby wasted staggered me. When I was thirteen I got glasses, and I often wonder if I was not, in fact, nearsighted when I was seven, such that this little commercial trick was especially cruel for me. To tell you the truth, after all these years it still amazes me to see, around the world, all of these false prices.

Self-deception was also in my consciousness at an early age, because my mother once gave me a Bible verse uniquely suited for me, a lifelong mediation, if you will: "Judge not that ye be not judged for with the judgment you pronounce shall you be judged" (Matthew 7: 1–5). The mote as a metaphor for self-deception, in the service of evil, distorting appreciation of reality (the beam). I feel certain my mother was wagging a finger in my face, as she said, "Now, Ludlow, judge not that ye be not judged!"

When I was a senior at Harvard, I briefly considered psychology as a possible area of study, but I quickly discovered that psychology was not, in fact, a real discipline, but rather a competing set of hypotheses about what was important in human behavior. Each school of thought typically specialized in the arguments for why the other schools of thought were erroneous. So, social psychologists had rat psychologists pinned to the board and vice versa—hardly a formula for self-improvement. Freudian theory beckoned from some dark corner as offering deep insights into self-deception and human personality development, but when I actually read Freud, I rapidly formed a contrary opinion. I had met Henry Murray in 1965, a then well-known emeritus professor of psychology at Harvard. I believe I told Dr. Murray that I was interested in understanding the psychology of dreams, and he suggested that I begin with Freud's *The Interpretation of Dreams*. I soon sought out the book but was perplexed with the style of procedure being taken. I was hoping, of course, for a book that laid out the evidence and theories pertaining to dreams in a coherent fashion. Instead, I was seeing what looked like argumentation by assertion—that is, a series of assertions were made without any supporting logic or evidence whatsoever (except for an occasional classical allusion or metaphor from everyday life)—and an edifice of argumentation was thereby created. I kept searching for the missing references. I believe I returned to Dr. Murray and asked whether, by chance, there were two editions, one with references and one without, and I had happened on the wrong one, but of course there was no such luck, so this route also seemed hopeless at the outset.

When I learned evolutionary theory, mechanisms of self-deception seemed counterintuitive, to put it mildly. How could natural selection simultaneously favor ever-more-refined sensory organs and perception only to favor systematically distorting the information once it had arrived in the brain?! I remember enjoying a couple of years of argumentation on the subject with a good friend, Malcolm Slavin, with whom I was writing children's books on animal behavior for a fifth-grade curriculum. Slavin had a psychotherapeutic bent and went on to become a psychotherapist (in Cambridge, Mass.), and he had a rich fund of alleged knowledge on self-deception (i.e., psychoanalytic findings). At some point it became clear to me, certainly by 1972, that the social nature of the human being could easily induce self-deception, that is, that we are selected to deceive ourselves the better to deceive others.

I know I had a brainstorm on the subject during a very memorable visit to East Africa with Irven DeVore in 1972. Observations of baboons made it clear to me that individuals were sometimes engaged in intensive psychological warfare. And I could easily imagine that complicated mechanisms of reality manipulation might be selected in such a context. Since I was also working on parent–offspring conflict at the time, I naturally tended to ally the two ideas and to think especially about deceit and self-deception generated in the context of the family. This was, of course, congenial to the emphasis that psychoanalysis and related disciplines placed on family interactions, and I thought I could easily see a dark side to parent–offspring relations in which self-deceptions might be induced in the offspring in the face of certain styles of parental domination and manipulation. In any case, the brainstorm eventually landed me in the Harvard infirmary for three weeks with a mild version of the breakdown I had suffered in 1964.

Although there are oblique references to human self-deception in my paper on reciprocal altruism, I chose the foreword to Richard Dawkins's book to first state the connection to self-deception. He had written me out of the blue to ask my comments on a paper he had written on the so-called Concorde Fallacy applied to parental investment (Dawkins and Carlisle 1976). I had argued in my 1972 parental investment paper that females who already invested more in their offspring should be less likely to desert so as not to waste the great investment they had put in each offspring, and this is a new instance of an old fallacy. In poker the advice is, "Don't throw good money after bad." That is, the fact that you have already contributed $40 to a pot on the table does not mean that you owe it to the $40 to put an additional $20 in. The money has already been invested. You should only calculate future investment on the basis of future returns. I wrote back accepting the argument on the spot, and he then wrote and said he really wanted me to write the foreword to a book he had just written called *The Selfish Gene*. When I received it I thought it was such a well-written and attractive account of the recent set of ideas in my discipline that I readily agreed to write a foreword. Since Dawkins had the wit to emphasize deception in his book, I took the opportunity to give the argument for self-deception: "if . . . deceit is fundamental to animal communication, then there must be strong selection to spot deception and this ought, in turn, to select for a degree of self-deception, rendering some facts and motives unconscious so as not to betray—by the subtle signs of self-knowledge—the deception being practised. Thus, the conventional view that natural selection favors nervous systems which produce ever more accurate images of the world must be a very naïve view of mental evolution" (see Foreword, Dawkins 1976).

I Meet Huey P.

One of the few benefits of moving to the University of California at Santa Cruz was meeting the legendary founder of the Black Panther Party, Huey

P. Newton. He was a graduate student in "History of Consciousness" (roughly equivalent to Western Civilization), who had the wit to see that "consciousness" started long before the Greeks and, in some form, certainly by the time of the insects. He had gotten his undergraduate degree from Santa Cruz in 1974 and had befriended Dr. Burney LeBoeuf, the celebrated student of elephant seals. Burney had been preaching the beauties of evolutionary biology—and my own work in particular—so I had the good fortune of meeting Huey after he had been primed by Professor LeBoeuf.

Dr. William Moore called in the fall of 1978 to say that Huey, who was then in prison charged with the beating of an Oakland tailor, wanted to take a reading course from me. I said that was fine with me but I wanted a paragraph from Mr. Newton on what he wanted to read about. Before he could reply he was released on bail, and he traveled to Santa Cruz to meet me. We met at Professor LeBoeuf's home, Huey accompanied by Dr. Moore, his aide-de-camp Mark Alexander, and his bodyguard Larry Henson. When he told me that he had spent three years in solitary confinement, I asked him if he had never feared a mental breakdown in all that solitude. He described a night when, indeed, his psyche seemed to fragment and tear apart and he had to struggle to hold himself together and fight the fragmentation. So we formed a strong personal connection right off the bat. Incidentally, if you had a feeling for "vibes" (as Californians called it), you could feel Huey shivering as he sat talking, a result, I believe, of just coming out of two months of the male–male hell that is U.S. prisons.

We soon decided to do a reading course on deceit and self-deception, a subject I was eager to develop and on which Huey turned out to be a master. He was a master at propagating deception, at seeing through deception by others and at beating your self-deception out of you. He fell down, as so many of us do, when it came to his own self-deception.

Huey Newton was a natural-born genius, one of five or six I have met in my lifetime. Each is unique, of course, but Huey was unique in the explosive and persuasive power of his personality. While you might have to lean into Bill Hamilton to catch exactly what he is saying, you rarely had that problem with Huey. Quite the contrary, you might have to hold tight onto your chair to keep from being blasted twenty feet backward by the force of his exposition. His particular forte was logic and, especially, aggressive logic. That is, he was a master logician but unlike some I have met, his special ability was to use his logical powers aggressively against you so as to quickly beat you out of the position you may have taken. The argument often had a double-or-nothing quality about it where, in effect, he was doubling the stakes for each logical alternative, giving you the unpleasant sensation that you were, more than likely, losing ever more heavily as the argument continued. He claims not to have read a book until he was eighteen and to have learned to read studying Plato, but I imagine that this was, at least, some-

thing of an exaggeration. Like many philosophers who discovered evolutionary logic relatively late in life (including my father), he often said he
would have become an evolutionist if he had known such a thing existed
when he was in college.

Huey had an extraordinary, intuitive understanding for animals and animal imagery. Although he was not the inventor of the following usages, he
seized upon them and popularized them: "Black panther" for black revo-
lutionary and "pig" for a police officer. He once mimicked a moose giving
its territorial defense call, when challenged—a deep rising moaning sound—
and said that he had brought out the moose call in me when, during an
argument over royalties, he had suggested that instead of one-third to him
and two-thirds to me, as we had originally agreed, it should perhaps be two-
thirds to him and one-third to me. And I do, indeed, remember the feeling
of territorial invasion that seized me, as if suddenly being attacked on two
sides simultaneously, and there was a "moose" quality to my voice, as I said,
"Wait one minute there, Huey!" When he stayed out all night and his wife
was haranguing him, he would sometimes hang his head and say, "I'm sorry,
dear, I can't help myself; it's the *dog* in me."

Dummying Up

We normally think of deception where self-image is concerned as involving
inflation of self: you are bigger, better, or better looking than you really are.
But there is a second kind of deception—deceiving down—in which the
organism is selected to make itself appear less large, less threatening, and
perhaps less attractive, thereby gaining an advantage. Being less threatening
to an individual may permit you to approach more closely, for example.
The most memorable instance of deceiving down that I learned about was
when Huey Newton taught me the Black-American term "dummying up."
This refers to a tendency to represent yourself as being less intelligent and
less conscious than you actually are, usually the better to minimize the work
you have to do. So an employee may dummy up in order to avoid being
required to do more difficult things. In Panama, I routinely saw many instances in which the Spanish-speaking people would misrepresent themselves as understanding less English than they did or as being considerably
less intelligent than they actually were, all in order to gain benefits from the
benighted *gringos* who all too easily believed the dummying up.

One day I asked Huey how he dealt with dummying up by others. As
head of a large organization he must have repeatedly faced this problem
among his own underlings. It took him a while to catch on to what I was
asking him, but when he did his whole face lit up and he became animated,
saying, "Oh, you want to know how I handle *that*?!" He then launched into

a brilliant verbal attack on an imagined example of dummying up. I only wish that one of Richard Nixon's aides had been on hand to silently activate a tape recorder so that none of this would have been lost. Unfortunately, I can only give a rough sketch of Huey's answer. If I remember correctly, he imagined a situation in which a waiter working for you always managed to position himself so as to avoid seeing you calling him and to otherwise appear to be working while not actually doing any work. His own monologue in response to this ran roughly as follows: "Oh, so you are so dumb that you happen to be looking the other way whenever I am trying to get your attention? And you are so dumb that when you know I *am* watching you, you decide to polish silverware that needs no polishing? And you are so dumb that you are always walking toward the pantry without ever reaching it? Well, you're not *that* damn dumb!" followed by either slapping the organism to the ground or continuing with some verbal assault. In short, Huey revealed to the actor the hidden logic behind his actions, and the final ironic punch line was that "you're not that damned dumb" to have arranged all this dumb-acting behavior, in highly coherent patterns.

The Crash of Air Florida Flight 90

In January 1982 I was due to fly from California to Atlanta, New York, and then Boston on a lecture tour. But a snow storm blanketed the entire East Coast and I had to cancel the trip. The same snow storm claimed Air Florida Flight 90, which took off from Washington, D.C.'s National Airport on the way to Tampa, Florida, but landed instead in the Potomic River, and only three or four people in the back of the plane were fished out of the river alive. I called Ernst Mayr at Harvard to tell him that I was canceling my trip, and he asked me if I had heard that Robert Silberglied had perished on that flight. Bob Silberglied and I had been assistant professors together at Harvard. He was a student of insects, especially butterflies, who did field-work at the Smithsonian Tropical Research Institute in Panama. He had a new girlfriend, Ernst told me, was excited about the path his work was taking, and in his eagerness to return to girlfriend and work had caught a flight earlier than the one on which he had a reservation. This earlier flight was Air Florida Flight 90.

Perhaps because of this personal connection to the tragedy, I was listening to the news with unusual care one evening when they played the cockpit conversation between pilot and copilot. You could hear at once that the copilot, who was the one actually flying the airplane, was frightened and was concerned with conflicting information on their instruments as they roared down the runway. He kept asking questions of the pilot, who should have been reading the instruments and interpreting them, but the pilot was silent. Then the plane took off and now the copilot shut up and the pilot started talking, as if to the airplane, begging it to "just climb, just climb . . ." before the conversation ended with the sound of impact. I do not believe Huey heard this transcript, but I told him about it, and we agreed that it seemed to be a paradigmatic case of the cost of self-deception. The copilot was frightened and relatively reality oriented, while the pilot seemed over-confident and inattentive to reality. All of this was illustrated by the striking role reversal in talking: the copilot talked while it still mattered (the flight could still be safely aborted); the self-deceptionist spoke up only when it was too late. We decided to send for the transcripts of the full cockpit conversation as provided by the National Transportation Safety Board. The report provided other evidence that we were dealing here with a timid but reality-oriented copilot paired with an overconfident self-deceiving pilot. The reversal in roles, we felt, also contributed to the tragedy, since it made it likely that neither was playing his assigned role correctly that day.

I always regretted the fact that I failed to dedicate this paper to the memory of Bob Silberglied. He was a student of deception in the insect world and had written, with his girlfriend, a beautiful description of a butterfly cocoon that resembled a snake's head so well as to frighten even those experienced biologists who could see that it was a mimic. I take this opportunity to dedicate the paper to the memory of Robert Silberglied.

The Crash Of Flight 90

Doomed by Self-Deception?

ROBERT TRIVERS AND HUEY P. NEWTON

Two biologists offer a startling interpretation: this tragic event, they say, was triggered by behavior that has evolved over eons as a survival tactic.

The following article is unsettling. The editors believe that its disturbing thesis may help us understand why some disasters occur and enable us to place what is sometimes labeled negligence in a larger context. While some of the language of the story may seem brutally frank the editors believe the story has a construc-tive purpose, and they feel only sympathy and respect for the aircrew and the others who perished.

Seven months after a Tampa-bound Air Florida 737, Flight 90, slammed into a bridge and plunged into the Potomac River, killing 78 people, the National Transportation Safety Board reached its verdict. It attributed the crash to several factors. First, the crew failed to activate the anti-ice system of the plane's engines before and during takeoff. This in turn caused an engine-pressure-ratio (EPR) sensor to give false readings that registered more thrust than in fact was there. Also implicated were the pilot's decision to take off with snow and/or ice on the plane's wings and his failure to abort takeoff after being informed by the copilot (who was at the plane's controls) that the EPR readings were inconsistent with other instrument readings. The board also concluded that the pilot could have averted the crash by applying full thrust seconds after lift-off.

Air Florida has disputed these findings, charging that the crash was caused by a flaw in the design of the 737 that makes it pitch sharply and by "undetected and undetectable" ice that formed on the leading edge of Flight 90's wing.

Drs. Robert Trivers and Huey P. Newton, biologists at the University of California—Santa Cruz have drawn a more startling conclusion after reviewing the available evidence. The roots of the disaster, they say, lie in evolutionary biology. The adaptive mechanism of self-deception, they say, doomed Flight 90. Their article offers a unique interpretation of a tragic event.

[—Editors]

The benefit of self-deception is the more fluid deception of others. The cost is an impaired ability to deal with reality. Ultimately we measure the cost of self-deception by its negative effects on reproductive success and survival, but we are often far from able to make this final connection. One approach is to begin with a disaster and work backward, looking for evidence of a pattern of self-deception leading up to the event.

Reality Evasion

Consider, for example, the crash of Flight 90 immediately after takeoff on January 13, 1982, during a heavy snow-storm. The transcript of the final 30 minutes of conversation between the pilot and copilot suggests a pattern of self-deception and reality evasion on the part of the pilot that contributed directly to the tragedy. By contrast, the copilot comes across as reality oriented, but insufficiently strong in the face of his captain's self-deception. These are relatively crude characterizations, but useful to bear in mind as we try to capture the complex way in which patterns of self-deception may generate a human disaster.

Let us begin as the airplane is cleared for takeoff and its engines are fired up to head down the runway. It will roar down the runway for 47 seconds before reaching the speed at which a final decision must be made about whether to go or not. At any moment during this time the pilot can abort the flight safely. Ten seconds after starting down the runway the copilot responds to instrument readings that are inconsistent.

Copilot: God, look at that thing.

Then, four seconds later:

Copilot: That don't seem right, does it?

Three seconds later:

Copilot: Ah, that's not right.

Two seconds later:

Copilot: Well . . .

Two seconds later:

Pilot: Yes, it is, there's 80.

It takes 11 seconds for the pilot to respond to the copilot. Apparently referring to an air speed of 80 knots, he seeks to explain away the instrument readings that are troubling the copilot. This fails to satisfy the copilot, and one second later:

Copilot: Naw, I don't think that's right.

Nine seconds later, having received no support from the pilot, the copilot wavers:

Copilot: Ah, maybe it is.

Two seconds later, the pilot states the speed at which they are traveling:

Pilot: 120.

Two seconds later:

Copilot: I don't know.

Caught between his own doubts and the pilot's certainty, the copilot finally lapses into uncertainty. Eight seconds later the pilot says, "V-1." This is the go/no go decision speed. After this point, the flight can no longer be aborted safely because it would run out of runway. Now we note a striking reversal in the roles of pilot and copilot. So far the copilot has done all the talking, the pilot only giving routine information. Now that they have passed the speed at which they are committed to their course, the copilot no longer speaks, and the pilot speaks repeatedly. Two seconds after V-1, the pilot says "Easy." Four seconds later he says, "V-2." This is the speed that you must maintain to clear the end of the runway if an engine fails. Two seconds later the sound of the stickshaker, a device that signals an impending stall, is heard in the cockpit. Six seconds later:

Pilot: Forward forward.

Two seconds later:

Speaker undetermined: Easy.

One second later:

Pilot: We only want 500.

Two seconds later:

Pilot: Come on, forward.

Three seconds later:
  Pilot: Forward.
Two seconds later:
  Pilot: Just barely climb.

Avert a Stall

The pilot is apparently urging the copilot to reduce the rate of climb to avert the stall. Before the pair were committed to the fatal flight, the pilot had little or nothing to say. Now that they have made their mistake, he comes out into the open and tries to reason. Four seconds later:
  Speaker undetermined: Stalling, we're falling.
One second later:
  Copilot: Larry, we're going down, Larry.
One second later:
  Pilot: I know it.
Almost simultaneously, the recorder picks up "sounds of impact."
The copilot did all his talking while it still mattered. At the end, he is only heard from telling his pilot what the pilot has been so reluctant to see: "Larry, we're going down, Larry." And the pilot finally says, "I know it."
The dichotomy between self-deceiver and reality-seeker was evident in earlier exchanges between pilot and copilot as they sat in the cockpit together prior to departure in extremely cold weather and a driving snowstorm. A half hour before takeoff the following exchange took place:
  Copilot: We're too heavy for the ice.
  Copilot: They get a tractor with chains on it? They got one right over here.
He is referring to the unsuccessful efforts of a tractor to push the plane from the deicing and anti-icing position back to its runway position. The tractor has failed because of icy ground.
  Copilot: I'm surprised we couldn't power it out of here.
  Pilot: Well, we could of if he wanted me to pull some reverse.
The copilot is suggesting using the plane's own power to get back into position. The pilot replies that it could be done with reverse thrust. They try, but the attempt fails and in the end a tractor with chains on does the job.
Just before takeoff, the condition of the wings is considered. Given the seating arrangements in the cockpit, each man checks the wing on his own side.
  Pilot: Get your wing now.
  Copilot: D'they get yours? Can you see your wing tip?
  Pilot: I got a little on mine.
  Copilot: This one's got about a quarter to half an inch on it all the way.

We see that the self-deceiver gives an imprecise and diminutive answer concerning a danger, while the copilot gives a precise description of the extent of the danger. The copilot also curses the snow, saying it is "probably the [expletive deleted] snow I've seen."

Seven minutes before takeoff:

Copilot: Boy, this is a losing battle here on trying to deice those things, it gives you a false feeling of security, that's all that does.

Pilot: That, ah, satisfies the Feds.

Copilot: Yeah—As good and crisp as the air is and no heavier than we are I'd . . .

Pilot: Right there is where the icing truck, they oughta have two of them, you pull right.

The pilot and copilot now explore a fantasy together on how the plane should be deiced just before takeoff on the runway. Note that the copilot begins this exchange with an accurate description of their situation; they have a false sense of security. The pilot notes that the arrangement satisfies the higher-ups, but then switches the discussion to the way the system *should* work. This is not without its value and may, indeed, lead to an improved system in the future, but in their immediate situation concentration on the general issue rather diverted attention from the difficulties at hand.

Just before takeoff the copilot asks the pilot for advice on their situation:

Copilot: Slushy runway, do you want me to do anything special for this or just go for it?

Pilot: Unless you got anything special you'd like to do.

Copilot: Unless just takeoff the nose wheel early like a soft field takeoff or something.

The pilot, to whose greater experience the copilot appears repeatedly to defer, has no help to offer on how to take off in these particular circumstances. This makes their final conversation all the more vivid. The copilot is at the controls of the plane. Having failed to give his copilot any advice and having failed to plan in the slightest for difficulty in takeoff, the pilot's only responsibility is to read the instruments and warn the copilot of any problem. Yet it is the copilot who first calls attention to the strange instrument readings. It is the copilot who refers to them three times before the pilot responds to him.

The transcript suggests how easily the disaster could have been averted. Imagine that earlier conversations about the snow on the wings, the heavy weight of the airplane and the slushy conditions underfoot had induced a spirit of caution in both pilots. How easy it would have been for the pilot to say, "Well, this is a somewhat tricky situation. I think we should take off with full speed but watch our instruments carefully, and if we fail to develop sufficient power, I think I should abort the takeoff." Yet the conversation never had a chance to turn in this direction, for every time the copilot approached the subject, the pilot chose either not to respond or to divert attention from the problem they faced. Mechanisms of self-deception, having deprived him of even the most rudimentary advance planning, offer him a quick fix for the disturbing instrument readings and, after the fateful decision is made, a 10-second illusion that he may be able to get the plane into the air safely.

Dr. Aaron Waters, a noted geologist and professor emeritus at the University of California, Santa Cruz, who has been a member of mountain rescue groups, responded to our account as follows (in a letter dated 2/23/82):

A Disturbed Feeling

Your example of the Flight 90 crash, however, left a disturbed feeling about the way you wrote it up. You correctly blame the pilot for the crash, but maybe you do not bring out clearly enough that it was the pilot's complete insensitivity to the copilot's doubts, and to his veiled and timid pleas for help, that was at the root of all this trouble. The pilot, with much more experience, just sat there completely unaware and without any realization that the copilot was desperately asking for friendly advice and professional help. Even if he [the pilot] had gruffly grunted, "If you can't handle it, turn it over to me," such a response would have probably shot enough adrenaline into the copilot so that he would either have flown the mission successfully or aborted it without accident.

From limited experience in mountain rescue work, and considerable experience with dangerous work in abandoned mines, I've found that the people who lead others into trouble are the hale and hearty insensitive jocks trying to show off. They cannot perceive that a companion is so terrified that he is about to "freeze" to the side of a cliff—and for very good reason. And once this has happened the one that led him into it becomes an even worse basket case, and the most difficult one to rescue. I think the copilot "froze" and immediately the pilot "froze" even worse and began talking to the airplane. However, the copilot is also at fault; left to himself he would have called the tower and not flown the mission, but in the presence of his companion he was guilty of self-deception.

The media have concentrated on the icing of the wings, but the master geologist sees a human parallel to the freezing weather. Each man, in turn, "freezes" in fright and the disaster is complete. The most recent evidence on the faulty instrument readings bears out Dr. Water's interpretation. It is now known that the airplane was getting 25 percent *less* thrust than its instrument readings showed! Takeoff consumed almost 17 seconds more time (and a greater length of the runway) than it should have. Had the pilot, in fact, aborted at the go/no go speed, he would have run out of runway.

If the copilot was cold prior to takeoff, the pilot was positively "cool." Nothing fazed him. The situation in which he found himself was nothing new to his industry nor to his company. In the previous September, for example, Air Florida's chief 737 pilot attached a 737 winter-flight note to the monthly Air Florida crew newsletter. He specifically warned of the dangers of winter flying at the more northerly airports. "Nobody can be *too* prepared for La Guardia, O'Hare, White Plains or Washington National." He told crews to look for snow and ice buildup and to arrange for as late an airframe deicing as practical: *"If heavy freezing precipitation exists, it may be necessary to get deiced again if significant ground delays occur"* (emphasis added). Nine airliners taking off before Flight 90 were deiced between 9 and 44 minutes before takeoff, but Flight 90 went 49 minutes between its last deicing and anti-icing and takeoff.

We now see the final deicing discussion between pilot and copilot in a new light. Both pilot and copilot know that their plane needs a second deicing, but instead of seeking it, the pilot leads them into a fantasy world in which they get their second deicing without losing their place in line waiting to take off.

The American Airlines maintenance chief whose men serviced the Air Florida plane said he twice told the pilot that he should wait until just before takeoff for deicing; otherwise, the deicing fluid could cause wet snow to collect, which is precisely what happened. A picture taken of the plane just after deicing shows snow already covering the upper fuselage.

Moving the Aircraft

The problem of snow and ice on the wings may have been compounded by the decision to use the plane's own power to try to move the aircraft back from the gate. This kind of casual incaution is exactly what one would expect from an "airplane jock." Certain types of adventurous men are especially prone to this form of self-deception. (Both pilot, age 34, and copilot, age 31, had been military pilots before turning to commercial work.)

The use of the reverse thrust could have pushed the slush to the leading edge of the wings. This is precisely where ice and snow do the greatest damage. Indeed, in a 1980 bulletin, Boeing, the plane's manufacturer, had already warned against using 737 reversers during snowfalls. If reverse thrust *is* used, Boeing advised, the wing's edges should be cleared of any ice or snow. It can cause the plane's nose to "pitch up" too far at takeoff and roll to the side, threatening a stall. This is what seems to have happened to Flight 90.

A second consequence of using reverse thrust is that it may have caused snow to swirl up and block the sensors that caused the false readings on the amount of the engine thrust and speed of forward movement.

Superimposed on all the detailed information stands one obvious fact.

On the mission in question the copilot was flying the plane. That is, he was playing the role of pilot and the pilot, meanwhile, was playing the role of copilot. This is intended to be educational for the copilot, since he thereby learns how to become a pilot, but the pilot is still in charge. In effect, he is to do two things at once: discharge the duties of copilot while remaining responsible for the flight itself.

Did this confusion of roles contribute to the disaster? We believe it did. Had the pilot been flying the plane that day, we believe the chances for survival would have been better. The copilot shows himself to be a careful man. In this flight, he even discharges some of his customary duties, such as reading the instruments. By contrast, the pilot handles the airplane the way one might handle a horse, by seat-of-the-pants control. The pilot himself might have ignored the instrument readings, heading down the runway at full speed as judged by his own body. In his split role he neither discharges the copilot's role nor assumes full responsibility for the flight. Indeed, he repeatedly seeks to convey to the copilot the message that this is a routine flight, requiring nothing more than the usual self-confidence.

On the Air Florida flight, a natural question is "What were the potential benefits to the copilot of acquiescing to the pilot's self-deception?" To answer this we would have to present detailed information on the way in which copilots are required to relate to their superiors. But we can speculate on the cost to the copilot of becoming known as a "chicken," someone too frightened to take on the role of a pilot when circumstances are adverse.

What are the benefits of the pilot's self-deception? An analogy to fights in nature may be beneficial. When two animals are evenly matched in a fight, each will attempt to convince the other that the fight will go in his favor. At this time, a convincing false front may succeed in frightening away one's opponent. By contrast, when two fighters are poorly matched, a display of bravado by the underdog will carry little weight. Thus, we imagine that presenting a falsely positive front may often have been advantageous to the pilot prior to Flight 90, giving him the illusion that skill plus overconfidence works in all encounters. *Put another way, a pattern of self-deception can become ingrained through many small instances of positive feedback, thereby lulling the self-deceiver into the comfortable illusion that self-deception will always work in his favor.*

Summary

We have tried to show that processes of self-deception, acting primarily in the pilot, contributed directly to the disaster of Flight 90. This pattern included insensitivity to numerous signals from the copilot and a confusion of roles between pilot and copilot. We conclude that the human element of self-deception is the main factor leading up to the disaster. This conclusion has implications for air safety and, by analogy, implications for our understanding of the way in which natural selection acts on processes of self-deception.

Postscript

There were two mildly interesting reactions to our piece on the airplane crash after it was published. One was a call from a lawyer representing Boeing Company. He was interested in the possibility that I could serve as an expert witness in the trials that would arise over assigning blame to the crash. Since our analysis placed some blame on the pilot and copilot, it tended to absolve Boeing of responsibility, or at least lessen any responsibility attributable to the aircraft itself. Unfortunately, I have no knack for the commercial world and I immediately told him something that probably ended my career as an expert witness on the spot: I would be happy to testify but he should understand that in my opinion a disaster like this required a number of things to go wrong simultaneously, including some possible design failures in the airplane. In any case, I never heard from the man again. It was also possible that someone told him who my coauthor was and he decided the whole thing was not worth the effort.

The second reaction was slightly more interesting. I received a letter one day from a Lieutenant Mudd writing from a U.S. Army post in Alabama. He said that they were putting together a manual for pilots called "Cockpit Conversations." He wanted to know if they had our permission to reprint the article and, if so, whether we wished the manual to give a citation to the original publication. I was surprised to learn that Lieutenant Mudd seemed to think that you could reprint something without giving the original citation, but I suppose if you are the U.S. Army you can probably do pretty much as you please. In any case, I wrote back to the lieutenant giving him our permission and, at the same time, asked if he would be kind enough to provide us with a copy. Knowing that the government prefers a one-way flow of information in its direction, I doubted that I would receive a copy of the manual. In any case, I never heard from Lieutenant Mudd again.

Huey Newton was later charged by the State of California with embezzling funds in the process of running an elementary school in Oakland. Among other claims made by the state was the assertion that the educational corporation of which he was head had done no useful educational work during the year 1982. The letter from Lieutenant Mudd suggested other-wise. I wrote the lieutenant asking for confirmation that our article had, in fact, been used as planned and once again asked, if possible, for a copy of the manual itself. As I have mentioned already, I never heard from Lieutenant Mudd again. I do not know whether such an entity actually ever existed.

⟡ ◯ ⟡

The Elements of a Scientific Theory of Self-Deception

ROBERT TRIVERS

Abstract.   An evolutionary theory of self-deception—the active misrepresentation of reality to the conscious mind—suggests that there may be multiple sources of self-deception in our own species, with important interactions between them. Self-deception (along with internal conflict and fragmentation) may serve to improve deception of others; this may include denial of ongoing deception, self-inflation, ego-biased social theory, false narratives of intention, and a conscious mind that operates via denial and projection to create a self-serving world. Self-deception may also result from internal representations of the voices of significant others, including parents, and may come from internal genetic conflict, the most important for our species arising from differentially imprinted maternal and paternal genes. Selection also favors suppressing negative phenotypic traits. Finally, a positive form of self-deception may serve to orient the organism favorably toward the future. Self-deception can be analyzed in groups and is done so here with special attention to its costs.

Introduction

An important component of a mature system of social theory is a sub-theory concerning self-deception (lying to oneself, or biased information flow within an individual, analogous to deception between individuals). This sub-theory can always be turned back on the main theory itself. There can be little doubt about the need for such a theory where our own species is concerned—and of the need for solid, scientific facts which bear on the theory. Whether through a study of one's own behavior and mentation (e.g., for a novelist's treatment[1]) or of societal disasters (e.g., in aviation[2,3] or misguided wars[4,5]), or a review of findings from psychology,[6-13] we know that processes of self-deception—active misrepresentation of reality to the conscious mind—are an everyday human occurrence, that struggling with one's own tendencies toward self-deception is usually a life-long enterprise, and that at the level of societies (as well as individuals) such tendencies can help produce major disasters (e.g., the U.S. war on Viet Nam). With potential costs so great, the question naturally arises: what evolutionary forces *favor* mechanisms of self-deception?

A theory of self-deception based on evolutionary biology requires that we explain how forces of natural selection working on individuals—and the genes within them—may have favored individual (and group) self-deception, where natural selection is understood to favor high inclusive fitness, roughly speaking, an individual's (or gene's)reproductive success (RS = number of surviving offspring) plus effects on the RS of relatives, devalued by the degrees of relatedness between actor and relatives.[14] There is ample evidence that this simple principle provides a firm foundation for a general theory of social interactions.[15] Deception between individuals who are imperfectly related may often be favored when this gives an advantage in RS to the deceiver (see Refs. 15 and 16 for some examples) but the argument for *self*-deception is not so obvious.

For a solitary organism, the prospects seem difficult, if not hopeless. In trying to deal effectively with a complex, changing world, where is the benefit in misrepresenting reality to oneself? Only in interactions with other organisms, especially con-specifics, would several benefits seem to arise. Because deception is easily selected between individuals, it may also generate *self*-deception, the better to hide ongoing deception from detection by others.[2,15,17] In this view, the conscious mind is, in part, a social front, maintained to deceive others—who more readily attend to its manifestations than to those of the actor's unconscious mind. At the same time, social processes, such as parent-offspring conflict[18] in a species with a long period of juvenile dependency—or, a more general group-individual conflict—may generate conflicting internal voices, representing parental and own self-interests (or group and self), with consequent reality-distortion within the individual.[18] For example, the parental view may be overstated internally (for example, via parental manipulation), requiring careful devaluation or counter-assertion.

A stronger force may arise from the fact that different sections of our genome (mtDNA, sex chromosomes, autosomes and, separately, the maternal and paternal chromosomes) often enjoy differing degrees of relatedness to others, with consequent internal conflict between the sections potentially generating deception within the individual, a kind of "selves-deception."[19] Internal conflict may occur for other reasons, as well, and may or may not involve biased information flow. For example, it is certain that all of us possess disadvantageous traits, both genetic and developmental, and, thus, natural selection may have favored super-ordinate mechanisms for spotting negative traits in the phenotype (perhaps especially behavioral ones) and then attempting to suppress them. This may be experienced sometimes as internal psychological conflict and may or may not involve biased information flow. Finally, a positive stance toward life may have intrinsic benefits (and not only for social species). A concentration on the future—and positive outcomes therein—may benefit from seeing past setbacks as blessings in disguise and the current path chosen as the best available option. In short, positive illusions may give intrinsic benefit.[8,9] Is this self-deception or merely optimism in the service of reproductive success?[20,21]

Self-Deception in the Service of Deceit

One model for internal fragmentation and conflict is represented in Figure 1. True and false information is simultaneously stored in an organism with a bias towards the true information's being stored in the unconscious mind, the false in the conscious. And, it is argued, this way of organizing knowledge is oriented towards an outside observer, who sees first the conscious mind and its productions and only later spots true information hidden in the other's unconscious. This is self-deception in the service of deception of others. It may be expected to flourish in at least the following five kinds of situations.

1. Denial of ongoing deception.   Being unconscious of ongoing deception may more deeply hide the deception. Conscious deceivers will often be under the stress that accompanies attempted deception. Evidence from other animals suggests that, as in humans, deception, when detected, may often be met with hostile and aggressive actions by others.[22-25] Thus, if I were in front of you now, lying to you about something you actually care about, you might pay attention to my eye movements, the quality of my voice, and the sweat on the palms of my hands (if you can reach them) as a means of detecting the stress accompanying deception, but if I am unconscious of the deception being perpetrated all these avenues will be unavailable to you.

2. Unconscious modules involving deception.   In the above example, the main activity—verbal persuasion directed at others—is deceptive, but there are also situations in which your dominant activity (say, lecturing) is honest, but a minor activity is deceitful (stealing the chalk). These can be thought of as directed by unconscious modules favored by selection so as to allow us to pursue surreptitiously strategies we would wish to deny to others. Naturally these will often remain unconscious to us. I will shortly describe in detail a deceitful little module in my own life which I have discovered

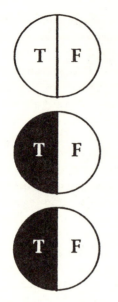

Figure 1.

primarily because my pockets fill up with contraband: hard, concrete objects that others may soon miss. What is the chance that I perform numerous unconscious selfish modules whose social benefits do not pile up in one place, where I can notice them (and others confirm them), e.g., ploys of unconscious manipulation of others (including, of course, as an academic, expropriating their *ideas*)?

I have discovered over the years that I am an unconscious petty thief. I steal small, useful objects: pencils, pens, matches, lighters and other useful objects easy to pocket. I am completely unconscious of this activity while it is happening. I am, of course, now richly aware of it in retrospect, but after at least 40 years of performing the behavior I am still unconscious ahead of time, during the action, and immediately afterwards. Perhaps because the trait is so unconscious, it appears now to have a life of its own and often seems to act directly against even my narrow interests. For example, I steal chalk from myself while lecturing and am left with no chalk with which to lecture (nor do I have a blackboard at home). I steal pencils and pens from my office and, in turn, from my home, so if I download my pockets at either destination, as I commonly do, I risk being without writing implements at the other end. Recently I stole the complete set of keys of a Jamaican school principal off of his desk between us. And so on.

In summary, noteworthy features of this module are that: (1) it is little changed over the course of my life; (2) increasing consciousness of the behavior *after* the behavior has done little or nothing to increase consciousness during or in advance of the behavior; and (3) the behavior seems increasingly to misfire, that is, to fail to steal *useful* objects.

What is the benefit of keeping this petty thievery unconscious? On the one hand, if challenged, I can act surprised and be confident in my assertion that nothing like this was ever my conscious intention (see below). On the other hand, unconsciousness ensures that my thievery will not interfere with ongoing behavior, while the piece of brain devoted to stealing can concentrate on the problem at hand, i.e., snatching the desired item undetected. Part of its consciousness has to be devoted to studying my *own* behavior since integrating its thievery into my other behavior will presumably make this harder to detect by others, including myself.

Incidentally, I believe I never, or almost never, pilfer from someone's office when it is empty. I have seen a choice pen and have seen my hand move toward it but I immediately stop myself and say, "but, Bob, that would be stealing," and I stop. Perhaps if I steal from you in front of your face I unconsciously imagine you have provided some acquiescence, if not actual approval. When I stole the principal's keys, I believe I was simultaneously handing him repayment of a small amount and wondering if I were slightly overpaying. Perhaps I reasoned to myself, "Well, this is for you, so *this* must be for me."

3. *Self-deception as self-promotion.*   Another major source of self-deception has to do with self-promotion, self-exaggeration on the positive side, denial on the negative, all in the name of producing an image that we are "benef-fective," to use Anthony Greenwald's[7] apt term, toward others. That is, we benefit others and are effective when we do it. If you ask high school seniors in the United States to rank themselves on leadership ability, fully 80% say they have better than average abilities, but for true feats of self-deception you can hardly beat the academic profession. When you ask professors to rate themselves, an almost unanimous 94% say they are in the top half of the profession![26] For many other examples, see Refs. 7 and 13. Tricks of the trade are biased memory, biased computation, changing from active to passive voice when changing from describing positive to negative outcomes, and so on.

4. *The construction of biased social theory.*   We all have social theories. We have a theory of our marriages. Husband and wife, for example, may agree that one party is a long-suffering altruist, while the other is hopelessly selfish, but they may disagree over which is which.[15] We each tend to have a theory regarding our employment. Are we an exploited worker, underpaid and underappreciated for value given (and fully justified in minimizing output and stealing company property)? We usually have a theory regarding our larger society as well. Are the wealthy unfairly increasing their own resources at the expense of the rest of us? Does democracy permit us to reas-

sert our power at regular intervals? Is the judicial system systematically biased against our kind of people (African-Americans for example)? The capacity for these kinds of theories presumably evolved in part to detect cheating in our relationships and in the larger system of reciprocal altruism.

Social theory is easily expected to be biased in favor of the speaker. Social theory inevitably embraces a complex array of facts and these may be very partially remembered and very poorly organized, the better to construct a consistent self-serving body of social theory. Contradictions may be far afield and difficult to detect. When Republicans in the House of Representatives bemoaned what the Founding Fathers would have thought had they known that a successor President was having sex with an intern, the Black American comedian Chris Rock replied that the Founding Fathers were not having intercourse with their interns, they were having intercourse with their *slaves*! This kind of undercuts the moral force of the argument given (for recent evidence supporting his assertion, see Ref. 27).

Alexander[17] was, I think, the first person to point out that group selection thinking—the mistaken belief that natural selection favors things that are good for the group or the species—is just the kind of social theory you would expect to be promulgated in a group-living species whose members are concerned to increase each other's group orientation.

5. *Fictitious narratives of intention.*   Just as we can misremember the past in a self-serving way, so we can be unconscious of ongoing motivation, instead experiencing a conscious stream of thoughts which may act, in part, as rationalizations for what we are doing, all of which is immediately available verbally should we be challenged by others: "But I wasn't thinking that at all, I was thinking such-and-such." A common form in myself is that I wish to go to point C, but can not justify the expense and time. I leap, however, at a chance to go to point B, which brings me close enough to point C so that, when there, I can easily justify the extra distance to C, but I do not think of C until I reach B. We may have much deeper patterns of motivation which may remain unconscious, or nearly so, for much longer periods of time, unconscious patterns of motivation in relationships, for example.

In summary, the hallmark of self-deception in the service of deceit is the denial of deception, the unconscious running of selfish and deceitful ploys, the creation of a public persona as an altruist and a person beneffective in the lives of others, the creation of self-serving social theories and biased internal narratives of ongoing behavior which hide true intention. The symptom is a biased system of information flow, with the conscious mind devoted, in part, to constructing a false image and at the same time being unaware of contravening behavior and evidence. The general cost of self-deception, then, is misapprehension of reality, especially social, and an inefficient, fragmented mental system. For a deeper view of these processes we must remember that the mind is not divided into conscious and unconscious, but into differing degrees of consciousness. We can deny reality and then deny the denial, and so on, *ad infinitum*. Consciousness comes in many, many degrees and forms. We can feel anxious and not know why. We can be aware that someone in a group means us no good, but not know who. We can know who, but not why, and so on.

The examples in this article are all taken from human life. While language greatly increases the possibilities for deceit and self-deception in our species, selection probably favored deception in social species for hundreds of millions of years and this may have selected for some mechanisms of self-deception. Two animals evaluating each other in an aggressive encounter (or even in courtship) will be selected to pay close attention to the other individual's apparent self-evaluation and level of motivation, both of which can be boosted by selective forgetting, as in humans.[28] In humans the major sex hormones (e.g., testosterone and estradiol) seem to be positively associated with degree of self-inflation.[28] Since testosterone is sometimes positively associated with aggression and aggression with self-deception (see below) such connection may make functional sense in both humans and other animals[5] (where it could easily be pursued experimentally).

Neuronal Times in Consciousness

It is common to imagine that our conscious mind occupies a central place in our life, where apprehension of reality and subsequent decision-making is concerned. It is easy to imagine that information reaching our brain is immediately registered in consciousness and likewise that signals to initiate activity originate in the conscious mind. Of course, unconscious processes go on at the same time and unconscious processes may affect the conscious mind but there is not a great deal of time, for example, for something like denial to operate, certainly not if this requires spotting a signal and then, before it can reach consciousness, shunting it aside. And, voluntary activity, of which we are conscious as we act, may be affected by unconscious factors, but nevertheless plays the overriding role in directing activity. This is the conventional (pre-Freudian) view.

Thirty years of accumulating evidence from neurophysiology suggests that this is an illusion (Table 1). The first and, perhaps, most startling fact is that while it takes a nervous signal only about 20 ms to reach the brain, it requires a full 500 ms for a signal reaching the brain to register in consciousness! This is all the time in the world, so to speak, for emendations, changes, deletions, and enhancements to occur. Indeed, neurophysiologists have shown that stimuli, at least as late as 100 ms before an occurrence reaches consciousness, can affect the content of the experience.[29] Some additional times are the following.

Table 1. Consciousness (neuronal times)

			Time
Finger	⇒	Brain	20 ms
Round trip			50 ms
Sensation	⇒	Consciousness	500 ms
Round trip	+	Cognitive processing	100–200 ms
Neuronal start of act	⇒	Conscious "intention"	350 ms
"Intention"	⇒	Action	200 ms

It takes only 50 ms for a signal from the finger to cause, via a round-trip to the brain stem, the finger to be moved. Additional cognitive processing may require another 150 ms, but all of this is achieved without consciousness. Finally, what do we make of the following fact? 350 ms *before* we consciously intend to do something the relevant neuronal activity begins and there is a further 200 ms delay after we "intend" to do something before we actually do it. It seems as if our conscious mind is more of an on-looker than a decision-maker.

The Logic of Denial and Projection

Denial and projection are basic psychological processes serving self-deception, though in slightly different ways. Sometimes we will wish to deny something, usually negative (e.g., that we have caused harm to others or "incriminating" personal facts regarding adultery, robbery or something shameful). At other times we may wish to project something onto others which is true of ourselves. In simple voice-recognition choice experiments (see below) denying one variable means choosing (or projecting) the other, and *vice versa*, but the two are distinguished by relevance to self (own voice more important than other). Projection and denial are likely to have different dynamics. Denial will easily engender denial of denial, the deeper to bury the falsehood. Denial may plausibly require a heightened level of arousal, the better to attend quickly to the facts needing denial and shunt them from consciousness. Projection, by contrast, may often be a more relaxed operation: it would be nice if the facts were true, but not critical if they are not.

These speculations are supported by the classic voice-recognition experiments of Gur and Sackeim,[6] where unconscious self-recognition is measured by a relatively large jump in galvanic skin response (GSR). Some people deny their own voices some of the time, while others project their voice some of the time. In each case, the skin (GSR) has it right. Furthermore, when interviewed afterwards, almost all deniers deny their denial, while half of those projecting their voices are conscious after the fact that they sometimes made mistakes of exactly this sort. A comparison of the overall levels of GSRs shows that deniers exhibit the greatest GSRs to all stimuli, while projectors show the more relaxed profile typical of those who make no mistakes, as well as the hopelessly confused (those who both deny and project, sometimes fooling their own skin). Finally, Gur and Sackeim showed that denial and projection were motivated in a logical fashion: individuals made to feel bad about themselves started denying their own voices, while people made to feel better about themselves started projecting their own voices—as if self-presentation was being contracted or expanded according to relevant facts.

A student could go a long way by devising a series of follow-up experiments requiring only a tape recorder and a machine for reading the galvanic skin response. Is denial really associated with greater arousal than projection or correct apprehension of reality? What is happening with those individuals who make both kinds of mistakes—are they really completely confused some of the time? And if so, why? What kinds of voices of yourself do you deny after failure and which kinds do you project when you succeed (or believe you have)? If it is really true, as Douglas and Gibbins[30,31] seem to show, that voices of familiar others evoke a GSR stronger than unfamiliar others, then what kinds of events cause us to deny familiar others? Is denial or projection more likely with increasing testosterone? And so on. Note that there is a large industry in the U.S. devoted to the use of lie-detector tests (which employ GSR as one if their measures), largely in the world of job interviews and job-related thefts. There is a parallel academic literature investigating the lie detector methodology in various settings. It should be easy to integrate the study of self-deception into these studies. Indeed, it may be possible to see under what conditions self-deception decreases detection by others, both those using a lie-detector machine and those not. It would also be possible, in principle, to adapt their methodology to the study of self-deception in animals: birds, for example, may also show greater physiological arousal to their own or close relative's voice than to others and they could be trained to peck when they "thought" they heard their own voice instead of another's.[25]

Denial of personal malfeasance may often strongly necessitate its projection onto others. Once years ago while driving, I took a corner too sharply and my one-year-old baby fell over in the back seat and started to cry. I heard myself harshly berating her nine-year-old sister (my stepdaughter)—as if she should know by now that I like to take my corners on two wheels, and brace the baby accordingly. But the very harshness of my voice served to warn me that something was amiss. Surely the child's responsibility in this matter was, at best, 10%. The remaining 90% belonged to me, but by denying my own role, someone else had to bear a greater burden. That is,

Table 2. Homophobia scale: sample questions[32]

---

1. I would feel comfortable working closely with a male homosexual.
4. If a member of my sex made a sexual advance toward me, I would feel angry.
5. I would feel comfortable knowing that I was attractive to members of my own sex.
12. I would deny to members of my peer group that I had friends who were homosexual.
14. If I saw two men holding hands in public I would feel disgusted.
17. I would feel uncomfortable if I learned that my spouse or partner was attracted to members of his or her own sex.

---

denial of my own responsibility required that responsibility be strongly projected onto someone else, to balance the "responsibility equation."

In a somewhat similar fashion, it has been argued that denying one's own homosexual tendencies will cause one to project these sexual tendencies onto others. It is as if we are aware that there is some homosexual content in the immediate neighborhood and, denying our own portion, we go looking for the missing homosexuality in others. Some striking experimental work has recently been produced in support of this possibility.[12] Fully heterosexual men (no homosexual behavior, no homosexual fantasies) are divided into those that are relatively homophobic and those who are not. Homophobic men are defined as those who are uncomfortable with, fearful of, and hostile toward homosexual men. Homophobia is measured by a series of 25 questions (Table 2). A rough analogue with the GSR was provided by a plethysmograph attached to the base of the penis which measures changes in circumference, while interviews provided information on conscious perception of tumescence and arousal. Of course, we are unconscious of our GSRs, but conscious of changes in penile circumference, at least beyond some threshold, so the analogy is not precise, but the methodology provides results of parallel interest to those of Gur and Sackeim.[6]

When the two groups of men are exposed to four-minute sexual videos (heterosexual, lesbian, and male homosexual), the plethysmograph shows that both sets of men respond with similar levels of arousal to the heterosexual and lesbian videos but that only the *homophobic* men show a significant response to the male homosexual video. Interviews afterwards show that both categories of men give accurate estimates of their degree of tumescence and arousal to all stimuli with one exception: the homophobic men deny their response to the male homosexual video!

The results make a certain kind of superficial sense. Those heterosexual males who are, according to their own account, fully heterosexual in behavior and in fantasy yet who will actually experience arousal to the sight of two men making love would be expected to be more uncomfortable in the vicinity of homosexual men. These men, after all, represent continual possible sources of arousal for the man's latent homosexual affect. Discomfort around homosexuals and disgust and anger at them may be expected to be larger where homosexual threat is greater. Note again a dynamic between denial and projection. Denying their own homosexual feelings may force the individual to project a greater danger of those same tendencies onto others.

Ramachandran[10,11] has recently produced very striking evidence that processes of denial—and subsequent rationalization—appear to reside preferentially in the left brain. People with a stroke on the right side of the body (damage to the left brain) never or very rarely deny their condition, while a certain, small percentage of those with left-side paralysis deny their stroke and, when confronted with strong counter-evidence, indulge in a remarkable array of rationalizations denying the *cause* of their inability to move (arthritis, a general lethargy, etc.). This is consistent with other evidence that the right hemisphere is more emotionally honest, while the left hemisphere is actively engaged in self-promotion. It goes without saying that we need much more evidence on the underlying physiology, neurobiology and anatomy of mechanisms of self-deception.

Interaction with Other Biological Systems

Internal conflict and biased information flow within the individual probably have multiple biological sources, self-deception evolving in the service of deceit being only one. The alternative sources are taken up here with particular attention to their interactions.

1. Parent-offspring conflict.   As is now well recognized, parents and offspring are expected to be in conflict in any outbred species since each will be related to self by 1 but to the other by only 1/2. This 1/2 degree of relatedness leads to a strong overlap in self-interest, but also an imperfect one, giving scope to various kinds of conflict.[18] Especially important in our own species is the fact that parent-offspring conflict extends to the behavioral tendencies of the offspring with the parent being selected to mold a more altruistic and less selfish offspring—at least as these behaviors affect other relatives—than the offspring is expected to act on its own. On the assumption that an internal representation of the parental voice is valuable to the child when their interests overlap closely, it can easily be imagined that selection has accentuated the parental voice in the offspring to benefit the parent and that some conflict is expected within an individual between its own self-interest and the internal representation of its parents' view of its self-interest.

It is easy to imagine that mechanisms of deceit and self-deception could be parasitized in this interaction. For example, low parental investment may coexist with exaggerated displays of parental affections, the latter serving as cover for the former. The offspring may be tempted to go along with the parental show since resistance and malaffection may lead to even less investment. Yes, mommy loves you and you love mommy too. But one can easily imagine that having to adopt this self-deception as one's own may have long-term negative consequences and may lead to later internal psychological conflict when you are no longer under your parent's immediate control (via investment).

A good clinical example of this is provided by Dori LeCroy.[33] A thirty-year-old woman arrived for therapy appointments in a hesitant and apprehensive manner which (when challenged) she explained by her desire to avoid intruding on another's "personal space." In a wispy, vacant style she described herself as a loving and "spiritual" person who put special value on kindness, tolerance, and forgiveness. She related events in her life with an emphasis on ill-treatment by friends and relatives, including physical abuse as a child from her alcoholic mother, but the complaints were accompanied by rationalizations which absolved others of blame (her mother was really "a beautiful person" with troubles of her own, for example). Most notably, she displayed no anger, no outrage, no desire for revenge. Instead, she worried about the well-being of the perpetrators! LeCroy speculates that abuse suffered as a child led to overidentification with the abuser: "Self-deception of his kind would have enabled her to behave devotedly as abused children frequently do, and thereby solicit nurture."[33] It is important to note that the woman had not forgotten the facts regarding the past, indeed she volunteered them, but she had apparently transmuted her anger and resentment into oversolicitous indulgence, first for her mother and then for others. To reconcile the facts of the matter—that the maternal abuse was just fine— she had to agree to a negative self-image: she became bad and her mother good. Recently her mother has come around and now provides some real investment, but the patient herself still seems saddled with an imposed self-deception going to the heart of her identity. We are not attracted to people with a negative self-image, too timid to intrude and displaying an otherworldly attachment to altruism, even in the face of mistreatment!

2. *Internal genetic conflict.*   A stronger potential source of internal conflict and biased information flow within the individual is internal *genetic* conflict, due to differing degrees of relatedness to others enjoyed by different parts of an individual's genome; these in turn are due to different rules of inheritance. For example, mitochondrial DNA (mtDNA) is passed only mother to offspring, while, of course, the autosomes are passed from each parent. One kind of conflict this can set up is over inbreeding. Autosomes enjoy an increased relatedness to offspring when they practice inbreeding, but this is not true for mtDNA, which is always related to progeny on the maternal side by 1. Put another way, since the mtDNA in any given individual is only coming from one parent, it does not increase relatedness for that parent to

be related to the second parent. An autosome deciding whether inbreeding would be advantageous has to set against the increase in relatedness a decrease in quality of the offspring due to inbreeding depression. But the mtDNA will only see the inbreeding depression; thus as long as there is any inbreeding depression (and there often is), mtDNA will oppose inbreeding that the autosomes may favor.

There is no evidence regarding such interactions in animals, but there is striking evidence of exactly this kind of conflict in plants (for a good review of the relevant theory, see ref. 34). Since most plants are hermaphrodites they can, in principle, practice "selfing" where the pollen and the ovules come from the same plant: this raises degree of relatedness to offspring from 1/2 to 1. About a 1/2 to 3/4 of all flowering plants are capable of selfing and in these species (but only rarely in obligately non-selfing species) one finds a most interesting conflict: mtDNA causes abortion of the male function or sterile pollen (cytoplasmic male sterility), while the nuclear DNA often acts to re-establish male function. This is exactly consistent with mtDNA's always opposing inbreeding when there is an outbred alternative.

The kind of conflict I have been describing pits a small part of the genotype against almost all the other genes: for this reason, such conflict is expected to be infrequent and resolved usually in favor of the dominant set of genes. A more important kind of internal genetic conflict for our own species pits one-half of the genotype against the other half. I refer to the phenomenon of genomic imprinting or parent-specific-gene expression (reviewed in Ref. 35). A small number of genes in us have the property that they are expressed only when inherited from one sex, the copy inherited from the opposite sex being silenced (or sometimes there is only a quantitative difference in gene expression depending on parent-of-origin).

The importance of genomic imprinting is that it allows imprinted genes to act on the basis of exact degrees of relatedness to each parent. This inevitably leads to conflict between paternal and maternal genes.[36] Possible psychological conflicts arising from imprinting are easy to describe.[37] Consider, for example, a contemplated act of inbreeding with your mother's sister's offspring. You are related on the maternal side and will thus enjoy an increase in relatedness to any resulting offspring by inbreeding on the maternal side, but the paternal genes will enjoy no increase in relatedness though they will suffer any inbreeding depression associated with the inbreeding. We can imagine your maternally active genes urging you to consider the inbreeding while your paternally active genes might take a moralistic posture and emphasize the biological defects thereby generated. Whether mtDNA has also been selected to decrease inbreeding, as in plants, is as yet unknown.

There is very intriguing indirect evidence suggesting that parts of the body may differ in the degree to which they express maternally active versus paternally active genes (reviewed in ref. 19). In mice chimeras which consist of a mixture of normal cells with cells that have either a double dose of maternal genes (and no paternal ones) or a double dose of paternal genes (and no maternal ones), it turns out that the two added kinds of cells survive and proliferate differentially according to tissue: thus, doubly maternal cells do well in the neocortex of the brain but do not survive and proliferate in the hypothalamus and vice versa for doubly paternal cells. By similar logic the tissue producing dentin appears to be more maternally active, while the tissue producing enamel is more paternally active. Thus, it is possible that there are conflicts at the level of tissues in which one can also imagine *selves*-deception, that is, deceitful signals sent out from one tissue, over-emphasizing one parent's interests whose signals are devalued by another tissue, overemphasizing the opposite sexed parent's interests. Where maternal kin are much more frequent in the social group than paternal kin, maternally active tissue in the neocortex may say, in effect, "Family is important, I like family, I believe in investing in family" while the hypothalamus may reply, "I'm hungry!"[37]

We can imagine interactions between genomic imprinting and other systems we have been discussing. For example, parental indoctrination will work better when it interacts with the appropriate imprinted genes: maternal manipulation with maternally active genes in the progeny and paternal manipulation with paternally active genes.[19] At the same time it is easy to imagine that mechanisms useful in self-deception to deceive others may prove useful in within-individual conflict. If selfish impulses are kept unconscious, the better to hide them from others, and they may also stay unconscious, the better not to be spotted by oppositely imprinted genes.

3. *Selection to suppress negative traits.* Everyone can expect to have some negative traits that are stuck in the phenotype either through misdevelopment or through genetic defect, and these are likely to have been such a regular part of our existence for so long that we may well wonder whether selection has not favored a mechanism which searches for such negative traits and attempts to suppress them. All genes in the individual would be in agreement with such a program, including the defective gene. Mutation will inevitably supply some negative traits,[38] but it is well to be aware of the fact that even in the absence of such a supply some selective factors by themselves generate some negative traits. For example, sex antagonistic genes are those that have opposite effects on reproductive success when found in the two sexes. As long as the net effect is positive, the gene will be favored, even though, when found in the opposite sex, it has negative effects on lifetime reproductive success.

William Rice's[39–41] beautiful experiments on *Drosophila* demonstrate clearly that sex antagonistic genes are a regular part of the *Drosophila* genome and by extension are expected in all sexually reproducing species that are not perfectly, life-long monogamous. This means that each sex is a partial compromise between the two sexes and contains numerous traits disadvantageous to that sex (but advantageous in the opposite sex).

Naturally, if a mechanism for suppressing negative traits does exist, one may well expect internal conflict, forces acting to maintain the negative trait being opposed by efforts at suppression. There is no selection to increase the resistance, but as suppression is selected to become more effective, more negative genes will remain in the genotype because the suppression has reduced or eliminated the cost. It is easy to imagine an interaction between this mechanism and parent-offspring conflict, since parents may help you locate—and encourage you to suppress—such negative traits, but due to imperfect overlap in self-interest, they may encourage you to think a trait negative to yourself when it is in reality only negative to themselves. Similarly, it is conceivable that paternally active genes (for example) may attempt to suppress maternally active ones (or vice versa) by pretending that it is an organism-wide negative phenotypic trait that needs to be suppressed.

Prayer and meditation are two widespread examples of people wrestling with their phenotypes, some of which may have been favored by selection to suppress negative phenotypic traits, including the negative phenotypic trait of self-deception! Many famous passages from the world's great religions, as well as rituals of prayer and meditation, are directed against self-deception, as in this loose translation of Matthew 7:1–5 in the New Testament of the Bible: "Judge not that ye be not judged, for you are projecting your faults onto others; get rid of your own self-deception first, then you will have a chance of seeing others objectively."

4. *Positive illusions?*   Another important possibility is that self-deception has intrinsic benefit for the organism performing it, quite independent of any improved ability to fool others. In the past twenty years an important literature has grown up[8,9] which appears to demonstrate that there are intrinsic benefits to having a higher perceived ability to affect an outcome, a higher self-perception, and a more optimistic view of the future than facts would seem to justify. It has been known for some time that depressed individuals tend not to go in for the routine kinds of self-inflation that we have described above. This is sometimes interpreted to mean that we would all be depressed if we viewed reality accurately, while it seems more likely that the depressed state may be a time of personal re-evaluation, where self-inflation would serve no useful purpose. While considering alternative actions, people evaluate them more rationally than when they have settled on one option, at which time they practice a mild form of self-deception in which they rationalize their choice as the best possible, imagine themselves to have more control over future events than they do, and see more positive outcomes than seem justified. What seems clear is that they gain direct benefits of functioning from these actions.[42] Life is intrinsically future-

oriented and mental operations that keep a positive future orientation at the forefront result in better future outcomes (though perhaps not as good as those projected). The existence of the placebo effect is another example of this principle (though it requires the cooperation of another person ostensibly dispensing medicine). It would be very valuable to integrate our understanding of this kind of positive self-deception into the larger framework of self-deception we have been describing.

Self-Deception and Human Disasters

There can be little doubt that self-deception makes a disproportionate contribution to human disasters, especially in the form of misguided social policies, wars being perhaps the most costly example. This is part of the large downside to human self-deception. Since the general cost of self-deception is the misapprehension of reality, especially social reality, self-deception may easily generate large social costs (everyone on the airplane dies, the entire nation is devastated by a war some of its members started).

Disasters are, of course, studied in retrospect so the evidence is not yet scientific for the connection to self-deception, but it is certainly suggestive. In the following examples, we also see how analysis of individual self-deception can easily be extended to groups: pairs of individuals, an organization and an entire society.

Two-party self-deception.   Trivers and Newton's[2] analysis of the crash of Air Florida's Flight 90 suggests that the pilot was practicing self-deception and the co-pilot acquiesced. The first clue comes from the cockpit conversation during take-off (Table 3). The co-pilot was flying the airplane, yet it was he who noticed contradictory information from the instrument panel and repeatedly spoke while it mattered (i.e., while they could still safely abort the flight). The pilot spoke only once, offering a false rationalization for the disturbing instrument readings. Only when it was too late—they were in the air—did the pilot start talking, while the co-pilot fell silent. An analysis of their conversation prior to take-off showed a consistent pattern of reality denial by the pilot (Table 4). His casual approach to reality, coupled with overconfidence, may have served him well in many minor situations, but proved fatal when real danger required close attention to reality, including the psychological state of his co-pilot.

When an organization practices deception toward the larger society, this may induce organizational self-deception.   Richard Feynman[3] analyzed the cause of the Challenger disaster and concluded that NASA's deceptive posture toward U.S. society had bred organizational self-deception. When NASA was given the assignment and the funds to travel to the moon in the 1960s, the society for better or worse, gave full support to the project: Beat the

Table 3. Crash of Air Florida flight 90[2]

	Co-pilot	Pilot
During take-off	Speaks	Silent[a]
After lift-off	Silent	Speaks

[a]Except for one rationalization.

Soviets to the moon (Table 5). As a result, NASA could design the moon vehicle in a rational way. The vehicle was designed from the bottom up, with multiple alternatives tried at each step, permitting maximum flexibility as the spacecraft was developed. Once the U.S. reached the moon, NASA was a five-billion-dollar bureaucracy with no work to do. Its subsequent history, Feynman argued, was dominated by a need to generate funds, and critical design features, such as manned flight versus unmanned flight, were chosen precisely because they were costly. In addition, manned flight had glamour appeal, which would generate enthusiasm for the funding. At the same time it was necessary to sell this project to Congress and the American people. The very concept of a reusable vehicle—the so-called Shuttle—was designed to appear inexpensive, while in fact it was very costly (more expensive, it turned out, than using brand new devices each time).

Means and concepts were chosen for their ability to generate cash-flow and the apparatus was then designed top-down. This had the unfortunate effect that when a problem surfaced, such as had with the O-rings, there was little parallel exploration or knowledge to solve the problem. Thus NASA chose to minimize the problem and the unit within NASA that was consigned to deal with safety became an agent of rationalization and denial, instead of one of rational study of safety factors.

Some of the most extraordinary mental gyrations in service of institutional self-deception occurred within the Safety Unit. Seven of twenty-four Challenger flights had shown O-ring damage. Feynman showed that if you merely plotted chance of damage as a function of temperature at time of take-off you got a significant negative relationship: lower temperature meant a higher chance of O-ring damage. To prevent themselves from seeing this, the Safety Unit performed the following mental operations. They said that

Table 4. Conversations during taxiing prior to take-off[2]

Co-pilot	Pilot
• Detailed description of snow on wings	• Diminutive description of snow on wings
• Calls attention to danger they face (too long since de-icing)	• Deflects attention to ideal world (de-icing machine on runway)
• Asks for advice on take-off	• Tells him to do what he wants

Table 5. Feynman's analysis of NASA's shift to self-deception[3]

1960s	Aim:	• Go to moon
		• No conflict with larger society
		• No internal conflict re facts
		• Built from bottom up
	Result:	• Success
1970s	Aim:	• Employ a $5 billion bureaucracy
		• Need to convince larger society—repeated manned flight via shuttle
		• Bottom splits from top, which does not wish to know true facts re safety
		• Built from top down
	Result:	• Challenger disaster

seventeen flights showed no damage and were thus irrelevant and could be excluded from further analysis. Since some of the cases of damage occurred during high-temperature take-offs, temperature at take-off could be ruled out as a causative agent. One of the O-rings had been eaten one-third of the way through. Had it been eaten all the way through, the flight would have blown up, as did the Challenger. But NASA cited this case of one-third damage as a virtue. They claimed to have built in a "threefold safety factor"! This is a very unconventional use of language. By law you must build an elevator strong enough so that the cable can support a full load with no damage. Then you must make it eleven times stronger. This is called an eleven-fold safety factor. NASA has the elevator hanging by a thread and calls it a virtue. They even used circular argumentation with a remarkably short radius: since manned flight had to be much safer than unmanned flight, it perforce was. In short, in service of the larger institutional deceit and self-deception, the Safety Unit was thoroughly corrupted to serve propaganda ends, that is, to create the appearance of safety where there was none.

There is thus a close analogy between self-deception within an individual and self-deception within an organization, both serving to deceive others. In neither case is information completely destroyed (all 12 engineers at Thiokol, which built the O-ring, voted against flight that morning). It is merely relegated to portions of the person or the organization that are inaccessible to consciousness (we can think of the people running NASA as the conscious part of the organization). In both cases the entity's relationship to others determines its internal structure of information. In a nondeceitful relationship information can be stored logically and coherently. In a deceitful relationship information will be stored in a biased manner the better to deceive others—but with serious potential costs. Note, however, that it is the astronauts who suffered the ultimate cost, while the upper echelons of NASA—indeed, the entire organization minus the dead—may have enjoyed a net benefit (in employment, for example) from their casual and self-deceived approach to safety.

Self-deception is especially likely in warfare.   Richard Wrangham has recently extended the analysis of self-deception to human warfare in a most revealing way.[5] Evolutionary logic suggests that self-deception is apt to be especially costly in interactions with outsiders, members of another group. In interactions with group members, self-deception will be inhibited by two forces: a partial overlap in self-interest gives greater weight to the opinion of others and within-group feedback provides a partial corrective to personal self-deception. In interactions between groups, every day processes of self-enhancement are uninhibited by negative feedback from others, nor by concern for their welfare, while derogation of the outsiders' moral worth, physical strength, and bravery is likewise unchecked by feedback and shared self-interest. These result in faulty mechanisms of assessment, and aggression will be more likely where each partner is biased in an unrealistic direction in self- and other-assessment, making conflicts more likely to occur and contests therefore more costly, on average, without any average gain in benefits.[5] Derogation of the moral status of your enemies only makes you underestimate their motivation (consider U.S. assessment of the Vietnamese). For an excellent analysis of this phenomenon, as applied to the Old Testament of the Bible, see Hartung.[43]

Processes of group self-deception only make matters worse: Within each group individuals are misoriented in the same direction, easily reinforcing each other and absence of contrary views is taken as confirming evidence (even silence is misinterpreted as support).[5] Tuchman[4] has frightening stories to tell of an individual leader and his cohorts whipping themselves into a frenzy of self-deception prior to launching an ill-advised, indeed disastrous, attack on neighbors.

Military incompetence—losing while expecting to win—is accompanied by four common symptoms: overconfidence, underestimation of the neighbor, ignoring intelligence reports, and wastage of manpower.[5] The latter two are noteworthy. The logic of self-deception preserves conscious illusion by becoming unconscious of contrary evidence, even when provided by one's own agents, whose very purpose it is to provide accurate information. Note in the Challenger disaster how the unit assigned to consider safety ended up being subverted to rationalize unsafety, even though its ostensible purpose was to view the matter objectively.[3] Wastage of manpower is a direct cost of self-deception since forces are deployed along illusory lines of attack, instead of rationally calibrated toward the real situation.

Wrangham makes an important distinction between raids and battles.[5] Lethal raids are attacks on a few neighbors, with numerical superiority being a key stimulus to attack. Raids have a long evolutionary history (chimpanzee males practice lethal raids)[44] and opportunities for self-deception are minimized by the ease of rational assessment (e.g., evidence of numerical superiority). Battles are set pieces between large opposing armies. They are a recent invention (within historical times, more or less), rational assessment is much more difficult, and a long evolutionary history of derogating others[13] makes misassessments especially likely. In short, we should be especially vigilant in guarding against self-deception when contemplating warfare.

Conclusion

Self-deception appears to be a universal human trait which touches our lives at all levels—from our innermost thoughts to the chance that we will be annihilated together in warfare. It affects the relative development of intellectual disciplines (the more social the content, the less developed the discipline: contrast physics and sociology) as well as the relative degree of consciousness of individuals (generally, more self-deceived, less conscious). An evolutionary analysis suggests that the root cause is social, including selection to deceive others, selection on others to manipulate and deceive oneself, and selection on competing sections of one's own genotype. There are undoubtedly complex and important interactions between these (and other) kinds of self-deception. The relevant evidence stretches from personal anecdote to historical analysis, but we especially need more biological evidence on the genetics, endocrinology, physiology and neuroanatomy of self-deception and we need to integrate very disparate findings from experimental, social and clinical psychology into the evolutionary analysis. We also need a detailed theory for the evolution of deception (many elements exist already) and a theory of consciousness based on our understanding of self-deception. Evolutionary theory promises to provide a firm foundation for a science of self-deception, which should eventually be able to predict both the circumstances expected to induce greater self-deception and the particular forms of self-deception being induced.

ACKNOWLEDGMENTS

I am grateful to Drs. Helena Cronin, Dori LeCroy, and Peter Moller for encouraging me to publish this article, to Mr. John Martin (Rockford, Illinois), and to the Ann and Gordon Getty Foundation for generous financial support. I am also thankful to Drs. David Haig, Dori LeCroy, David Smith, and especially Richard Wrangham for a series of detailed and valuable comments on the manuscript.

REFERENCES

1. McEwan, I. 1997. Enduring Love. Jonathan Cape. London.
2. Trivers, R. L. & H. P. Newton. 1982. The crash of Flight 90: doomed by self-deception? Science Digest (November): 66, 67 and 111.

3. Feynman, R. 1988. What Do You Care What Other People Think? Further Adventures of a Curious Character. Norton. New York.

4. Tuchman, B. 1988. The March of Folly: From Troy to Viet Nam. Ballantine, New York.

5. Wrangham, R. 1999. Is military incompetence adaptive? Evol. Hum. Behav. **20:** 3–12.

6. Gur, R. & H., A. Sackeim. 1979. Self-deception: a concept in search of a phenomenon. J. Pers. Soc. Psychol. **37:** 147–169.

7. Greenwald, A. G. 1980. The totalitarian ego: fabrication and revision of personal history. Am. Psychol. **35:** 603–618.

8. Taylor, S. E. & D. A. Armor. 1996. Positive illusions and coping with adversity. J. Pers. **64:** 873–898.

9. Taylor, S. E. 1998. Positive illusions. *In* Encyclopedia of Mental Health, Vol. 3. H. S. Friedman, Ed.: 199–208. Academic Press, San Diego, CA.

10. Ramachandran, V. & D. Rogers-Ramachandran. 1996. Denial of disabilities in anosognosia. Nature **382:** 501.

11. Ramachandran, V. 1997. The evolutionary biology of self-deception, laughter, dreaming and depression: some clues from anosognosia. Med. Hypotheses **47:** 347–362.

12. Adams, H. E., L. W. Wright, Jr & B. A. Lohr. 1996. Is homophobia associated with homosexual arousal? J. Abnorm. Psychol. **105:** 440–445.

13. Krebs, D. L. & Denton, K. 1997. Social illusions and self-deception: the evolution of biases in person perception. *In* Evolutionary Social Psychology. J. A. Simpson & D. T. Kenrick, Eds.: 21–48. Erlbaum Associates. Mahwah, NJ.

14. Hamilton, W. D. 1994. The genetical evolution of social behaviour. I, II. J. Theor. Biol. **7:** 1–52.

15. Trivers, R. 1985. Social Evolution. Benjamin/Cummings. Menlo Park, CA.

16. Krebs, J. R. & R. Dawkins. 1984. Animal signals: mindreading and manipulation. *In* Behavioural Ecology, 2nd ed. J. R. Krebs & N. B. Davies, Eds.: 380–402. Sinauer. Sunderland, MA.

17. Alexander, R. D. 1979. Darwinism and Human Affairs. University of Washington Press: Seattle, WA.

18. Trivers, R. L. 1974. Parent-offspring conflict. Am. Zool. **14:** 249–264.

19. Trivers, R. & A. Burt. 1999. Kinship and genomic imprinting. *In* Genomic Imprinting. An Interdisciplinary Approach. R. Ohlsson, Ed.: 1–23. Springer. Heidelberg, Germany.

20. Lewis, M. 1997. Altering Fate: Why the Past Does Not Predict the Future. Guilford Press. New York, NY.

21. Tiger, L. 1979. Optimism: The Biology of Hope. Simon and Schuster. New York.

22. Rohwer, S. 1977. Status signalling in Harris sparrows: some experiments in deception. Behaviour **61:** 107–129.

23. Rohwer, S. & F. A. Rohwer. 1978. Status signalling in Harris sparrows: experimental deception achieved. Anim. Behav. **26:** 1012–1022.

24. Møller, A. P. & J. P. Swaddle. 1987. Social control of deception among status signalling house sparrows *Passer domesticus*. Behav. Ecol. Sociobiol. **20:** 307–311.

25. Trivers, R. 1991. Deceit and self-deception: the relationship between communication and consciousness. *In:* Man and Beast Revisited. M. Robinson & L. Tiger, Eds.: 175–191. Smithsonian. Washington, DC.

26. Mele, D. 1997. Real self-deception. Behav. Brain Sci. **20:** 91–136.

27. Foster E, M. Jobling, P. Taylor, P. Donnelly, P. de Knijff, R. Mieremet, T. Zerjal & C. Tyler-Smith. 1998. Jefferson fathered slave's last child. Nature **396:** 27–28.

28. Cashdan, E. 1995. Hormones, sex and status in women. Horm. Behav. **29:** 354–366.

29. Libet, B. 1996. Neuronal time factors in conscious and unconscious mental functions. *In* Toward a Science of Consciousness: The First Tucson Discussion and Debates. S. R. Hameroff, A. W. Kaszniak & A. Scott, Eds.: 337–347. MIT Press. Cambridge, MA.

30. Douglas, W. & K. Gibbins. 1983. Inadequacy of voice recognition as a demonstration of self-deception. J. Pers. Soc. Psychol. **44:** 589–592.

31. Sackeim, H. A. & R. C. Gur. 1985. Voice recognition and the ontological status of self-deception. J. Pers. Soc. Psychol. **48:** 213–215.

32. Hudson, W. W. & W. A. Ricketts. 1980. A strategy for the measurement of homophobia. J. Homosexuality **5:** 356–371.

33. LeCroy, D. 1998. Darwin in the clinic: an evolutionary perspective on psychodynamics found in a single case study. Ascap **11:** 6–12.

34. Frank, S. 1989. The evolutionary dynamics of cytoplasmic male sterility. Am. Naturalist. **133:** 345–376.

35. Ohlsson, R. Ed. 1999. Genomic Imprinting. An Interdisciplinary Approach. Springer. Heidelberg.

36. Haig, D. 1997. Parental antagonism, relatedness asymmetries, and genomic imprinting. Proc. Roy. Soc. Lond. B. **264:** 1657–1662.

37. Trivers, R. 1997. Genetic Basis of intra-psychic conflict. *In* Uniting Psychology and Biology: Integrative Perspectives on Human Development. N. Segal, G. E. Weisfeld & C. C. Weisfeld, Eds.: 385–395. Am. Psychol. Assoc. Washington, DC.

38. Eyre-Walker, A. & P. D. Keightley. 1999. High genomic deleterious mutation rates in hominids. Nature **397:** 344–347.

39. Rice, W. 1992. Sexually antagonistic genes: experimental evidence. Science **256:** 1436–1439.

40. Rice, W. 1998. Male fitness increases when females are eliminated from gene pool: implications for the Y chromosome. Proc. Natl. Acad. Sci. USA **95:** 6217–6221.

41. Holland, B. & W. R. Rice. 1999. Experimental removal of sexual selection reverses intersexual antagonistic coevolution and removes a reproductive load. Proc. Natl. Acad. Sci. USA **96:** 5083–5088.

42. Krebs, D., K. Denton & N. C. Higgins. 1988. On the evolution of self-knowledge and self-deception. *In* Sociobiological Perspectives on Human Development. K. B. MacDonald, Ed. Springer. New York.

43. Hartung, J. 1995. Love thy neighbor: the evolution of in-group morality. Skeptic **3:** 86–99.

44. Wrangham, R. & D. Peterson. 1996. Demonic Males: Apes and the Origins of Human Violence. Houghton Mifflin. Boston, MA.

Postscript

In 1985 I reviewed deceit and self-deception in chapter 16 of my book and in 1991 I published a casual essay on the topic, notable mostly for suggesting a novel way of communicating with animals (Trivers 1991). The preceding paper is my best understanding of self-deception to date. It grew out of a conference organized at Hunter College in New York by Peter Marler and Dori LeCroy. As the paper suggests, there is an enormous amount of work waiting to be done on all aspects of self-deception, from its neural substrate, to an integration of the psychological findings on the subject, to the development of a deeper theory. People from a wide range of disciplines can contribute. For an interesting study of lie detection and deception, see Etcoff et al. (2000).

9

GENOMIC IMPRINTING

Genomic imprinting refers to a surprising discovery in the 1980s that in mammals (and flowering plants, such as corn) some genes in an individual have their pattern of expression affected by the sex of the parent that contributed the gene. Typically, there are paternally active genes, in which the copy from the father is active while the copy from the mother is inactive, and conversely, there are maternally active genes, in which the paternal copy is silenced. This difference does not reside in the DNA but is some kind of extragenetic piece of information (involving, e.g., how methylated the genes are). I was aware of this new discovery by the late 1980s and felt intuitively that it must have something to do with intragenomic conflict, a topic I was then beginning to work on. Intragenomic conflict refers to situations in which the genes within an individual are being selected in different directions (due, e.g., to differing degrees of relatedness to others, e.g., Y chromosomes vs. mtDNA vs. autosomes). But I could not see a solution. Then in late 1991 Jon Seger sent me a two-page paper from *Cell* written by David Haig and Chris Graham that solved the problem in one stroke and provided a very striking fact in its favor. Haig and Graham (1991) pointed out that genomic imprinting had immediate and strong effects on degrees of relatedness.

An unimprinted gene, that is, a gene whose activity was identical no matter which parent donated it, was a gene without information on parental origin. Thus, when calculating degree of relatedness of offspring to parent for any given gene, one begins by saying there is a ½ chance it came from the mother and a ½ chance it came from the father. But an imprinted gene had exact information on where it came from: a paternally active gene in an individual came from the father; its silent homolog must have come from the mother. In modeling the effects of the active allele, we see that if it acts as if it assumes that it came from the father with certainty and is hence unrelated to the mother and relatives related only through her, then it will spread. The opposite is true of a maternally active gene. Given that sibships in mice (and many other species) inevitably consist of half-siblings related through different fathers as well as full siblings related through both parents, paternally active genes are expected to be associated with more selfish growth patterns and maternally active genes to have opposite effects. Haig and Graham then drew attention to the first two imprinted genes discovered from mice, *Igf2* and *Igf2r*. *Igf2*, or *insulinlike growth factor 2*, is a paternally active gene and is associated with an increase in size at birth. *Igf2r*, *insulinlike growth factor 2 receptor*, binds to *Igf2* and carries it into lysosomes, where it is degraded. It reduces size of offspring at birth, and it is maternally active! This was the most exciting development in kinship theory since Hamilton's (1964) original work. It revealed a large potential conflict early in development with one half of the genotype pitted against the other, and it promised the possibility of later conflict as well, in adult behavior, for example, when behavior is preferentially directed toward maternal or paternal kin.

I Meet David Haig

It was spring of 1993 and I was eagerly anticipating my first meeting with David Haig. He was then a Junior Fellow at Harvard when I visited to give a talk. He did not disappoint. We spent nearly nine hours together the first day, and my joke later was that if I only could have written fast enough I would have finished my half of my current book (coauthored with Austin Burt). David Haig knows genetics. At one point he said, "You know, Bob, most biologists do not even know how meiosis and mitosis work." I raised my hand. He then volunteered to teach me meiosis and mitosis, and later in my room, with his long fingers twirling around, he created in front of me the three-dimensional cell undergoing mitosis and then meiosis.

The word "egghead" was invented for someone like David, to capture both the beautiful egglike shape of his head, bald on top, orange hair elsewhere, and also his completely intellectual approach to things. He enjoys thinking and he is very good at it. David had all but figured out genomic imprinting before it was discovered. Deriving degree of relatedness in plants, he was used to keeping these values separate for maternal and paternal genes, since in many plants there is a haploid gametophytic phase before sex cells are united in a diploid genome. Thus, when he learned of genomic imprinting, he said to himself, "I have a theory for that!" This has given him a rather unfair advantage in the interpretation of imprinting, and he has used this advantage very effectively. What we have seen is a situation in which almost every new molecular study is soon greeted by an appropriate evolutionary explanation, almost always supplied by David himself. A recent review of the kinship approach to imprinting bears this out (Trivers and Burt 1999).

Austin Burt Looms Up Out of Nowhere

When I finished my textbook in 1985, I cast around for something new to do and it occurred to me that there was an area of genetics which was in many ways analogous to social behavior in that some genes within individuals were seen to be in conflict with others over representation in the next generation and also over preferred actions toward relatives. Transmission ratio distortion encompasses a very wide range of phenomena, from classic meiotic drive elements to B chromosomes, paternal genome loss systems, DNA transposons, retrotransposons, and homing endonucleases, all of which usually increase in frequency solely because they give themselves a selfish advantage in propagating themselves into the next generation. The resulting intragenomic conflict, I felt certain, was likely to have profound implications for genetic processes (such as meiosis and recombination), as well as the genome itself.

Although at first I naively assumed I could whip the literature into shape in two or three years, it soon became clear that the subject was far too vast and complicated for me to handle on my own. Fortunately, since 1992 I have had the good fortune of working on the subject with Austin Burt, a brilliant young Canadian biologist now working at Imperial College London. We are finishing a major book on *Selfish Genetic Elements* (Harvard University Press), the first review of this whole subject and this article, published in 1999 is a small part of our larger chapter of genomic imprinting.

I first met Austin in 1988 when I delivered a talk on a theory I had concerning sex differences in recombination at a conference in Vancouver (Trivers 1988). He came up after the talk and said in a somewhat diffident style that he had recently been reviewing the evidence regarding sex differences in recombination and the evidence looked different than my summary of it. I had a sinking feeling. If he had said the *theory* was wrong, then the theory could always be readjusted, but if I had the evidence wrong, then the theory, which was made to fit the evidence, was likely to be wrong as well. He said he was a graduate student working with Graham Bell at McGill University in Montreal. My heart sunk further. Graham was known for the care with which he reviewed evidence so it was most unlikely that this quiet-spoken graduate student of his had gotten the facts wrong. Later, his correction to my paper appeared (Burt, Bell, and Harvey 1991) and to this day no one has figured out why there should be sex differences in recombination nor why they show the pattern they do; for example, in mammals, placentals such as ourselves show greater recombination in female gametogenesis, while in marsupials female gametogenesis usually shows tighter linkage. This problem remains a major unsolved problem in sexual selection with important implications for the evolution of the sex chromosomes.

From this interaction began a friendship that was cemented shortly thereafter when Austin arrived at the University of California at Santa Cruz to take up a postdoc under William Rice, the brilliant student of sex-antagonistic effects and the evolution of the sex chromosomes. Austin and I taught a course in 1992 on selfish genes, and at that time I saw that the argument for degrees of relatedness under genomic imprinting did not apply to the genes involved in establishing the initial imprint in the parental generation. That is, one could imagine a new kind of parent-offspring conflict over the process of imprinting itself. I slowly developed the argument in my usual style, laboriously deriving degrees of relatedness and looking for points of conflict. For several years Austin paid little attention to the argument until one day it caught his attention, and he rapidly developed the idea in a more general form, with additional implications, the resulting work being this article.

Genetic Conflicts in Genomic Imprinting

AUSTIN BURT AND ROBERT TRIVERS

Abstract.   The expression pattern of genes in mammals and plants can depend upon the parent from which the gene was inherited, evidence for a mechanism of parent-specific genomic imprinting. Kinship considerations are likely to be important in the natural selection of many such genes, because coefficients of relatedness will usually differ between maternally and paternally derived genes. Three classes of gene are likely to be involved in genomic imprinting: the imprinted genes themselves, *trans*-acting genes in the parents, which affect the application of the imprint, and *trans*-acting genes in the offspring, which recognize and affect the expression of the imprint. We show that coefficients of relatedness will typically differ among these three classes, thus engendering conflicts of interest between Imprinter genes, imprinted genes, and imprint-recognition genes, with probable consequences for the evolution of the imprinting machinery.

Introduction

For some genes in mammals and plants, maternally and paternally derived alleles have different patterns of expression (Barlow 1995; Reik & Surani

1997). In the usual case, one allele is silent and the other active; sometimes this difference is seen in some tissues and not in others, and sometimes there is only a quantitative difference in gene expression. This parent-specific gene expression is presumably due to differential imprinting of the alleles in the maternal and paternal germ lines. Why would such a system evolve? One likely explanation is that maternally and paternally derived genes have different coefficients of relatedness to many relatives, and so have different optimal levels of expression. For example, paternally derived genes will be less related to an individual's mother than will maternally derived genes, and so will usually be selected to extract more maternal investment (Haig & Westoby 1989; Moore & Haig 1991). Indeed, Haig (1992) has shown that the optimal level of maternal investment will generally differ between maternal genes, maternally derived offspring genes, paternally derived offspring genes, and unimprinted offspring genes (see also Queller 1994; Haig 1996). In this paper we show that these differing optima will also apply to the evolution of the imprinting machinery itself.

Three different classes of gene are likely to be involved in any particular instance of genomic imprinting: the imprinted genes themselves, *trans*-acting genes in the parents, which affect the application of the imprint, and *trans*-acting genes in the offspring, which recognize and affect the expression of the imprint (Efstratiadis 1994). We show that coefficients of relatedness between individuals will typically differ among these three classes, thus engendering conflicts of interest between Imprinter genes, imprinted genes, and imprint-recognition genes.

As a disproportionate number of imprinted genes in mice and humans are involved in placental and juvenile growth (Barlow 1995), we will again consider the example of interactions between mother and offspring. In addition, we use Hamilton's Rule (Hamilton 1963, 1964a,b), restricting ourselves to the simplest case of panmictic populations with weak selection, for which coefficients of relatedness can be simply derived from genealogical relationships (Grafen 1985). Conflicts between different classes of genes are demonstrated by showing that different conditions for the spread of a new mutation apply to the different classes (Trivers 1974).

Conflicts in the Maternal Germ Line

Maternally derived genes in a juvenile (or placenta) are, by definition, found in the mother with probability 1. Therefore, for a locus (e.g. a growth promoter) that is initially silent, a new mutant that is active if inherited from the mother will be selected for only if the benefit it brings to the offspring expressing it (b_0) is greater than the cost it incurs to the mother (c_m): $b_0 > c_m$, (figure 1a) (benefits and costs refer to changes in the reproductive value

(RV) of individuals (Hamilton 1966; Charlesworth & Charnov 1981); in populations of constant size, this is equivalent to changes in the expected number of offspring). The same condition applies for a locus (e.g. a growth suppressor) that is biallelically expressed, and a new mutant that is silent if inherited from the mother. However, now consider an Imprinter gene that acts in *trans* in the maternal germ line to apply the imprint. The probability of it being inherited along with the imprint is ½ (that is, the offspring expressing the imprint is related to the Imprinter allele by a coefficient of ½), and so a new mutant will spread only if the benefit to the offspring is more than twice the cost to the mother ($b_0 > 2c_m$). Therefore, if the benefit of a maternally imprinted gene to the offspring expressing it is in the range $c_m < b_0 < 2c_m$, then there will be a conflict of interest, with the target gene being selected to acquire the imprint and the Imprinter gene being selected not to apply it.

Conflicts between Imprinter genes and target genes can also arise in the opposite direction, for altruistic mutations that benefit the mother at the expense of the offspring expressing them (figure 1*b*). That is, for a locus (e.g. a growth suppresser) that is initially silent, a new mutant that is active if inherited from the mother will increase in frequency only if the benefit to the mother (b_m) is more than the cost to the offspring (c_0): $b_m > c_0$ (and similarly for a growth promoter locus that is biallelically expressed and a new mutant that is silent if maternally inherited). However, a *trans*-acting Imprinter gene will be selected to apply the imprint as long as $b_m > c_0/2$. If $c_0/2 < b_m < c_0$, then again there will be a conflict of interest, except that now the Imprinter gene will be selected to apply the imprint, and the target gene selected to avoid it.

Conflicts in the Paternal Germ Line

Such conflicts between Imprinter loci and their targets may also arise in the paternal germ line. Here, selection on genes affecting maternal investment will depend critically upon how costs borne by the mother affect the father. Suppose a unit change in maternal RV has effect k on the father's value. Then, for a growth promoter that is initially silent, a new mutation that is active if inherited from the father will increase in frequency if $b_0 > kc_m$. However, a *trans*-acting paternal Imprinter gene will only be selected to apply the imprint if $b_0 > 2kc_m$. Thus if $kc_m < b_0 < 2kc_m$, then there will be a conflict, with the growth gene being selected to acquire the imprint and the Imprinter gene being selected not to apply it. Again, conflicts with the opposite orientation arise for altruistic mutations, when $c_0/2 < kb_m < c_0$.

Note that conflicts only arise in the paternal germ line when $k > 0$ (figure

1). What is this parameter? As defined, it measures how fitness effects on mothers affect fathers. Thus, its value will clearly depend on the mating system: $k = 0$ with complete promiscuity and $k = 1$ with lifetime monogamy. With polygyny, $k = 1$ if there is no interference between females; it will be less than 1 if there is interference (so that if one female suffers a cost, another female will replace at least part of the loss), and greater than 1 if there are synergistic or cooperative effects between females in a harem. Note that in the latter case, one would expect the usual imprinting asymmetries to reverse, with growth suppressers being paternally expressed. Note too that our general approach of attributing the costs of offspring selfishness directly to parental RV is apparently novel, contrasting with the more usual approach in which these costs are borne by maternal siblings, either extant or future, which may or may not have the same father ($r = 1/2$ or $1/4$, respectively; see, for example, Trivers 1974). However, this approach can be misleading, or at least difficult to apply correctly (Mock & Parker 1997, pp. 151–154). Thus, Haig (1992, 1996) suggests that the conflict between maternally and paternally derived genes disappears if females only mate with one male in their lifetime, but this is unlikely to be generally true. Even if females are monandrous, if offspring selfishness causes reduced maternal survival, and if a male is able to replace a dead mate, then $k < 1$ and paternally derived genes will be selected to extract more maternal investment than maternally derived genes. In an analysis of parent-offspring relations, our approach seems the more direct and less likely to lead to errors. (Even so, the relation between maternal and paternal RV may be more complicated than assumed here: for example, decrements in maternal survival will have less effect on paternal RV than decrements in maternal fertility if a dead mate is replaced more easily than a subfertile mate. However, these complications need not concern us here.)

Conflicts in the Offspring

Supposing that the maternally and paternally derived genes in a newly formed zygote are differentially imprinted, this does not guarantee differential expression, for the imprint must be inherited through the many mitoses in offspring development and must have some effect on gene expression. Both requirements will involve the action of *trans*-acting genes in the offspring, to maintain and read the imprint. Assuming that these genes are not themselves imprinted, they will have yet another threshold for the spread of a new mutation, namely, that $b_0 > (1 + k) c_m/2$ for a growth promoter, and $b_m > 2 c_0/(1 + k)$ for a growth suppresser, assuming they are autosomal (figure 1). Thus, even if a gene is selected to acquire an imprint, and Imprinter genes selected to apply it, offspring genes may nonetheless

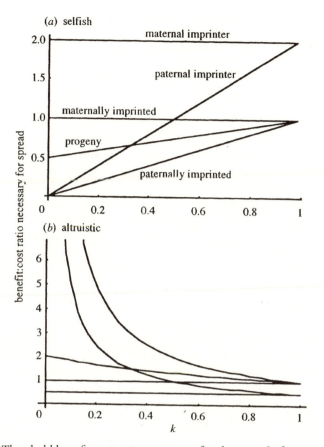

Figure 1. Threshold benefit-cost ratios necessary for the spread of a gene as a function of k, the effect on paternal reproductive value of a unit change in the maternal value, for different types of genes. (a) Genes that increase offspring growth (benefits to offspring, costs to mother); (b) genes that reduce offspring growth (benefits to mother, costs to offspring; labelling of lines is in inverse order to that above). Values in (a) also give the stopping rule for maternal investment for the different types of genes. Note that for $k = 1$, all conflict is parent–offspring, and that as k gets smaller, the conflict gradually becomes more maternal–paternal.

be selected to remove the imprint or to ignore it. Imprints might easily be removed by changing patterns of methylation (Chaillet *et al.* 1995), and they can be ignored by, for example, starting transcription in a different place, as occurs for human *IGF2* when it switches from paternal expression in foetal liver to biallelic expression in adult liver (Vu & Hoffman 1994). This switch is presumably under the control of *trans*-acting transcription factors. Thus, intragenomic conflicts may arise both over the application of an imprint in the parental germ lines and over the expression of an imprint in the offspring.

Discussion

Patterns of relatedness will often differ for maternally and paternally derived genes, and this asymmetry is a likely source of natural selection for parent-specific gene expression (Haig & Westoby 1989; Moore & Haig 1991). This maternal-paternal conflict was first described as depending upon females having more than one mate; as we have noted above, it more accurately depends upon decrements in the female's RV being, from the male's point of view, replaceable. Previous models of this kinship theory of imprinting have considered the evolution of *cis*-acting control regions affecting the level of imprinting (Haig 1992, 1996; Mochizuki *et al.* 1996; Spencer *et al.* 1998). We have extended these analyses to the other sorts of genes likely to be involved in genomic imprinting, *trans*-acting genes in the parental germ lines and *trans*-acting genes in the offspring, and have demonstrated that there will be well-defined conflicts of interest between these different classes of gene. Spencer & Williams (1997) have previously presented models for the evolution of imprinting with *cis*- and *trans*-acting germ line modifiers, but did not consider traits in which kinship is important, and so did not find a conflict. The existence of conflicts between Imprinter and imprinted loci, and between imprinted and imprint-recognition loci, could have a number of important consequences for the evolution of genomic imprinting.

First, conflicts of interest of the sort described here could lead to a constantly dynamic pattern of perpetual selection, as evolutionary change in one component of the imprinting machinery selects for an evolutionary response by another. As noted above, conflicts in the parental germ lines can be in either direction, each with its expected evolutionary dynamic. First, a locus may be selected to acquire an imprint while an Imprinter is selected to not apply it. In this case, the target gene may, for example, be selected to mimic other imprinted genes, by acquiring their recognition sequences, and the imprinting apparatus therefore selected to make ever finer discriminations between genes it wants to imprint and those it does not. Alternatively, an Imprinter may be selected to apply an imprint and the target selected to avoid it, in which case the imprinting apparatus will be chasing the target locus through sequence space. Antagonistic coevolution may also occur between imprinted genes and imprint-recognition genes. One possible manifestation of such perpetual selection would be a breakdown of the normal pattern of parent-specific gene expression in species hybrids and backcrosses. Vrana *et al.* (1998) have shown that expression of some imprinted genes differs between reciprocal F1 hybrids of *Peromyscus maniculatus* and

P. polionotus, indicating there has been recent evolution of these genes and/ or of the imprint recognition machinery since the species diverged. Interestingly, not all imprinted genes showed the same pattern in the reciprocal hybrids; this result indicates that parent-specific expression of different loci is under at least somewhat separate control.

Second, not all genes whose evolution might be affected by asymmetric coefficients of relatedness have parent-specific expression (e.g. *Igf1*, a gene affecting foetal growth). Why not? Haig (1997) suggests that this may simply be due to the absence of appropriate mutations, Mochizuki *et al.* (1996) suggest that it may be due to the increased expression of deleterious recessives at imprinted loci, and Spencer *et al.* (1998) suggest that it may be because the costs to the mother outweigh the benefits to the offspring. As an alternative, we suggest that the different components of the imprinting machinery will not always be selected in the same direction (especially for maternal imprinting; see below), and parent-specific expression may not evolve because the 'nays' have won.

Third, conflicts within the imprinting machinery may also help explain why imprinted genes tend to occur in clusters (reviewed by Reik & Maher (1997)) as follows: if there is a conflict between a gene selected to acquire an imprint and Imprinter genes selected not to apply it, perhaps the former can evolve to make use of mechanisms operating at other nearby loci where Imprinters and targets are both positively selected. Consistent with this idea, there appear to be complex interconnected causal pathways acting both in *cis* and in *trans* within these clusters (Buiting *et al.* 1995; Dittrich *et al.* 1996; Forné *et al.* 1997; Webber *et al.* 1998). Previous explanations for clustering have posited a lack of genetic variation for becoming imprinted (Haig 1997) or an imprinting process that is costly (Mochizuki *et al.* 1996).

Fourth, Moore & Reik (1996) have suggested that such conflicts could account for the complex pattern of de- and remethylation observed at some imprinted loci. For example, at a particular cytosine of *Igf2r*, the maternal copy is methylated at the zygote and two-cell stage, is unmethylated at the four-cell stage, then reacquires methylation at the eight-cell stage (Razin & Shemer 1995; Shemer *et al.* 1996). Perhaps even the genome-wide demethylation that occurs early in mouse development (Li 1997) is the organism's attempt to reduce the frequency of unwanted imprints, both maternal and paternal.

Fifth, recognition of conflicts in the imprinting machinery gives reasons for thinking that paternal imprinting may be more common than maternal imprinting, as follows. Each of the five classes of gene we have discussed will have a different optimal level of maternal investment: in particular, each class of gene will be selected to continue investment until the marginal benefit-to-cost ratio falls below the corresponding value in figure 1*a*. The actual amount of investment at any point in evolutionary time seems likely to be intermediate between the various optima, perhaps closest to the maternal Imprinter optimum because maternal genes have so much more control than other genes over maternal investment. If so, then it will be between the optima for maternal Imprinter genes and maternally imprinted genes, and so these will be selected to change investment in opposite directions, maximizing the conflict of interest. On the other hand, both paternal Imprinter genes and paternally imprinted genes will often be selected to increase the level of maternal investment, particularly when k is small, and so are less likely to disagree. Imprints arising in the paternal germ line are also more likely to be in the offspring's interest, and so are more likely to be maintained and used. Such reasoning suggests that imprinting may be more stable over evolutionary time in the paternal germ line than in the maternal germ line, and that paternally imprinted genes should therefore come to outnumber maternally imprinted genes.

Other lines of reasoning lead to the same prediction: (i) mothers have many ways to influence maternal investment other than via imprinting, whereas fathers are much more limited in their options; and (ii) the actual level of investment is likely to be further away from the paternal optima than from the maternal optima, and so selection for change will be stronger. Against these considerations must be weighed the greater influence maternal genes can have over imprinted loci, through cytoplasmic RNAs in the oocyte (Latham & Sapienza 1998), which could lead to a preponderance of maternally imprinted genes.

Unfortunately, we do not know for most imprinted genes whether the imprinting is maternal or paternal. Expression patterns cannot easily be used to decide because imprinting may be the parent-specific silencing of an allele that used to be active, or the activation of an allele that used to be silent. For example, *Ins1* and *Ins2* in mice are paternally expressed in the yolk sac, but biallelically expressed in the pancreas, and it is not yet clear whether the evolutionary innovation was silencing of the maternal allele (maternal imprinting; the ancestral state was biallelic expression in the yolk sac) or activation of the paternal allele (paternal imprinting; the ancestral state was no expression in the yolk sac). Similarly, methylation patterns cannot be used to decide because imprinting may be the addition of methyl groups that are usually absent, or the removal of ones that are usually present (Chaillet *et al.* 1995). Rather, comparative and genetic studies are required to determine what type of gene changed, and in which germ line the change occurred, in the evolution of parent-specific expression.

The five classes of gene discussed thus far do not exhaust the possibilities. For example, Imprinter genes might themselves be imprinted and work in a parent-specific manner, in which case one would have to consider relatedness over three generations. This is perhaps not too far-fetched, as genes that affect the methylation of artificially constructed transgenes have been shown to work in a parent-specific manner (Allen & Mooslehner 1992). Sex-linked genes can also have optima for kin-selected traits that are different from those of autosomal genes, and hence their own set of benefit—cost thresholds (Hamilton 1972). This will almost certainly be the case if the effects on relatives are sex-specific (e.g. foetal testosterone production, which is good for brothers and bad for sisters [Clark & Galef 1995]). Considering only the case where litter-mates have the same father, autosomes and maternally derived Xs in a female foetus are related to all litter-mates by a factor of 1/2, but the paternally derived X is related to sisters by a factor of 1 and to brothers by a factor of 0. Thus, paternally derived Xs will be selected to produce a relatively female-beneficial, male-detrimental uterine environment, as will X-linked Imprinter genes active in the male germ line, whereas maternally derived Xs and all autosomes will be selected to produce a more gender-neutral foetal environment. The result will be conflicts between different components of the imprinting apparatus, and between different components of the imprint-recognition apparatus, if some of the genes involved are sex-linked and others autosomal.

Conflicts of interest over imprinting do not depend upon maternal—paternal asymmetries, for even if $k = 1$, parental Imprinter loci may be selected to apply an imprint (e.g. inactivate a growth enhancer, or activate a growth suppresser), while target loci and imprint-recognition loci are selected to lose it (figure 1). Similarly, they need not be limited to foetal characters; indeed, it is difficult to think of any class of interaction between relatives that will not produce such conflicts, including alarm calls, dispersal, dormancy, inbreeding and inbreeding avoidance, helping to raise siblings, etc. (Trivers & Burt 1998). On the other hand, imprinting itself need not be limited to kin-selected traits: if there is selection for a change in gene dosage or tissue-specificity and the first appropriate mutation happens to work in a parent-specific manner, then it may be selected for and go to fixation. One possible example is the demethylation of *Xist* in the paternal germ line of mice, which apparently marks the X chromosome for inactivation in the offspring trophectoderm (Norris *et al.* 1994; Ariel *et al.* 1995; Zuccotti & Monk 1995). By inactivating the X chromosome, the male is making it match the degenerate Y chromosome transmitted in the other half of his gametes, a simple form of dosage compensation available only to taxa with methylation (so, for example, not *Drosophila* or *Caenorhabditis*). Imprinting of genes unrelated to kin selection will not lead to intragenomic conflicts.

NOTE

We thank Charles Godfray, David Haig, Paul Vrana and anonymous referees for useful discussions and/or comments on a previous draft. A.B. is supported by the NERC (GR3/10626).

REFERENCES

Allen, N. D. & Mooslehner, K. A. 1992 Imprinting, transgene methylation and genotype-specific modification. *Semin. Dev. Biol.* **3**, 87–98.

Ariel, M., Robinson, E., McCarrey, J. R. & Cedar, H. 1995 Gamete-specific methylation correlates with imprinting of the murine *Xist* gene. *Nature Genet.* **9**, 312–315.

Barlow, D. P. 1995 Gametic imprinting in mammals. *Science* **270**, 1610–1613.

Buiting, K., Saitoh, S., Gross, S., Dittrich, B., Schwartz, S., Nicholls, R. D. & Horsthemke, B. 1995 Inherited micro-deletions in the Angelman and Prader-Willi syndromes define an imprinting centre on human chromosome 15. *Nature Genet.* **9**, 395–400.

Chaillet, J. R., Bader, D. S. & Leder, P. 1995 Regulation of genomic imprinting by gametic and embryonic processes. *Genes Devel.* **9**, 1177–1187.

Charlesworth, B. & Charnov, E. L. 1981 Kin selection in age-structured populations. *J. Theor. Biol.* **88**, 103–119.

Clark, M. M. & Galef, B.G.J. 1995 Prenatal influences on reproductive life history strategies. *Trends Ecol. Evol.* **10**, 151–153.

Dittrich, B. (and 10 others) 1996 Imprint switching on human chromosome 15 may involve alternative transcripts of the *SNRPN* gene. *Nature Genet.* **14**, 163–170.

Efstratiadis, A. 1994 Parental imprinting of autosomal mammalian genes. *Curr. Opin. Genet. Devel.* **4**, 265–280.

Forné, T., Oswald, J., Dean, W., Saam, J. R., Bailleul, B., Dandolo, L., Tilghman, S. M., Walter, J. & Reik, W. 1997 Loss of the maternal *H19* gene induces changes in *Igf2* methylation in both cis and trans. *Proc. Natl. Acad. Sci. USA* **94**, 10 243–10 248.

Grafen, A. 1985 A geometric view of relatedness. *Oxf. Surv. Evol. Biol.* **2**, 28–89.

Haig, D. 1992 Genomic imprinting and the theory of parent–offspring conflict. *Semin. Dev. Biol.* **3**, 153–160.

Haig, D. 1996 Placental hormones, genomic imprinting, and maternal—fetal communication. *J. Evol. Biol.* **9**, 357–380.

Haig, D. 1997 Parental antagonism, relatedness asymmetries, and genomic imprinting. *Proc. R. Soc. Lond.* B **264**, 1657–1662.

Haig, D. & Westoby, M. 1989 Parent-specific gene expression and the triploid endosperm. *Am. Nat.* **134**, 147–155.

Hamilton, W. D. 1963 The evolution of altruistic behavior. *Am. Nat.* **97**, 354–356.

Hamilton, W. D. 1964*a* The genetical evolution of social behaviour. I. *J. Theor. Biol.* **7**, 1–16.

Hamilton, W. D. 1964*b* The genetical evolution of social behaviour. II. *J. Theor. Biol.* **7**, 17–52.

Hamilton, W. D. 1966 The moulding of senescence by natural selection. *J. Theor. Biol.* **12**, 12–45.

Hamilton, W. D. 1972 Altruism and related phenomena, mainly in social insects. *A. Rev. Ecol. Syst.* **3**, 193–232.

Latham, K. E. & Sapienza, C. 1998 Localization of genes encoding egg modifiers of paternal genome function to mouse chromosomes one and two. *Development* **125**, 929–935.

Li, E. 1997 Role of DNA methylation in mammalian development. In *Genomic imprinting* (ed. W. Reik & A. Surani), pp. 1–20. Oxford University Press.

Mochizuki, A., Takeda, Y. & Iwasa, Y. 1996 The evolution of genomic imprinting. *Genetics* **144**, 1283–1295.

Mock, D. W. & Parker, G. A. 1997 *The evolution of sibling rivalry.* Oxford University Press.

Moore, T. & Haig, D. 1991 Genomic imprinting in mammalian development: a parental tug-of-war. *Trends Genet.* **7**, 45–49.

Moore, T. & Reik, W. 1996 Genetic conflict in early development: parental imprinting in normal and abnormal growth. *Rev. Reprod.* **1**, 73–77.

Norris, D. P., Patel, D., Kay, G. F., Penny, G. D., Brockdorff, N., Sheardown, S. A. & Rastan, S. 1994 Evidence that random and imprinted *Xist* expression is controlled by preemptive methylation. *Cell* **77**, 41–51.

Queller, D. C. 1994 Male—female conflict and parent—offspring conflict. *Am. Nat.* **144**, S84–S99.

Razin, A. & Shemer, R. 1995 DNA methylation in early development. *Hum. Molec. Genet.* **4**, 1751–1755.

Reik, W. & Maher, E. A. 1997 Imprinting in clusters: lessons from Beckwith—Wiedemann syndrome. *Trends Genet.* **13**, 330–334.

Reik, W. & Surani, A. (eds) 1997 *Genomic imprinting.* Oxford University Press.

Shemer, R., Birger, Y., Dean, W. L., Reik, W., Riggs, A. D. & Razin, A. 1996 Dynamic methylation adjustment and counting as part of imprinting mechanisms. *Proc. Natl. Acad. Sci. USA* **93**, 6371–6376.

Spencer, H. G. & Williams, M.J.M. 1997 The evolution of genomic imprinting: two modifier-locus models. *Theor. Popul. Biol.* **51**, 23–35.

Spencer, H. G., Feldman, M. W. & Clark, A. G. 1998 Genetic conflicts, multiple paternity and the evolution of genomic imprinting. *Genetics* **148**, 893–904.

Trivers, R. L. 1974 Parent—offspring conflict. *Am. Zool.* **14**, 249–264.

Trivers, R. & Burt, A. 1998 Kinship and genomic imprinting. In *Genomic imprinting* (ed. R. Ohlsson), pp. 1–21. Heidelberg: Springer.

Vrana, P. B., Matteson, P., Guan, X.-J., Ingram, R. S. & Tilghman, S. M. 1998 Genomic imprinting is disrupted in interspecific *Peromyscus* hybrids. *Nature Genet.* (In the press.)

Vu, T. H. & Hoffman, A. R. 1994 Promoter-specific imprinting of the human insulin-like growth factor-II gene. *Nature* **371**, 714–717.

Webber, A. L., Ingram, R. S., Levorse, J. M. & Tilghman, S. M. 1998 Location of enhancers is essential for the imprinting of *H19* and *Igf2* genes. *Nature* **391**, 711–715.

Zuccotti, M. & Monk, M. 1995 Methylation of the mouse *Xist* gene in sperm and eggs correlates with imprinted *Xist* expression and paternal X-inactivation. *Nature Genet.* **9**, 316–320.

Postscript

A good review of genomic imprinting from an evolutionary, as well as molecular, standpoint can be found in Ohlsson (1999). This includes an introductory chapter on the kinship (i.e., Haig's) approach to imprinting (Trivers and Burt 1999). For a collection of Haig's papers, with commentary, see Haig (2002).

10

FLUCTUATING ASYMMETRY AND THE 2ND : 4TH DIGIT RATIO IN CHILDREN

I have always found it advantageous to keep an empirical project going alongside whatever theoretical work I am trying to do, fieldwork, for example, in Jamaica. It provides several kinds of relief from the theoretical work and gives you something to do when you are not otherwise particularly creative. In any case, as an early graduate student at Harvard, I watched pigeons in my spare time and then studied lizards as part of my thesis (and as a means of visiting Jamaica frequently). In more recent years, most of my academic position being in an anthropology department, I have tried to study our own species. In particular, in January 1996 with a team of scientists and students, we measured nearly three hundred school children in Jamaica from head to toe for degree of bodily symmetry. We are especially interested in each individual's degree of fluctuating asymmetry and differences between traits in degrees of fluctuating asymmetry. My interest in this topic came about in the following way.

In 1990 Bill Hamilton visited the University of California at Santa Cruz to deliver a talk. Since he had done the most exciting piece of work on sexual selection in the 1980s—namely, work suggesting that a key aim of mate choice in many species is to identify relatively parasite-resistant genes via degree of development of such secondary traits as bright color and complex song—I said to him, "What's new in sexual selection, Bill?" Hamilton replied, "Fluctuating asymmetry." I said, "Fluctuating who?" and he repeated, "Fluctuating asymmetry." I had never heard the term before.

Fluctuating asymmetry refers to minor deviations from perfect bilateral symmetry (in bilaterally symmetrical species) that vary randomly from right to left and tend in a population to cancel out and give true symmetry. In other words, the trait lacks and *directional* bias in the population. Hamilton directed me to a paper just published by the great Danish ornithologist

Anders Moller (now at the Université de Louis Pasteur) showing that female swallows prefer males whose tail feathers are symmetrical. He also told me to get hold of a soon-to-be-published work by Randy Thornhill (University of New Mexico) on fluctuating asymmetry and sexual selection in scorpion flies.

I like to do whatever Bill Hamilton tells me to do, so within a day or two I had read Moller's paper and I sent for Thornhill's manuscripts. Moller's paper was an eye-opener. By experimentally cutting and glueing sections of the outer tail feathers of male swallows, he artificially created asymmetrical individuals whom he then released into nature in competition with control males (same sections cut and reglued for no net effect). He found that females in nature tended to settle more often—and reproduce more often—with the symmetrical males.

Thornhill and Sauer (1992) showed that degree of fluctuating asymmetry (FA) was also an important variable in insects, predicting reproductive success in both sexes of scorpion flies but more strongly in males. Furthermore, he showed that FA had a large effect on another component of sexual selection—in this case, male–male combat. Symmetrical males won fights over less symmetrical ones, and degree of symmetry was a stronger predictive variable than body size (which also made a positive contribution). Incidentally, Thornhill (1992) also showed that female scorpion flies prefer the smell of more symmetrical males. While smell is produced by paired scent glands and females could, in principle, be measuring asymmetry of scent production, this seemed very unlikely since they were choosing between the smells of two males each of which was sent down a relatively long tube, the scents from the two glands probably mixing fairly well. In any case, Thornhill was to duplicate this feat ten years later when he showed that women who are not taking a contraceptive pill prefer the smell of symmetrical men and do so more strongly the closer they are to their own time of maximum fertility (Gangestad and Thornhill 1998).

Furthermore, both authors made reference to a forty-year literature on fluctuating asymmetry and its correlates, a literature that, though overlooked by most biologists, had developed nicely on its own. It showed that degree of FA was a measure of (1) stress during development and (2) genetic inability to compensate for stress. It is presumably the latter that made FA an important parameter in mate choice.

I remember well the sensation I had when reading Moller's paper on swallows. A kind of bias I share says that if a human thinks symmetry is important it may or may not be, but if a bird thinks symmetry is important, it very likely is! Moller's swallows seemed to tell me that I would be overlooking symmetry at my own peril. Of course, symmetry has an almost visceral attractiveness of its own, a powerful concept in mathematics, physics, chemistry, art, and so on. Now there was a mate choice/sexual selection angle to symmetry that gave it renewed importance in biology as well.

Degree of symmetry had a couple of other things going for it as a generally useful measure of genetic quality, both for members of a species evaluating each other and for scientists studying the species. One was that symmetry in principle could easily be observed by the organism in front of you. The dimensions of the two sides were subject to visual inspection. Second, the optimal value remained always the same, perfect symmetry itself. One can easily imagine relatively strong selection for symmetry—for example, on wing length in birds and insects—and relatively weaker selection for symmetry (leg dimensions in the same species) but as long as the trait showed no directional bias, it would usually be safe to assume that symmetry itself was the optimal value. Subsequent work has indeed borne out the supposition that degree of FA may be of broad significance. Six or seven species of birds have now been shown to prefer symmetry in mate choice. Its importance has been demonstrated in numerous insects, in some fish, in the antlers and horns of ungulates, and in flowers, and it is now known to be an important variable mediating attractiveness for both sexes in our own species (review in Moller and Swaddle 1997).

I immediately checked to see if it was important in lizards, using dewlap coloration in a Jamaican lizard as the object of study. I was already studying parasite load and dewlap coloration in *Anolis lineatopus*. When deciding to measure dewlap coloration, like a good conventional biologist, I always picked one side to measure in case there was any variation, but, of course, by measuring only one side I was not gathering data about symmetry. So I started comparing the two sides for symmetry of the pattern of coloration, but as I ended up showing no significant correlations for color itself, it is perhaps not surprising that I also failed to find correlations for degree of asymmetry in color. In any case, I soon turned my attention to humans, and this really because of the suggestion of a friend.

Fluctuating Asymmetry in Jamaican Children

Dr. Michael McGuire is a psychiatrist at the Neuropsychiatric Institute at UCLA in Los Angeles who has worn many hats in his career. He has practiced psychiatry, has done extensive research on neurotransmitters in vervet monkeys, and is one of the first to publish a detailed approach to psychiatry based on evolutionary logic (McGuire and Troisi 1998). As a friend, he has urged several propositions on me. First is his seven-year rule, which states that if you screw up, it takes at least seven years for people to forgive you and therefore it takes seven years to fund new work. Since that is roughly the frequency with which I have breakdowns, that leaves me permanently outside the loop where financial research support is concerned.

Second, he urged on me the value of a long-term project, especially in Jamaica, for the overall health of my life and suggested that fluctuating asymmetry was an almost ideal topic. It was clear that asymmetry was likely to develop into a major variable in the study of human life, and if so, it needed long-term studies that, by their very nature can provide certain kinds of evidence unavailable by other means.

I decided to do as he said and eventually Rutgers University underwrote the beginning measurements of two hundred and eighty-five children in rural Jamaica. Incidentally, you cannot get permission to do this kind of work in the United States. The general rule is that if your research requires touching children, you will not get permission. Even were you to find a school system willing to let you make a series of measurements and endless re-measures, the number of parents which would agree to this would be much smaller still. By contrast, with the payment of very modest sums, (about ten dollars per child per measurement session) we have been able to get over 95% participation in a well-defined rural community. Let me briefly review the major findings of the work to date.

The Jamaican children are more symmetrical than a similar group of English children. When bodily symmetry of the Jamaican children is compared with the bodily symmetry of a comparable set of children in Liverpool, England, the Jamaican children are found to show an average difference of 1.9% between the two sides of the body while the English children showed a 2.5% difference (Trivers et al 1999). We do not know why this is true, though it may be due to greater genetic variability in the Jamaican population or more intense selection for developmental stability (e.g., because of less medical attention).

Boys are more symmetrical than girls. The effect is very small but it is real (that is, highly significant). On average Top Hill area boys are about 1/10th of 1% more symmetrical than are Top Hill girls (1.7% difference between the sides compared to a 1.8% difference). This is a very interesting result but, again, we have no idea why it may be true. It is primarily caused by boys having much more symmetrical elbows than girls. (Elbows, in turn, are the most asymmetrical part of the human body.) The British children likewise showed a tendency for boys to be more symmetrical than girls but the difference was not significant.

As children age they become more symmetrical. In our data there is a trend in this direction but it is not significant. On the other hand, comparable British data make it clear that, from early on in life, we become more symmetrical as we age, at least until the time of puberty when there is a brief decrease in symmetry followed by a further increase in symmetry until early adulthood (Wilson and Manning 1996). From a practical standpoint, the trend means that we must "correct" all our data by reference to age, so as to make sure we are measuring an effect of only symmetry itself.

The lower body is more symmetrical than the upper body. This has never been shown before for the human being because no one had previously measured the body from head-to-toe. But our data clearly show that the lower body—the knees, the ankles, and the feet—are much more symmetrical than the upper body—the wrists, the fingers, the ears, and so on. The average difference between the two sides of the upper body is 2.1%, while that for the lower body is only 0.8% (Trivers et al 1999). We think it is likely that the lower body is more symmetrical than the upper because it is used much more symmetrically, and symmetry is probably more important for efficient walking and running than it is for upper body activities. The elbows are the least symmetrical part of the body, differing on average by 3.5%.

Symmetry in one part of the body may be correlated with some traits and not others. This is a general rule for which we now have many examples. For example, in our study academic achievement is not related to symmetry of the child, in fact, slightly less symmetrical children seem to do better in school, but this is a small effect, and not significant. By contrast, children whose *palm* prints are more symmetrical do score significantly higher in academic achievement (unpublished data).

Boys whose lower bodies are more symmetrical are more aggressive. This was a surprising finding but has now been confirmed on British children (Manning and Wood 1998), and on U.S. men as well (Furlow, Gangestad and Armijo-Prewitt 1998). Boys whose lower bodies are more symmetrical are more likely to be involved in aggressive acts and to escalate these acts to higher levels. Why should lower body symmetry be associated with aggressiveness? In some animals we know that symmetrical males win fights against asymmetrical males, so perhaps the lower body is especially important in winning fights and those boys who know that they can win fights may be more likely to start them and to escalate them.

Girls whose ears are more symmetrical are more likely to cradle a doll on the left side of the body. This was a very interesting finding because John Manning (University of Liverpool) had discovered a similar finding in British women. He interpreted left-side cradling as being the preferred position (as it is around the world) because it facilitates communication with the *right* brain, itself specialized for emotional interactions, particularly in women. (The body is cross-wired: sounds in the left ear go first to the right brain.) Thus, we were eager to see if Jamaican girls resembled English women and they did (Manning et al 1997). Symmetry in other body parts seemed unrelated to left-side cradling but ear symmetry was highly predictive. The Jamaican boys were just like the English men: neither showed much of a tendency to cradle the baby (or doll) on the left side and there was no correlation between ear symmetry and tendency to left-side cradle.

Symmetrical boys are better dancers. In boys, but not in girls, symmetry was associated with greater dancing ability, as judged by other people viewing video-tapes of the children dancing (unpublished data). Curiously enough, it was not lower-body symmetry that was the best predictor of dancing ability, but rather upper-body symmetry.

Symmetrical girls are judged more attractive by others, while in boys the reverse is true. Whether adults rate children's pictures for attractiveness or the children rate the pictures themselves, we find that symmetrical girls are found to be more attractive but *asymmetrical* boys are seen as more attractive (unpublished data). Each effect is significant but relatively weak. The effect in boys is opposite of expectation, since many studies have shown elsewhere in the world that symmetrical *men* are seen as more attractive.

The Jamaican children have highly symmetrical palm- and finger-prints. Like African and mostly African derived populations around the world, the Jamaican children have relatively symmetrical palm- and fingerprints, much more so than those of Europeans and Asians, and their derived populations (Jantz, Trivers, and Strange unpublished manuscript). Why African and African-derived peoples have more symmetrical finger and palm prints is completely unknown. African Brazilians have higher dental symmetry than European-Brazilians and Jamaican children are more symmetrical in their bodies than English children (our study), so there may be a general tendency for African and African-derived people to enjoy greater symmetry, though a general tendency across many populations is know only for palm and fingerprints (Jantz and Webb 1982).

The 2nd : 4th Digit Ratio

While this project was underway, John Manning rediscovered a subtle sexual dimorphism in humans known since the nineteenth century: the 2nd : 4th digit ratio is about 2% higher in women than in men. That is, compared to the 4th finger, the 2nd is somewhat longer in women than in men. This fact had laid dormant until Manning had the bright idea of studying variation *within* each sex in this trait to see whether it was correlated with variation in other traits such as sex hormones (Manning et al 1998). He soon discovered a series of interesting correlates, some of which used data from our symmetry study, since we had already measured finger lengths in our children and it was easy to add measurements of their parents, as well.

The 2 : 4th finger ratio in a woman is positively related to the number of her children. The 2nd : 4th finger ratio has been discovered to be associated with fertility, as measured by number of children, for both men and women elsewhere in the world. In women, the relatively longer the 2nd finger is compared to the 4th, the greater her number of children, while for men, the smaller the 2nd : 4th ratio, the greater his number of children. These findings were tested in the mothers of our children by measurements taken from photocopies of the hands made in June 1998. The mother of the Jamaican children are like women elsewhere; higher 2nd : 4th ratio is associated with more children (Manning et al. 2000). In men, there is a weak tendency for men with smaller 2nd : 4th finger ratios to have more children. The average 2nd : 4th finger ratio in Jamaican women is considerably lower than that for British women. The latter have a ratio of 1 (that is, the two fingers are equally long) while in Jamaican women the ratio is 0.95 (that is, the 2nd finger is only 95% as long as the 4th). In both countries, men have a 2nd : 4th finger ratio that is about 2% smaller; within each sex, there is variation of about 10% from the average ratio in either direction. In a world-wide sample of peoples, only Finnish people had 2nd : 4th digit ratios as low as Jamaicans. What Finnish people and Jamaican people have in common otherwise would be an interesting thing to study!

Mothers with higher waist/hip ratio have had more sons; their children of both sexes have smaller 2nd : 4th digit ratios. The waist/hip ratio was measured for the mothers of our children. Like women in England and in the United States, the Jamaican women with thicker waists (and a higher waist/hip ratio) appear to have produced more sons (but not more daughters); cause and effect is unknown but it may be that having sons thickens a mother's waist more than having daughters (Manning et al. 1999). Also, for the Jamaican women, thicker waists and a higher waist/hip ratio are associated with children of both sexes who have smaller 2nd : 4th digit ratios. This may occur because women with higher waist/hip ratios have more circulating testosterone in their blood (a well-established fact) and high testosterone *in utero* is believed to result in relatively short 2nd digits.

For both men and women with low 2nd : 4th digit ratios, offspring produced are relatively male-biased. There is evidence that high testosterone in either sex at the time of conception is associated with a higher proportion of sons (James 2000). It follows from this that low 2nd : 4th digit ratios may be associated with a higher proportion of sons and this appears to be true for a sample which includes adults from our Jamaican study (Manning et al. unpublished manuscript).

In the children, a low 2nd : 4th finger ratio is associated with left-handedness. Because one's 2nd : 4th finger ratio appears to be set early in life, that is, before birth, we thought there might be a connection between left-handedness and low 2nd : 4th ratios since both are apparently partly affected by high testosterone levels inside the mother. Sure enough, in both boys and girls, low 2nd : 4th finger ratios (especially in the right hand) are associated with left-handedness. Degree of handedness was measured by using a peg-moving device, in which the speed at which pegs are moved from a back row to a front row with one hand is compared with the same speeds for the other hand. This paper is reproduced here. It was conceived and written by John Manning, using data generated from the Jamaican project.

The 2nd : 4th Digit Ratio and Asymmetry of Hand Performance in Jamaican Children

J. T. MANNING, R. L. TRIVERS, R. THORNHILL, AND D. SINGH

Abstract.   Testosterone, particularly prenatal testosterone, has been implicated in the aetiology of many extragenital sexually dimorphic traits. It is
difficult to test directly for the effect of prenatal testosterone in humans. However, Manning, Scutt, Wilson, and Lewis-Jones (1998b) have recently shown
that the ratio of the length of the 2nd and 4th digits (2D : 4D) in right hands
negatively predicts testosterone levels in men. As digit ratios are fixed *in utero*
it may be that the 2D : 4D ratio is associated with many prenatally determined
sexually dimorphic traits. We tested this for one case by examining the relationship between lateralised hand performance (LHP), as measured by an Annett peg board, and 2D : 4D ratio in rural Jamaican children. 2D : 4D ratio
was measured from photocopies and X rays of hands. A low 2D : 4D ratio in
the right hand of boys and girls (photocopies) and the right hand of boys only
(X rays) was associated with a reduction in rightward performance asymmetry.
In both samples the difference in 2D : 4D ratio between the hands (2D : 4D
left hand–2D : 4D right hand) showed the strongest relationship with LHP,
i.e., high ratio in the left and low in the right correlated with a tendency
towards a fast performance with the left hand. It is suggested that the 2D :
4D ratio may be associated with the expression of other sexually dimorphic
behavioural traits.

Prenatal testosterone has been implicated as an important factor in the development of extragenital sexual dimorphism including the differentiation
of the nervous system (Bardin & Caterall, 1981; McEwen, 1981; MacLusky
& Naftolin, 1981). One such dimorphism may be seen in the expression of
hand preferences (Geschwind & Behan, 1982; Geschwind & Galaburda,
1985; Hassler & Gupta, 1993). Geschwind and Galaburda (1985) have hy-
pothesised that testosterone may slow growth within some areas of the left
hemisphere and promote growth of certain areas in the right hemisphere.
Such a process may mean that high levels of testosterone in utero would be
associated with left-handedness and this left-preference could be seen in
higher frequencies in males.

The Geschwind and Galaburda model is controversial. The model predicts association between left-handedness and lateralised hand performance
and such things as auto-immune disorders, autism, and dyslexia. Not all of

these predictions are convincingly supported by available data (Bryden, McManus, & Bulman-Fleming, 1994). A more powerful test of the model would be to relate the expression of left-handedness and other traits directly to prenatal testosterone.

It is very difficult to test directly for the effect of prenatal testosterone on left-handedness and lateralised hand performance in humans. However, in utero exposure to testosterone may leave morphological markers which could be used to test the relationship indirectly. Prenatal testosterone comes from maternal testosterone, which may pass across the placenta and enter the foetal bloodstream, and from the foetus itself, which secretes increasing amounts from about 8 weeks (the time of Leydig cell differentiation) to mid-gestation (George, Griffin, Leshin, & Wilson, 1981; Jamison, Meier, & Campbell, 1991). From that point levels slowly decrease until a few months after birth when it has reached the low level characterising childhood (Tanner, 1990). The foetal source is dependent on the differentiation of the testes (Lording & Dekretser, 1972). Manning et al. (1998a,b) have pointed out that the differentiation of the gonads and the digits and toes is under the common control of the *Hox* genes and particularly the posterior-most *Hoxd* and *Hoxa* genes (Kondo, Zakamy, Innis, & Duboule, 1997; Peichel, Prabhakaram, & Vogt, 1997). Therefore patterns of formation of the digits would relate to the function of the gonads. In support of this Manning et al. (1998a) have found that digit asymmetry is negatively related to sperm numbers per ejaculate and to measures of sperm viability such as the Sperm Migration Test. In addition the ratio between the length of the 2nd digit or index finger and that of the 4th digit or ring finger (2D : 4D ratio) was sexually dimorphic. Men tended to have a lower 2D : 4D ratio than women, and in men 2D : 4D was negatively related to sperm numbers and testosterone levels and positively related to oestrogen and LH levels. In women 2D : 4D was positively correlated with oestrogen and LH. They also found that all relationships were stronger for the 2D : 4D ratio of the right hand compared to that of the left (Manning et al., 1998a). Relative digit length is determined early in foetal growth. Garn, Burdi, Babler, & Stinson (1975) have found that adult bone-to-bone ratios of the phalanges are established by 13 weeks. The correlations between 2D : 4D ratios and adult hormone levels are therefore likely to relate also to foetal hormone levels.

The purpose of this work was to examine the relationship between 2D : 4D ratio and measures of lateralised hand preference. The work was carried out within a large long-term study, the Jamaican Symmetry Project, which is concerned with patterns of developmental instability in rural Jamaican children (Manning et al., 1997; Trivers, Manning, Thornhill, & Singh, 1999). Our prediction was that low 2D : 4D ratio would be related to an increase in the speed of performance with the left hand relative to the right.

Method

Our Jamaican study population is described in detail in Trivers et al. (1999) and Manning et al. (1997). Our total sample was 285 children (156 boys and 129 girls) drawn from Southfield in the parish of St Elizabeth. The children were recruited from three schools in the area (Top Hill Primary, Mayfield All/Age, and Epping Forest All/Age) and the sample had an age range of 5–11 years. Subjects' age, height, and weight were recorded.

We assessed relative hand performance by the Annett peg-moving test (Annett, 1985, 1987). Trials were carried out in January 1996. A peg board with two rows of ten holes was used. Participants moved, with one hand, ten pegs from a row or holes to an empty row of holes situated about five inches in front. Each trial was timed from the moment the hand touched the first peg until the last peg was placed in its hole. There were ten trials in all, five for each hand. Mean left and right hand times were then calculated. Lateralised hand performance (LHP) was calculated by subtracting the mean left hand time from the mean right hand time, i.e. the greater the LHP the greater the tendency to perform faster with the left hand relative to the right hand. Subjects who had a faster time with the left hand compared to the right hand had an LHP > 1. There were 250 participants with a mean LHP of − 1.77 ± 1.14SD seconds and a range of −5.27 to 2.80 seconds. Boys had a significantly higher mean LHP than girls (boys, n = 134, $\bar{x}$ = 11 1.58 ± 1.08 seconds and girls, n = 116, $\bar{x}$ = − 1.99 ± 1.17 seconds, unpaired t test, t = 2.89, P = 0.004).

The 2D : 4D ratio was measured in two ways:

1. Photocopies were made of the right and left hands of 152 children (78 boys and 74 girls) in June 1998. The subjects placed their hands palm down on the centre of the glass plate and one photocopy per hand was made. Care was taken to ensure that details of major creases could be seen on the hands. When quality was poor a second photocopy was made. Vernier callipers measuring to 0.01 mm were used to measure the length of the 2nd and 4th photocopied digits from the ventral basal crease of each digit to the tip. Repeatabilities of similar measurements made directly on digit length have been high in previous studies (Manning, 1995; Trivers et al., 1999). We calculated our repeatabilities (r_1) of 2D : 4D ratios in the form of intra-class correlation coefficients using Model II single factor ANOVA tests as follows:

$$r_1 = \frac{\text{groups mean squares} - \text{error mean squares}}{\text{groups mean squares} + \text{error mean squares}}$$

In addition we used repeated measures ANOVA tests to calculate the ratio (F) between groups mean squares (i.e. the real differences between individuals) and the error mean squares (i.e. the error in our repeated measures; Zar, 1984).

Repeatabilities were obtained by measuring the 2nd and 4th digits twice from 30 hands. The r_1 values of the 2D : 4D ratios were: right hand $r_1 = 0.77$ and left hand $r_1 = 0.65$. The between-individual variance was greater than the measurement error for both hands (right hand $F = 8.00$, $P = .0001$, left hand $F = 4.62$, $P = .0002$). We concluded that our measurements reflected real differences between the 2D : 4D ratios of different subjects.

2. X rays of the right and left hands of 244 participants (135 boys and 109 girls) were taken in January 1996. The X rays were taken from the dorsal surface of the hand at 70kVp, 10mA and 1 second at a collimator distance of 70cm. Measurements of the 2nd and 4th digits were made from the proximal tip of the shaft of phalanx 1 to the distal tip of phalanx 3. Measurements were made with vernier callipers measuring to 0.01mm. The hands of 30 children were measured twice. Repeatabilities of the 2D : 4D ratios were: right hand $r_1 = 0.77$ and left hand $r_1 = 0.79$. The F ratios were highly significant (right hand $F = 7.87$, $P = -.0001$, left hand $F = 8.52$, $P = .0001$). We concluded our measurements reflected real differences in 2D : 4D ratios between subjects.

There were photocopies and X rays from 135 children. The 2D : 4D ratio calculated from the photocopies was significantly correlated with the ratio from the X rays (right hand, $r_1 = 0.46$, $F = 34.89$, $P = .0001$; left hand, $r_1 = 0.47$, $F = 38.90$, $P = .0001$). As the photocopies were made approximately 2.5 years after the X rays this indicates the 2D : 4D ratio is stable over time and our two different modes of measurement were recording real differences between subjects.

Results

Study 1: Photocopies of the hands

Mean 2D : 4D ratios were right hand x = 0.935 ± 0.035SD, left hand x̄ = 0.934 ± 0.036SD. The ratios of left and right hands were significantly correlated ($r = 0.68$, $F = 124.79$, $P = .0001$). As in the English sample (Manning et al., 1998b) there was evidence of sexual dimorphism in the ratio with boys showing a lower ratio than girls. The difference was significant for the right hand but not for the left (right hand, boys x̄ = 0.929 ± 0.037SD and girls, x̄ = 0.941 ± 0.032SD, unpaired t test, t = 2.12, P = .03; left hand, boys, 0.932 ± 0.36SD and girls, 0.936 ± 0.036SD, t = 0.78, P = 0.43).

Age, height (cm), and weight (kg) were not strong predictors of 2D : 4D with only one relationship, that of age and left hand 2D : 4D, showing significance (simple linear regressions, age; right hand, $b = 0.003$, $F = 1.75$, $P = .19$, left hand, $b = 0.006$, $F = 8.82$, $P = .004$: weight; right hand, $b = 0.0002$, $F = 1.59$, $P = .21$, left hand, $b = 0.0002$, $F = 0.93$, $P = .34$: height, right hand, $b = 0.001$, $F = 0.47$, $P = .49$, left hand, $b = 0.001$, $F = 1.3$, $P = .25$). Manning et al. (1998b) found no relationship between age and ratio in their English data.

There were 130 subjects (63 boys and 67 girls) with LHP scores and photocopies of their hands. Descriptive statistics (means and standard divisions) of the sample were as follows; 2D : 4D ratio, right hand $x = 0.936 \pm 0.036$, left hand $\bar{x} = 0.936 \pm 0.036$: LHP $\bar{x} = 0.87 \pm 0.07$: Age $\bar{x} = 7.82 \pm 1.45$ years: weight $\bar{x} = 24.79 \pm 5.35$kg: height $\bar{x} = 128.98 \pm 9.37$cms. Table 1 shows the results of simple linear regressions of LHP on 2D : 4D calculated from photocopies. The 2D : 4D ratio of the right hand was a significant negative predictor of LHP (Fig. 1) i.e. subjects with low 2D : 4D ratio had a greater tendency towards low left hand times than subjects with high 2D : 4D ratios. Left hand 2D : 4D ratio did not predict LHP. A regression of LHP on 2D : 4D left hand—2D : 4D right hand showed a significant positive association ($b = 9.68$, $F = 7.83$, $P = .006$, Fig. 2). This means that a low 2D : 4D ratio in the right hand relative to a high ratio in the left hand correlates with a tendency towards faster performance with the left hand relative to the right.

We found some evidence that LHP was associated with asymmetry of the 2nd and 4th digits. Absolute asymmetry was calculated by subtracting the length of the right side of the trait from the length of the left side (L-R). Descriptive statistics of asymmetries are in Table 2. Dermatoglyphic asymmetry has been reported to show a curvilinear relationship with hand preference (Yeo, Gangestad, & Daniel, 1993). We found a similar and significant U-shaped relationship between composite asymmetry of the 2nd and 4th digit and LHP (Table 2).

Table 1. Linear Regression Analyses: Photocopies

Trait	beta	F	P
2D:4D Right hand (boys and girls)	−6.24	4.65	0.03
2D:4D Right hand (boys only)	−6.05	2.51	0.11
2D:4D Right hand (girls only)	−4.04	0.88	0.35
2D:4D Left hand (boys and girls)	−0.73	0.06	0.80
2D:4D Left hand (boys only)	−5.30	1.74	0.19
2D:4D Left hand (girls only)	4.62	1.40	0.24

The results of simple linear regression analyses of LHP (lateralised hand preference) regressed on 2D:4D ratio for right hands and left hands measured from photocopies (boys and girls $n = 130$, boys $n = 60$, girls $n = 67$).

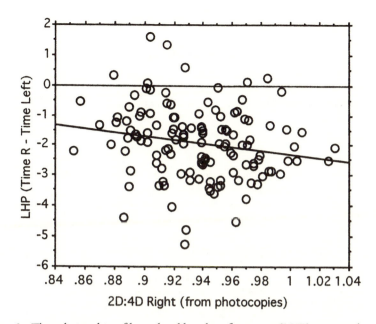

Figure 1. The relationship of lateralised hand performance (LHP) regressed on 2D : 4D (measured from photocopies) of the right hand of 130 boys and girls (y = − 6.24 × 1.21). Higher LHP values indicate an increased speed of left hand performance.

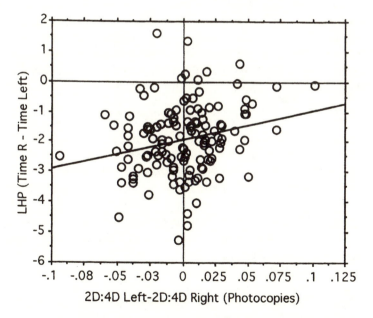

Figure 2. The relationship of lateralised hand performance (LHP) regressed on 2D : 4D left hand—2D : 4D right hand (measured from photocopies) of 130 boys and girls (y = 9.68 × 1.92). Higher LHP values indicate an increased speed of left hand performance.

Table 2. Descriptive Statistics: Photocopies

Tests of Directional Asymmetry (one-sample t tests, mean at zero)
2nd digit,	$x = -0.08$mm, $t = 0.65$, $P = .52$.
4th digit,	$x = -0.05$mm, $t = 0.32$, $P = .75$.

No evidence of directional asymmetry.

Tests of Skewness (g_1) and kurtosis (g_2)
2nd digit	$g_1 = 0.25$, $Z = 1.27$, $P = .10$.
	$g_2 = 0.09$, $Z = 0.23$, $P = .41$.
4th digit	$g_1 = 0.45$, $Z = 2.26$, $P = .01$.
	$g_2 = 3.47$, $Z = 8.89$, $P = .0001$.

There was evidence that the asymmetries of the 4th digit were skewed and leptokurtotic.

2nd Order Polynomial Regressions of Digit Asymmetry on LHP
2nd digit	$r^2 = 0.03$, $F = 2.03$, $P = .14$.
4th digit	$r^2 = 0.03$, $F = 1.98$, $P = .14$.
Composite Asymmetry	$r^2 = 0.06$, $F = 3.58$, $P = .03$.

Descriptive statistics of absolute asymmetries calculated (L–R) from the lengths of the 2nd and 4th digits measured from photocopies.

Study II: X rays of the Hands

Mean 2D : 4D ratios calculated from the X rays were lower than those from the photocopies (X rays, right hand $\bar{x} = 0.907 \pm 0.02$, left hand $\bar{x} = 0.903 \pm 0.02$). Also there was no evidence of sexual dimorphism in the data (right hand, boys $\bar{x} = 0.907 \pm 0.02$, girls $\bar{x} = 0.907 \pm 0.02$: left hand, boys $\bar{x} = 0.905 \pm 0.02$, girls $\bar{x} = 0.901 \pm 0.02$). These differences presumably reflect the different types of measurement i.e. bone to bone compared to soft tissue measurements.

As with the ratios calculated from the photocopies we found that age, height, and weight were not strong predictors of 2D : 4D ratio (simple linear regressions; age, right hand, $b = 0.0001$, $F = 0.02$, $P = .89$, left hand, $b = 0.0002$, $F = 0.05$, $P = .83$: height, right hand, $b = 0.0001$, $F = 0.58$, $P = .45$, left hand, $b = 0.0002$, $F = 2.01$, $P = .16$: weight, right hand, $b = 0.0002$, $F = 0.39$, $P = .53$, left hand, $b = 0.0002$, $F = 0.91$, $P = .34$).

There were 215 subjects (116 boys and 99 girls) with LHP scores and X rays of their hands. Descriptive statistics (means and standard deviations) of the sample were as follows; 2D : 4D ratio, right hand $\bar{x} = 0.907 \pm 0.023$, left hand $\bar{x} = 0.904 \pm 0.023$: LHP $\bar{x} = 0.87 \pm 0.07$: age $\bar{x} = 8.30 \pm 1.72$ years: weight $\bar{x} = 25.85 \pm 6.10$: height $\bar{x} = 129.99 \pm 10.33$cms) (Fig.3). Table 3 shows the results of simple linear regressions of LHP on 2D : 4D calculated from X rays. The 2D : 4D ratio of the right hand in boys was a significant negative predictor of LHP. That is, in boys those individuals with

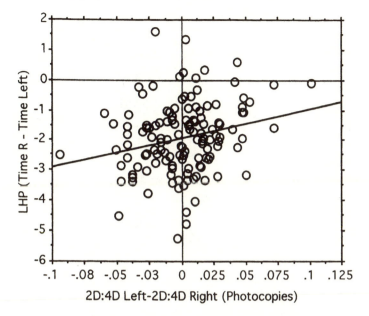

Figure 3. The relationship of lateralised hand performance (LHP) regressed on 2D : 4D right hand (measured from X rays) of 116 boys (y = − 7.99 × 1.39). Higher LHP values indicate an increased speed of left hand performance.

low 2D : 4D ratio in the right hand had a higher tendency towards left handedness than boys with high ratios. The remaining relationships were not close to significance. A regression of LHP on L-R 2D : 4D again showed a positive and significant association (b = 10.03, F = 4.69, P = .03, Fig. 4). That is a low 2D : 4D in the right hand and high 2D : 4D in the left hand gave a tendency towards a faster performance with the left hand compared to the right.

Table 3. Linear Regression Analyses: X Rays

Trait	beta	F	P
2D:4D Right hand (boys and girls)	−3.59	1.23	0.27
2D:4D Right hand (boys only)	−7.99	4.12	0.04
2D:4D Right hand (girls only)	1.70	0.11	0.74
2D:4D Left hand (boys and girls)	1.19	0.14	0.71
2D:4D Left hand (boys only)	−3.82	1.02	0.32
2D:4D Left hand (girls only)	6.58	1.60	0.21

The results of simple linear regression analyses of LHP (lateralised hand performance) regressed on 2D:4D ratio (measured from X rays) for right hands and left hands (boys and girls n = 215, boys n = 116 and girls n = 99).

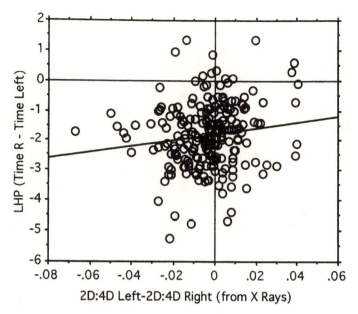

Figure 4. The relationship of lateralised hand performance (LHP) regressed on 2D : 4D left hand—2D : 4D right hand (measured from X rays) of 215 boys and girls (y = 10.03 6–1.78). Higher LHP values indicate an increased speed of left hand performance.

Table 4. Descriptive Statistics: X Rays

Tests of Directional Asymmetry (one-sample t tests, mean at zero)

2nd digit	$x = 0.31mm, t = 5.51, P = .0001.$
4th digit	$x = 0.11mm, t = 1.87, P = .07.$

Evidence of directional asymmetry in the asymmetries of digit 2.

Tests of Skewness (g_1) and kurtosis (g_2)

2nd digit	$g_1 = 0.22, Z = 1.37, P = .09.$
	$g_2 = 0.08, Z = 0.25, P = .40.$
4th digit	$g_1 = 0.55, Z = 3.43, P = .003.$
	$g_2 = 4.15, Z = 13.21, P = .0001.$

There was evidence that the asymmetries of the 4th digit were skewed and leptokurtotic.

2nd Order Polynomial Regressions of Digit Asymmetry on LHP

2nd digit	$r^2 = 0.01, F = 1.30, P = .27.$
4th digit	$r^2 = 0.01, F = 0.9, P = .40.$
Composite Asymmetry	$r^2 = 0.001, F = 0.10, P = .91.$

Descriptive statistics of absolute asymmetries calculated (L-R) from the lengths of the 2nd and 4th digits measured from X rays.

In this sample we found no evidence that LHP was related to asymmetry of the 2nd and 4th digits as measured from X rays. Absolute asymmetry was calculated as previously (Table 4) and there was no U-shaped relationship between composite asymmetry of the 2nd and 4th digit and LHP.

Discussion

We have found evidence that low 2D : 4D ratios in the right hand correlate with a reduction in left-hand performance times. This effect is seen in our male and female sample combined in Study I and in males only in Study II. Left hand 2D : 4D ratios alone did not predict LHP in either study. The strongest correlate of LHP in both studies was L-R 2D : 4D ratio.

There are two possible interpretations of our results. (1) They could arise from a direct effect of relative 2nd and 4th digit lengths on manual dexterity. For example, a hand with a 2D : 4D ratio = 1 may be more dextrous than a hand with a ratio of say 0.90. If this was so, an individual with a low ratio in the right hand and a high ratio in the left might favour the use of the latter hand over the former. We do not think this is a correct interpretation of our data. If low ratios were correlated with preferred use of the other hand then simple linear regressions of LHP on 2D : 4D ratio of the left hand would show strong positive relationships, in addition to the negative relationships of the right hand. An explanation involving both left and right hand 2D : 4D ratios is necessary. (2) An alternative explanation is the following model: (a) it is known that adult testosterone levels are negatively correlated with 2D : 4D in the right hand, this suggests that foetal levels of testosterone show similar association; (b) foetal testosterone may inhibit the growth of the left hemisphere (Geschwind & Galaburda, 1985) and may also lower the 2D : 4D ratio of the right hand (Manning et al., 1998B); (c) a reduction in growth of the left hemisphere could be associated with an enhanced growth rate of the right hemisphere but perhaps the effect of testosterone on the 2D : 4D ratio of the left hand is less marked; (d) an increase in dexterity of the left hand is facilitated by this relative effect on right and left hemisphere growth and is negatively correlated with 2D : 4D ratio of the right hand and positively with L-R 2D : 4D ratio. In support of this model it is known that dermatoglyph asymmetry, which is determined in the first trimester of pregnancy, is related to hand preference (Yeo et al., 1993) and to testosterone levels in adults (Jamison et al., 1993). Against the model is the lack of direct evidence that foetal testosterone is related to cerebral lateralisation (Grimshaw, Bryden, & Finnegan, 1995). We are at present investigating the relationship between foetal testosterone and 2D : 4D ratio.

Testosterone production may be ultimately dependent on *Hox* gene function. The posterior-most *Hoxd* and *Hoxa* genes are essential for limb and genital development (Herault, Fradeau, & Zakany, 1997; Peichel et al., 1997). In humans the hand-foot-genital syndrome is marked by anatomical defects in digits and genitalia, and is the result of mutation within *Hoxa* (Mortlock & Innis, 1997). Polymorphisms within *Hoxd* and *Hoxa* may therefore result in variation in gonad and digit form and function. The former could then affect Leydig cell differentiation and therefore the production of testosterone.

If the 2D : 4D ratio (particularly the ratio for the right hand) is a marker of *in utero* testosterone levels it may be useful as a predictor of other extra-genital traits. Geschwind and Galaburda (1985) have implicated testosterone in the aetiology of autism, dyslexia, migraine, stammering, autoimmune disease, sexual preferences, and spatial, language, music, and mathematical abilities. We suggest that these traits will also correlate with 2D : 4D ratios of the right hand and L-R 2D : 4D ratios.

NOTES

For permission to work, we thank the Jamaican Ministry of Education and Culture. For their help, we thank the principals and teachers, and we thank especially Mr Vernon Cameron. For help with the measurements, we are especially grateful to Mirit Cohen and Allison Sievwright. For their participation we thank the children and parents of the Top Hill area. For support we are grateful to Rutgers University, The Ann and Gordon Getty Foundation, and the Biosocial Research Foundation.

REFERENCES

Annett, M. (1985). *Left, right, hand and brain: The right shift theory.* Hove, UK: Lawrence Erlbaum Associates Ltd.

Annett, M. (1987). Handedness as chance or as species characteristic. *Behavioural and Brain Sciences, 10,* 263–264.

Bardin, C. W., & Caterall, C. F. (1981). Testosterone: A major determinant of extragenital sexual dimorphism. *Science, 211,* 1285–1294.

Bryden, M. P., McManus, I. C., & Bulman-Fleming, M. B. (1994). Evaluating the empirical support for the Geschwind—Behan—Galaburda model of cerebral lateralisation. *Brain and Cognition, 26,* 103–167.

Garn, S. M., Burdi, A. R., Babler, W. J., & Stinson, S. (1975). Early prenatal attainment of adult metacarpal—phalangeal rankings and proportions. *American Journal of Physical Anthropology, 43,* 327–332.

George, F. W., Griffin, J. E., Leshin, M., & Wilson, J. D. (1981). *Endocrine control of sexual differentiation in the human.* In M. J. Novy & J. A. Resko (Eds.), *Fetal endocrinology* (pp. 341–357). New York: Academic Press.

Geschwind, N., & Behan, P. (1982). Left-handedness: Association with immune disease, migraine and developmental learning disorder. *Proceedings National Academy of Sciences USA, 79,* 5097–5100.

Grimshaw, G. M., Bryden, M. P., & Finegan, J. K. (1995). Relations between prenatal testosterone and cerebral lateralization in children. *Neuropsychology, 9,* 68–79.

Hassler, M., & Gupta, D. (1993). Functional brain organisation, handedness, and immune vulnerability in musicians and non-musicians. *Neuropsychologia, 31,* 655–660.

Herault, Y., Fradeau, N., & Zakany, J. (1997). *Ulnaless (Ul),* a regulatory mutation inducing both loss-of-function and gain-of-function of posterior *Hoxd* genes. *Development, 124,* 3493–3500.

Jamison, C. S., Meier, R. J., & Campbell, B. C. (1993). Dermatoglyphic asymmetry and testosterone levels in normal males. *American Journal of Physical Anthropology, 90,* 185–198.

Kondo, T., Zakany, J., Innis, J. W., & Duboule, D. (1997). Of fingers, toes and penises. *Nature, 390,* 185–198.

Lording, D. W., & Dekretser, D. M. (1972). Comparative ultrastructural and histochemical studies of the interstitial cells of the rat testis during fetal and postnatal development. *Journal Reproduction and Fertility, 29,* 261–269.

MacKlusky, N. J., & Naftolin, F. (1981). Sexual differentiation of the nervous system. *Science, 211,* 1294–1303.

Manning, J. T. (1995). Fluctuating asymmetry and bodyweight in men and women: Implications for sexual selection. *Ethology and Sociobiology, 16,* 145–152.

Manning, J. T., Scutt, D., & Lewis-Jones, D. I, (1998a). Development stability, ejaculate size and sperm quality in men. *Evolution and Human Behaviour, 19,* 273–282.

Manning, J. T., Scutt, D., Wilson, J., & Lewis-Jones, D. I. (1998b). The ratio of 2nd to 4th digit length: A predictor of sperm numbers and concentrations of testosterone, luteinizing hormone and oestrogen. *Human Reproduction, 13,* 3000–3004.

Manning, J. T., Trivers, R. L., Thornhill, R., Singh, D., Denman, J., Eklo, M. H., & Anderton, R. H. (1997). Ear asymmetry and left-side cradling. *Evolution and Human Behaviour, 18,* 1–14.

McEwen, B. S. (1981). Neural gonadal steroid actions. *Science, 211,* 1303–1311.

Mortlock, D. P., & Innis, J. W. (1997). Mutation of Hoxa 13 in hand—foot—genital syndrome. *Nature Genetics, 15,* 179–180.

Peichel, C. L., Prabhakaran, B., & Vogt, T. F. (1997). The mouse *Ulnaless* mutation deregulates posterior *Hoxt* gene expression and alters appendicular patterning. *Development, 124,* 3481–3492.

Tanner, J. M. (1990). *Foetus into man: Physical growth from conception into maturity.* Cambridge, MA: Harvard Press.

Trivers, R. L., Manning, J. T., Thornhill, R., & Singh, D. (1999). The Jamaican Symmetry Project: A long-term study of fluctuating asymmetry in rural Jamaican children. *Human Biology, 71,* 419–432.

Yeo, R. A., Gangestad, S. W., & Daniel, W. F. (1993). Hand preference and development stability. *Psychobiology, 21,* 161–168.

Zar, J. H. (1984). *Biostatistical analysis.* New Jersey: Prentice-Hall.

Postscript

For a very detailed review of a fascinating series of correlations in humans between the 2nd:4th digit ratio and a wide range of variables, see Manning (2002). For a very useful review of the literature on fluctuating asymmetry, see Moller and Swaddle (1997).

REFERENCES

Agrawal, A. 2001. Sexual selection and the maintenance of sexual reproduction. *Nature* 411: 692–695.

Agrawal, A, Browdie, E, and Brown, J. 2001. Parent-offspring coadaptation and the dual genetic control of maternal care. *Science* 292: 1710–1712.

Alexander, R. 1974. The evolution of social behavior. *Annu. Rev. Ecol. Syst.* 5: 325–383.

Alexander, R, and Sherman, P. 1977. Local mate competition and parental investment in social insects. *Science* 196: 494–500.

Andersson, M. 1994. *Sexual Selection*. Princeton, NJ: Princeton University Press.

Arevalo, E, Strassmann, J, and Queller, D. 1998. Conflicts of interest in social insects: male production in two species of *Polistes*. *Evolution* 52(3): 797–805.

Aron, S, Passera, L, and Keller, L. 1999. Evolution of social parasitism in ants: size of sexuals, sex ratio and mechanisms of caste determination. *Proc. Roy. Soc. Lond. B.* 266: 173–177.

Aureli, F, Cozzolino, R, Cordischi, C, and Scucchi, S. 1992. Kin-oriented redirection among Japanese macaques: an expression of a revenge system? *Anim. Behav.* 44: 283–291.

Aureli, F, and van Schaik, C. 1991. Post-conflict behaviour in long-tailed macaques (*Macaca fascicularis*). *Ethology* 89: 89–100.

Axelrod, R, and Hamilton, W. 1981. The evolution of cooperation. *Science* 211: 1390–1396.

Baer, B, and Schmid-Hempel, P. 1999. Experimental variation in polyandry affects parasite loads and fitness in a bumble-bee. *Nature* 397: 151–154.

Bateman, A. 1948. Intrasexual selection in *Drosophila*. *Heredity* 2: 349–368.

Beukeboom, L, de Jong T, and Pen, I. 2001. Why girls want to be boys. BioEssays 23: 477–480.

Bendor, J, and Swistak, P. 1995. Types of evolutionary stability and the problem of cooperation. *Proc. Natl. Acad. Sci. USA* 92: 3596–3600.

Bendor, J, and Swistak, P. 1997. The evolutionary stability of cooperation. *Am. Pol. Sci. Rev.* 91: 290–307.

Betzig, L., and Turke, P. 1986. Parental investment by sex on Ifaluk. *Ethology and Sociobiology* 7: 29–37.

Bianchi, D, Zickwolf, G, Weil, G, Sylvester, S, and DeMaria, M. 1996. Male fetal progenitor cells persist in maternal blood for as long as 27 years postpartum. *Proc. Natl. Acad. Sci USA*. 93: 705–708.

Binmore, K. 1998. The evolution of fairness norms. *Rationality and Society* 10: 275–301.

Boesch, C, 1997. Evidence for dominant wild female chimpanzees investing more in sons. *Animal Behavior* 54: 811–815.

Bourke, A, and Franks, N. 1995. *Social Evolution in Ants*. Princeton: Princeton University Press.

Burt, A, Bell, G, and Harvey, P. 1991. Sex differences in recombination. *J. Evol. Biol.* 4: 259–277.

Cameron, E, Linklater, W, Stafford, K, and Veltman, C. 1999. Birth sex ratios relate to mare condition at conception in Kaimanawa horses. *Behav. Ecol.* 10: 472–475.

Clark, M, and Galef, B. 1995. Prenatal influences on reproductive life history strategies. *TREE* 10: 151–153.

Clutton-Brock, T, Albon, S. and Guinness, F. 1984. Maternal dominance, breeding success and birth sex ratios in red deer. *Nature* 308: 358–360.

Clutton-Brock, T., Albon, S., and Guinness, F. 1985. Parental investment and sex differences in juvenile mortality in birds and mammals. Nature 313: 131–133.

Clutton-Brock, T, and Parker, G. 1995. Punishment in animal societies. *Nature* 373: 209–216.

Connor, R, Heithaus, M, and Barre, L. 1999. Superalliance of bottlenose dolphins. *Nature* 397: 571–572.

Crozier, R, and Pamilo, P. 1996. *Evolution of Social Insect Colonies: Sex Allocation and Kin Selection*. New York: Oxford University Press.

Cronk, L. 1991. Preferential investment in daughters over sons. *Human Nature* 2: 387–417.

Cronk, L. 2000. Female-biased parental investment and growth performance in Mukugodo children. In L. Cronk, N. Chagnon and W. Irons (eds.) *Adaptation and Human Behavior: An Anthropological Perspective*. Hawthorne, NY: Aldine de Gruyter, pp. 203–221.

Day, T, and Taylor, P. Chromosomal drive and the evolution of meiotic nondisjunction and trisomy in humans. *Proc. Natl. Acad. Sci USA* 95: 2361–2365.

Dawkins, R. 1976. *The Selfish Gene*. New York: Oxford University Press.

Dawkins, R, and Carlisle, T. 1976. Parental investment, mate desertion and a fallacy. *Nature* 262: 131–132.

de Vos, H, and Zeggelink, E. 1997. Reciprocal altruism in human social evolution: the viability of reciprocal altruism with a preference for 'old-helping-partners'. *Evolution and Human Behavior* 18: 261–278.

de Waal, F. 1997a. The chimpanzee's service economy: food for grooming. *Evol. Hum. Behav.* 18: 1–12.

de Waal, F. 1997b. Food transfer through mesh in brown capuchins. *J. Comp. Psychol.* 111(4) 370–378.

de Waal, F, and Berger, M. 2000. Payment for labour in monkeys. *Nature* 404: 563.

Drury, W. 1998. *Chance and Change: Ecology for Conservationists* (ed. J. Anderson). Berkeley: University of California Press.

Elnum, S. 2000. Highly fecund mothers sacrifice offspring survival to maximize fitness. *Nature* 405: 565–567.

Emlen, S. and Wrege, P. 1992. Parent–offspring conflict and the recruitment of helpers in bee-eaters. *Nature* 356: 331–333.

Etcoff, N, Ekman, P, Magee, J, and Frank, M. 2000. Lie detection and language comprehension: people who can't understand words are better at picking up lies about emotions. *Nature* 405: 139.

Field, J, Shreeves, G, Sumner, S, and Casiraghi, M. 2000. Insurance-based advantage to helpers in a tropical hover wasp. *Nature* 404: 869–871.

Foster, K, and Ratnieks, F. 2001. Paternity reproduction and conflict in vespine wasps: a model system for testing kin selection predictions. *Behav. Ecol. Sociobio.* 50: 1–8.

Frank, S. 1995. Mutual policing and repression of competition in the evolution of cooperative groups. *Nature* 377: 520–522.

Frean, M. 1994. The prisoner's dilemma without synchrony. *Proc. R. Soc. Lond. B.* 257: 75–79.

Freed, L. 2000. Ornamentation, mate choice and sexual selection. *TREE* 15: 471.

Freeman, D, Wachocki, B, Goldschlage, D, and Michaels, H. 1994. Seed size and sex ratio in spinach: Application of the Trivers-Willard hypothesis to plants. *Ecoscience* 1: 54–63.

Furlow, B, Gangestad, S, and Armijo-Prewitt, T. 1998. Developmental stability and human violence. *Proc. Roy. Soc. Lond. B.* 265: 1–6.

Gangestad, S, and Thornhill, R. 1998. Menstrual cycle variation in women's preferences for the scent of symmetrical men. *Proc. Roy. Soc. Lond. B.* 265: 927–933.

Gaulin, S, and Robbins, C. 1991. Trivers/Willard effects in contemporary North American society. *Amer. J. Phys. Anthro.* 85: 61–69.

Godfray, H. 1992. The evolution of forgiveness. *Nature* 355: 206–207.

Godfray, H. 1995. Evolutionary theory of parent-offspring conflict. *Nature* 376: 133–138.

Godin, J, and Davis, S. 1995. Who dares, benefits: predator approach behaviour in guppy (*Poecilia reticulata*) deters predator pursuit. *Proc. R. Soc. Lond. B.* 259: 193–200.

Haig, D. 1993. Genetic conflicts in human pregnancy. *Q. Rev. Biol.* 68(4): 495–532.

Haig, D. 1996. Placental hormones, genomic imprinting, and maternal-fetal communication. *J. Evol. Biol.* 9: 357–380.

Haig, D. 1997. Parental antagonism, relatedness asymmetries, and genomic imprinting. *Proc. Roy. Soc. Lond. B.* 264; 1657–1662.

Haig, D. 2002. *Genomic Imprinting and Kinship.* New Brunswick, NJ: Rutgers University Press.

Haig, D, and Graham, C. 1991. Genomic imprinting and the strange case of the insulin-like growth factor II receptor. *Cell* 64: 1045–1046.

Hamilton, W. 1964. The genetical evolution of social behavior. *J. Theor. Biol.* 7: 1–52.

Hamilton, W. 1967. Extraordinary sex ratios. *Science* 156: 477–488.

Hamilton, W, and Zuk, M. 1982. Heritable true fitness and bright birds: a role for parasites? *Science* 218: 384–387.

Hansen, D, Moeller, H, and Olsen, J. 1999. Severe periconceptional life events and the sex ratio in offspring: follow up study based on five national registers. *BMJ* 319: 548–549.

Harris, J. 1995. Where is the child's environment? A group socialization theory of development. *Psychol. Rev.* 102: 458–489.

Hastings, M, Queller, D, Eischen, F, and Strassmann, J. 1998. Kin selection, relatedness, and worker control of reproduction in a large-colony epiponine wasp, *Brachygastra mellifica. Behav. Ecol.* 9: 573–581.

Herbers, J, and Stuart, R. 1998. Patterns of reproduction in slave-making ants. *Proc. R. Soc. Lond. B* 265: 875–887.

Hicks, R, and Trivers, R. 1983. The social behavior of *Anolis valencienni.* In A. Rhodin and K. Miyata (eds.) *Advances in Herpetology and Evolutionary Biology.* Cambridge, Mass.: Museum of Comparative Zoology. pp. 570–595.

Hrdy, S. 1999. *Mother Nature: A History of Mothers, Infants and Natural Selection.* New York: Pantheon.

James, W. 2002. The variation of the probability of a son within and across couples. *Hum. Repro.* 15: 1184–1188.

Jantz, R, and Webb, R. 1982. Interpopulation variation in fluctuating asymmetry of the palmar a-b ridge count. *Amer. J. Phys. Anthro.* 57: 253–259.

Jones, N. 1978. Natural selection and birthweight. *Ann. Hum. Biol.* 5(5): 487–489.

Kirkpatrick, M. 1982, Sexual selection and the evolution of female choice. *Evolution* 36: 1–12.

Koziel, S, and Ulijaszek, S. 2001. Waiting for Trivers and Willard: do the rich really favor sons? *Amer. J. Phys. Anthro.* 115: 71–79.

Kruuk, L, Clutton-Brock, T, Albon, S, Pemberton, J, and Guinness, F. 1999. Population density affects sex ratio variation in red deer. *Nature* 399: 459–461.

Leigh, E. 1984. Egg trading in the chalk bass, *Serranus toruagarum,* a simultaneous hermaphrodite. *Z. Tierpsychol* 66: 143–151.

Liersch, S, and Schmid-Hempel, P. 1998. Genetic variation within social insect colonies reduces parasite load. *Proc. R. Soc. Lond. B.* 265: 221–225.

Manning, J. 2002. *Digit Ratio: A Pointer to Fertility, Behavior and Health.* New Brunswick, NJ: Rutgers University Press.

Manning, J., Anderton, R., and Shutt, M. 1997. Parental age gap skews child sex ratio. *Nature* 389: 344.

Manning, J., Trivers, R., Thornhill, R., Singh, D., Denman, J., Eklo, M., and Anderton, R. 1997. Ear asymmetry and left-side cradling. *Evol. Hum. Behav.* 18: 327–340.

Manning, J., and Wood, D. 1998. Fluctuating asymmetry and aggression in boys. *Human Nature* 9: 53–65.

Manning, J., Scutt, D., Wilson, J., and Lewis-Jones, D. 1998. The ratio of 2nd: 4th digit length: a predictor of sperm numbers and concentrations of testosterone, luteinizing hormone and oestrogen. *Hum. Repro.* 13; 3000–3004.

Manning, J., Trivers, R., Singh, D., and Thornhill, R. 1999. The mystery of female beauty. *Nature* 399: 214–215.

Manning, J., Barley, L., Walton, J., Lewis-Jones, D., Trivers, R., Singh, D., Thornhill, R., Rohde, P., Bereczkei, T., Henzi, P., Soler, M., and Szwed, A. 2000. The 2nd: 4th digit ratio, sexual dimorphism, population differences and reproductive success: evidence for sexually antagonistic genes. *Evol. Hum. Behav.* 21: 163–183.

Margulis, S. 1997. Inbreeding-based bias in parental responsiveness to litters of old-field mice. *Behav. Ecol. Sociobiol.* 41: 177–184.

Marler, P. 1955. The characteristics of certain animal cells. *Nature* 176: 6–7.

McGuire, M., and Troisi, A. 1998. *Darwinian Psychiatry*. New York: Oxford University Press.

McNamara, J., Gasson, C., and Houston, A. 1999. Incorporating rules for responding to evolutionary games. *Nature* 401: 368–371.

Mock, D., and Parker, G. 1997. *The Evolution of Sibling Rivalry*. Oxford: Oxford University Press.

Moller, A. 1992. Female swallow preference for symmetrical male sexual ornaments. *Nature* 357: 238–240.

Moller, A. 1997 Developmental selection against developmentally unstable offspring and sexual selection, *J. Theor. Biol.* 185: 415–422.

Moller, A, and Swaddle, J. 1997. *Asymmetry, Developmental Stability, and Evolution*. Oxford: Oxford University Press.

Mueller, U. 1993. Social status and sex. *Nature* 363: 490.

Nowak, M., and Sigmund, K. 1998a. Evolution of indirect reciprocity by image scoring. *Nature* 393: 573–577.

Nowak, M, and Sigmund, K. 1998b. The dynamics of indirect reciprocity. *J. Theor. Biol.* 194: 561–574.

Ohlsson, R. 1999. *Genomic Imprinting*. Heidelberg: Springer-Verlag.

Packer, C. 1977. Reciprocal altruism in *Papio anubis*. *Nature* 265: 441–443.

Pugesek, B. 1990. Parental effort in the California gull: tests of parent-offspring conflict theory. *Behav. Ecol. Sociobiol.* 27: 211–215.

Queller, D., Peters, J., Solís, C, and Strassmann, J. 1997. Control of reproduction in social insect colonies: individual and collective relatedness preferences in the paper wasp, *Polistes annularis*. *Behav. Ecol. Sociobiol.* 40: 3–16.

Roberts, G. 1998. Competitive altruism: from reciprocity to the handicap principle. *Proc. R. Soc. Lond. B.* 265: 427–431.

Roberts, G, and Sherratt, T. 1999. Development of cooperative relationships through increasing investment. *Nature* 394: 175–179.

Rodriguez-Giroues, M., Zuniga, J., and Redondo, T. 2001. Effects of begging on growth rates of nest living chicks. *Behav. Ecol.* 12: 269–274.

Schall, J, and Staats, C. 1997. Parasites and the evolution of extravagant male characters: *Anolis* lizards as a test of the Hamilton-Zuk hypothesis. *Oecologia* 116: 543–548.

Schmid-Hempel, P. 1994. Infection and colony variability in social insects. *Philos. Trans. R. Soc. Lond. B.* 346: 313–321.

Schwabl, H. 1993. Yolk is a source of maternal testosterone for developing birds. *Proc. Natl. Acad. Sci. USA* 90: 11446–11450.

Seger, J. 1983. Partial bivoltinism may cause alternating sex-ratio biases that favour eusociality. *Nature* 301: 59–62.

Siller, S. 2001. Sexual selection and the maintenance of sex. *Nature* 411: 689–692.

Solís, C., Huges, C., Klinger, C., Strassmann, J., and Queller, D. 1998. Lack of kin discrimination during wasp colony fission. *Behav. Ecol.* 9(2): 172–176.

Stamps, J., Losos, J., and Andrews, R. 1997. A comparative study of density and sexual size dimorphism in lizards. *Am. Nat.* 149: 64–90.

Strassmann, J., Solís, C., Hughes, C., Goodnight, K., and Queller, D. 1997. Colony life history and demography of a swarm-founding social wasp. *Behav. Ecol. Sociobiol.* 40: 71–77.

Thornhill, R. 1992. Female preference for the pheromone of males with low fluctuating symmetry in the Japanese scorpionfly (*Panorpa japonica*: Mecoptera). *Behav. Ecol.* 3: 277–283.

Thornhill, R., and Sauer, K. 1992. Genetic sire effects on the fighting ability of sons and daughters and mating success of sons in the scorpionfly (*Panorpa vulgaris*). *Anim. Behav.* 43: 255–264.

Trivers, R. 1985. *Social Evolution*. Menlo Park, CA.: Benjamin/Cummings.

Trivers, R. 1988. Sex differences in rates of recombination and sexual selection. In R. Michod and D. Levin (eds.) *The Evolution of Sex: An Examination of Current Ideas*, pp. 260–276. Sunderland, MA.: Sinauer.

Trivers, R. 1991. Deceit and self-deception: The relationship between communication and consciousness. In: M. Robinson and L. Tiger (eds.) *Man and Beast Revisited*, pp 175–191. Washington, DC: Smithsonian.

Trivers, R., and Burt, A. 1991. Kinship and genomic imprinting. In R. Ohlsson (ed.) *Genomic Imprinting*. Heidelberg: Springer-Verlag.

Trivers, R., Manning, J. T., Thornhill, R., Singh, D., and McGuire, M. 1999. The Jamaican symmetry project: a long-term study of fluctuating asymmetry in rural Jamaican children. *Hum. Biol.* 71: 419–432.

Turner, P., and Chao, L. 1999. Prisoner's dilemma in an RNA virus. *Nature* 398: 441–443.

van Schaik, C., and Hrdy, S. 1991. Intensity of local resource competition shapes the relationship between maternal rank and sex ratios at birth in cercopithicine primates. *Am. Nat.* 138: 1555–1562.

Voland, E. 1988. Differential infant and child mortality in evolutionary perspective: data from the late 17th to 19th century Ostfriesland (Germany). In L. Betzig, M. Borgerhoff Mulder, and P. Turkey (eds.), *Human Reproductive Behavior: A Darwinian Perspective*, pp. 253–262. Cambridge: Cambridge University Press.

Westerdahl, H., Bensch, S., Hansson, B., Hasselquist, D., and von Shantz, T. 2000. Brood sex ratios, female harem status and resources for nestling provisioning in the great reed warbler (*Acrocephalus arundinaceus*). *Behav. Ecol. Sociobiol.* 47: 312–318.

Whitman, C. 1919. *The Behaviour of Pigeons*. Posthumous works of Charles Otis Whitman, vol. 3. Carr (ed.). Washington: Carnegie Institution.

Williams, G. 1966. *Adaptation and Natural Selection*. Princeton: Princeton University Press.

Williams, G. 1979. The question of adaptive sex ratio in outcrossed vertebrates. *Proc. Roy. Soc. Lond. B.* 205: 567–580.

Wilson, J., and Manning, J. 1996. Fluctuating asymmetry and age in children: evolutionary implications for the control of developmental stability. *J. Hum. Evol.* 30: 529–537.

INDEX

335

Television Studies After TV

Television ... now addr... ... ... hange
has been v... ... and ... ... vary
significantly ... ... the
multiplicatio... ... ... of
new conten... ... ... the
fragmentation ... ... all changing the ... ... on
today: its con... ... how it is consum...

Television Stud... ... ... developing ... of
understanding ... ...
leading interna... ... ...
now implicated ... ...
Anglophone ma... ...such as Asi... ...
reflect the wide ... ...ys of structures, form...
television around...

Contributors: M... ... ...
Michael Curtin, ... Anthony
Hartley, P. David ... ...ler, Albert Moran
Sinclair, Wanning ... ...
and Yuezhi Zhao.

Graeme Turner is a...
Director of the Cer...
Cultural Studies at th...
published widely in ...
include *British Cultur...*
standing Celebrity (2004...
sion in Australia (2005),

Jinna Tay is a Resear...
Studies at the Universit...
Asia, fashion magazines a...
of the *Creative Industries...* ...es section
... ...rtley.